MATHEMATICAL OPTIMIZATION AND ECONOMIC ANALYSIS

Springer Optimization and Its Applications

VOLUME 36

Aims and Scope
Optimization has been expanding in all directions at an astonishing rate during the last few decades. New algorithmic and theoretical techniques have been developed, the diffusion into other disciplines has proceeded at a rapid pace, and our knowledge of all aspects of the field has grown even more profound. At the same time, one of the most striking trends in optimization is the constantly increasing emphasis on the interdisciplinary nature of the field. Optimization has been a basic tool in all areas of applied mathematics, engineering, medicine, economics and other sciences.

The series *Optimization and Its Applications* publishes undergraduate and graduate textbooks, monographs and state-of-the-art expository works that focus on algorithms for solving optimization problems and also study applications involving such problems. Some of the topics covered include nonlinear optimization (convex and nonconvex), network flow problems, stochastic optimization, optimal control, discrete optimization, multiobjective programming, description of software packages, approximation techniques and heuristic approaches.

For other titles published in this series, go to
www.springer.com/series/7393

MATHEMATICAL OPTIMIZATION AND ECONOMIC ANALYSIS

By

MIKULÁŠ LUPTÁČIK
Vienna University of Economics and
Business Administration, Vienna, Austria

Springer

Mikuláš Luptáčik
Department of Economics
Vienna University of Economics
and Business Administration
1090 Vienna
Austria
mikulas.luptacik@wu-wien.ac.at

ISSN 1931-6828
ISBN 978-1-4614-2464-2 e-ISBN 978-0-387-89552-9
DOI 10.1007/978-0-387-89552-9
Springer New York Dordrecht Heidelberg London

Mathematics Subject Classification (2000): 91B02, 91B38, 91B66, 91B76, 90C05, 90C29, 90C30, 90C46, 90C90

Printed on acid-free paper

Springer is part of Springer Science+Business Media (www.springer.com)

To my wife Anni
my daughter Andrea,
and my sons Martin and Peter

Contents

Part II Multiobjective Optimization

Preface

The problem of allocating scarce resources among competing ends is central to economic analysis. Resources are not sufficiently available to produce all of the goods and services to satisfy human wants and therefore choices must be made concerning how resources will be used. Particularly in its "neoclassical" phase, since about 1870 economic analysis tends to presuppose that the economic agents are optimizing. Production units, or firms, maximize profit and households maximize their utility or well-being. In general, there exist a variety of objectives, besides profit maximization, high sales revenue or market share, environmental goals or the different goals followed by economic policy. The scarcity of resources acts as a bottleneck in the furthering of the objectives and represents the opportunity set from which the choices can be made. The problem of optimal allocation of scarce resources can thus be summarized as the optimization of some objective(s) subject to constraints. Constrained optimization, referred to as a mathematical programming model, is useful in economic analysis for providing deeper insights into the behavior of economic agents as well as for preparing of decision support systems for businessmen and policymakers.

This book is intended to offer the reader a systematic exposition of both single- and multiobjective optimization models with the focus on their use for economic analysis. The emphasis is given to the exposition of mathematical optimization as an instrument for qualitative analysis and to a wide range of applications in economics, including efficiency analysis, industrial economics (with focus on regulatory economics), international economics, input–output economics, quantitative economic policy and environmental economics.

Part I of the book is devoted to single-objective optimization and starts with the notion of scarcity and efficiency and with the formulation of different economic problems leading to optimization models (Chapter 1). Kuhn–Tucker conditions as the necessary optimality conditions for the general mathematical programming problem are explored and their application as an instrument of qualitative economic analysis is presented in Chapter 2.

Chapter 3 deals with convex programming and with the economic implications of the convexity property. For an economist, the problem of optimal allocation of scarce resources is immediately related to the pricing problem, referred to in the language of mathematical programming as the dual problem. Therefore, the basic duality theory is presented with the focus on its economic interpretation.

In Chapter 4, the theory of linear programming as the simplest and most widely spread class of convex programming is developed. The chapter concentrates on the implications of a linearity assumption for economic analysis and the applications of linear programming in economics.

Data envelopment analysis (DEA) as one of the most important recent applications of linear programming in economics is treated in Chapter 5. DEA represents a widely used approach for measuring efficiency and productivity even when dealing with multiple inputs and outputs without the need to assign prespecified weights to either.

Chapter 6 completes the first part of the book with geometric programming as a special class of nonlinear programming focusing on various applications in economics and management science.

In Part II of the book, multiobjective optimization is presented as an instrument of economic analysis providing a deeper insight into the trade-off choices that have to be made with respect to the objectives. Chapter 7 deals with the extension of the Kuhn–Tucker conditions and with the duality theory for multiobjective optimization. Examples from different fields of economics and the analysis of the behavior of a firm facing a bicriteria objective under regulatory constraint demonstrate the possibilities for the application of multiobjective optimization in economic analysis.

As in single-objective optimization, the most developed part of multiobjective optimization is multiobjective linear programming, treated in Chapter 8.

The extension of geometric programming from the first part of the book to problems with multiple objectives is the subject of Chapter 9. A list of references is added to each chapter separately with the aim of providing references for more detailed study and further reading related to particular topics.

Because of the increasing complexity of recent economic problems, the use of mathematical techniques including optimization plays a very important part in economics education and in applied economic research. The book is intended for university economists, graduate and postgraduate students and for quantitative oriented economists in applied research who want to expand the array of mathematical techniques at their disposal. Students of mathematics and operations research interested in economic applications of mathematical programming can also benefit from using this book. As a prerequisite to follow the text, the basics of calculus and linear algebra are needed. Definitions, theorems, and propositions are stated rigorously, but due to the mathematical prerequisite and to emphasize an economic interpretation, most of the proofs are omitted and referred to in the literature.

Following not only the principle of the division of labor by the classical economist Adam Smith, optimization under uncertainty referred to as stochastic programming and questions of choice in dynamic economic models (there are some excellent monographs in this field) are not discussed in the book. These problems and models require essentially higher mathematical background and I aimed to provide a not-too-voluminous text.

A number of students and colleagues have contributed to this book directly or indirectly. The book is an outgrowth of many courses in optimization and mathematical economics that I have taught at Vienna University of Economics and Business Administration, Vienna University of Technology and Comenius University Bratislava, Slovakia. Inspiring questions that students have raised in my courses have often helped me both to clarify and to deepen my own perception of particular topics. I am indebted to numerous authors and researchers who contributed to the study of mathematical optimization and economics. Relevant literature sources are listed at the end of each chapter.

I owe much to Gustav Feichtinger for his unending encouragement and support during the long and fruitful time I shared with him at the Vienna University of Technology. I wish to express my gratitude to my first teachers at the University of Economics in Bratislava, Juraj Fecanin, Milan Hamala, Jozef Sojka, and Ladislav Unčovský, who introduced me to optimization and mathematical economics. I thank Bernhard Böhm and František Turnovec for permission to include part of the research outcome published in our joint papers into the book.

I am grateful to Clemens Hödl, Wolfgang Katzenberger, Carl-Louis Sandblom, Susanne Warning, Wendy Williams, and Michael Weber who read the manuscript (or its parts) and suggested many improvements.

The book went through several drafts, and I am deeply indebted to Viera Zajačiková for her patience and excellent typing of the manuscript. I thank Daniel Ševčovič and Robert Zvonár for preparing the figures.

I want to thank the publisher's two anonymous referees for their very helpful comments. Any errors or omissions in the book are the responsibility of the author only and I will be grateful if they are pointed out to me.

Finally, I would like to thank the publisher for constructive cooperation and patience, understanding and encouragement during the years it took to complete the book.

Last but not least, I wish to express my thanks to my family—my wife Anni and children Martin, Andrea, and Peter—for their encouragement and understanding during the time-consuming task of preparing this book.

Mikuláš Luptáčik
Vienna, January 2009

Part I

Single-Objective Optimization

The very name of my subject, economics, suggests economizing or maximizing.

Paul A. Samuelson
Nobel Memorial Lecture
December 11, 1970

1

Scarcity and Efficiency

Scarcity is a fundamental problem faced by all economies. Not enough resources are available to produce all of the goods and services to satisfy human wants. According to a frequently cited definition of economics by Robbins [41, p. 16]: "Economics is the science which studies human behavior as a relationship between ends and scarce means which have alternative uses." In the *Concise Encyclopedia of Economics*,[1] this definition of economics is still used to define the subject today. Although this definition is simplified and "it cannot be understood as a complete list of topics belonging to economics, the scarcity principle certainly plays *some* role in all economic studies" [40, p. 13; translated by the author]. Scarce commodities are those which are both desired and not freely available. The scarcity of resources cannot be eliminated; rather, choices must be made about how resources will be used. The answer given by economists is based on the notion of *efficiency*. In a very simple formulation, it means that the desired end is achieved by minimal use of resources or under the given amount of resources the desired end is maximized. Efficiency implies the recognition of scarcity and at the same time the best possible use of the disposable resources.

The problems of optimal allocation of scarce resources are of particular interest in microeconomics: the neoclassical theory of the household and the neoclassical theory of the firm are two principal areas of study. Quantitative economic policy, the theory of optimal economic growth, international economics and environmental economics are other fields of application of the idea of best allocation of scarce resources and therefore of the optimization models which "have come to occupy a prominent position in modern economic theory" [36, p. 7]. Optimization models are useful for economists not only for understanding the behavior of economic agents (that means in a positive sense), but also in the preparation of decision support systems for businessmen and policymakers (that means in a normative sense). In this first chapter, some optimization models in economics will be formulated in order to illustrate the variety of economic problems and to provide a good starting point for the analysis in subsequent chapters.

[1] http://www.econlib.org/library/CEE.html.

M. Luptáčik, *Mathematical Optimization and Economic Analysis*,
Springer Optimization and Its Applications 36, DOI 10.1007/978-0-387-89552-9_1,

1.1 The Mathematical Programming Problem

The basic economic problem of allocating scarce resources among competing ends has three components. First, there are the *instruments* whose values can be chosen by the economic agent (such as a consumer or a producer). These are the *decision variables* in the problem. Second, the scarcity of the resources is represented by the *opportunity set* or the set of feasible values from which to choose. Finally, the competing ends are described by some criterion function, called the *objective function*, which gives the value attached to each of the alternative decisions. The mathematical problem in the language of economics is the following: how to choose the instruments within the opportunity set so as to maximize or minimize the objective function.

For this purpose, let us denote the decision variables by an n-dimensional vector $\mathbf{x}' = (x_1, x_2, \ldots, x_n)$ and the m constraints reflecting the availability of the resources by the inequalities

$$f_i(x_1, x_2, \ldots, x_n) \leqq 0 \quad (i = 1, 2, \ldots, m).$$

The set $K = \{\mathbf{x} | \mathbf{x} \in R^n,\ f_i(x_1, x_2, \ldots, x_n) \leqq 0\ (i = 1, 2, \ldots, m)\}$ is the set of feasible solutions or the opportunity set, which is a subset of the Euclidean n-space (the space of all n-tuples of real numbers), $K \subset E^n$. It will generally be assumed that K is not empty, that is, that there exists a feasible vector $\mathbf{x}$, where $\mathbf{x}$ is feasible if and only if $\mathbf{x} \in K$.

A mathematical programming model reflecting the basic problem of the allocation of scarce resources among alternative uses can be formulated as follows:

$$\begin{aligned} &\text{maximize (minimize)} && f_0(x_1, x_2, \ldots, x_n) \\ &\text{subject to} && f_i(x_1, x_2, \ldots, x_n) \leqq 0 \quad (i = 1, 2, \ldots, m), \end{aligned} \tag{1.1}$$

where $f_0(\mathbf{x})$ denotes the objective function. The term "programming" has been used initially to denote that the aim of the calculation is to find a program.

1.2 Mathematical Programming Models in Economics

Mathematical programming has been applied to a wide variety of problems in economics. According to [12, p. 9]—one of the groundbreaking books in this field—the first economic problem solved by the explicit use of mathematical (linear) programming was the diet problem. It was formulated and approximately solved by Jerome Cornfield in an unpublished memorandum. It was solved by Stigler [43] without obtaining the optimal solution and in 1947 by G. B. Dantzig and J. Laderman using linear programming.

1.2.1 The Diet Problem

Each type of food, such as potatoes, meat, bread, apples, grapes, etc., contains a specific quantity of nutrients, e.g., carbohydrates, protein, fat, vitamin D, and others.

We consider n different foods available on the market, denoted by the index $j = 1, 2, \ldots, n$, and m nutritional ingredients, denoted by the index $i = 1, 2, \ldots, m$. The quantity of the ith nutrient contained in one unit of food of type j is described by the coefficient a_{ij}. For healthy nourishment b_i units of the ith nutrient $(i = 1, 2, \ldots, m)$ as the minimum requirements per day or per unit of time should be consumed. The price per unit of food j $(j = 1, 2, \ldots, n)$, denoted by c_j, is given and independent of the purchased quantity of this food. In other words, no discount for a higher quantity of purchased food j is provided. What foods, and how much of each, should the hard-pressed consumer buy in order to provide a healthy diet for his family at minimum cost?

Denoting the number of units of food j in the diet by x_j, the constraints concerning nutritional requirements can be formulated in the following way:

$$\sum_{j=1}^{n} a_{ij}x_j \geqq b_i \quad (i = 1, 2, \ldots, m). \tag{1.2}$$

The total amount of the nutrient i $(i = 1, 2, \ldots, m)$ contained in the diet $\mathbf{x} = (x_1, x_2, \ldots, x_n)'$ should be at least b_i units.

The housewife or the houseman can only buy the food j or not. Therefore the nonnegative conditions on the decision variables $x_1, x_2, \ldots, x_n$ must be imposed:

$$x_j \geqq 0 \quad (j = 1, 2, \ldots, n). \tag{1.3}$$

The system of inequalities (1.2)–(1.3) describes the set of feasible solutions or the opportunity set and the mathematical programming problem consists of the minimization of the objective function (the total cost of the diet) $z = \sum_{j=1}^{n} c_j x_j$ under the constraints (1.2)–(1.3). In matrix notation, we can simply write the mathematical programming problem of the consumer as

$$\begin{aligned} \text{minimize} \quad & z = \mathbf{c}'\mathbf{x} \\ \text{subject to} \quad & A\mathbf{x} \geqq \mathbf{b}, \\ & \mathbf{x} \geqq \mathbf{0}, \end{aligned} \tag{1.4}$$

where $\mathbf{c} \in R^n$, $\mathbf{x} \in R^n$, $\mathbf{b} \in R^m$, and A is an $m \times n$ matrix.

There are some modifications and extensions of the basic diet model (1.4). A family food planning model taking into account changes in the eating habits, food prices, and dietary allowances is described in [2].

The Consumer and Food Economics division of the Agricultural Research Service of the U.S. Department of Agriculture has developed a system of specifying the food quantities "to help families to plan nutritionally adequate and satisfying meals for the money they can afford" [2, p. 327]. The foods are defined as food groups such as milk products, eggs, meat, vegetables, fruits, etc. for a number of sex–age groups and budget levels.

The goal was "to adjust the food plans as close as possible to the food consumption patterns established by the 1965–1966 household survey data" [2, p. 328] so that they would be nutritionally adequate. The problem was formulated as the minimization of the weighted total squared deviations of each food (group) quantity in the food plan relative to the 1965 consumption levels for each sex–age group and income level under consideration of the (new) nutritional constraints.

Let q_j denote the past consumption of food quantity j and x_j the corresponding quantity in the new food plan. To find the optimal food plan the following mathematical programming model was formulated:

$$\begin{aligned} \text{minimize} \quad & \sum_{j=1}^{n} w_j^2 (q_j - x_j)^2 \\ \text{subject to} \quad & A\mathbf{x} \geqq \mathbf{b}, \\ & R\mathbf{x} \geqq \mathbf{d}, \end{aligned} \tag{1.5}$$

where w_j are the weights for the relative contributions of deviations and A is the matrix of food cost and nutrient composition data. The R matrix represents a set of upper and lower bounds as well as proportionality constraints imposed on the components of the solution vector to assure strictly positive and acceptable food quantities.

For further developments of the diet model and for applications of mathematical programming to optimize human diets, the reader is referred to [2]. As shown in this paper, the menus planned by mathematical programming "are superior to those planned by conventional methods.... In the four hospitals cost savings of 10, 15% to 15, 30% were achieved relative to the conventional plans" [2, p. 333].

1.2.2 The Neoclassical Theory of the Household

The household and the firm are the two most important economic agents. The households consume commodities and supply their capacity to work; the firms produce commodities (and services) and demand labor.

Assuming that there are n goods (and services) available, let $\mathbf{x}$ be the column vector of the goods purchased and consumed by the household,

$$\mathbf{x}' = (x_1, x_2, \ldots, x_n)'.$$

The preferences of the household with respect to the consumption of particular goods (and services) are described by the utility function $U(\mathbf{x})$ of the household,

$$U(\mathbf{x}) = U(x_1, x_2, \ldots, x_n),$$

giving utility as a function of consumption levels. It fulfills the usual neoclassical properties:

(i) $\frac{\partial U}{\partial x_j} > 0$—the marginal utility (utility received from the consumption of the last "small" amount) of the good j ($j = 1, 2, \ldots, n$) is positive and
(ii) decreasing, $\frac{\partial^2 U}{\partial x_j^2} < 0$ ($j = 1, 2, \ldots, n$).

Marginal utility decreases as consumption increases. This principle is known as Gossen's first law.

Let $\mathbf{p}$ be the row vector of (positive) given prices of the goods,

$$\mathbf{p} = (p_1, p_2, \ldots, p_n),$$

and M the (positive) given income available to the household.

The household is assumed to behave so as to maximize utility subject to a budget constraint. Thus the household chooses nonnegative amounts of goods $\mathbf{x}$ (it will consume the particular good or not) so as to maximize the utility function $U(\mathbf{x})$ under the constraint that expenditure on all n goods cannot exceed disposable income M. The problem of the household is then

$$\begin{aligned} \text{maximize} \quad & U(\mathbf{x}) \\ \text{subject to} \quad & \mathbf{px} = \sum_{j=1}^{n} p_j x_j \leq M, \\ & \mathbf{x} \geqq \mathbf{0}. \end{aligned}$$

For a comparison of the diet problem with the neoclassical theory of household, see [12, Chapter II].

1.2.3 The Neoclassical Theory of the Firm

The firm is assumed to behave so as to maximize profit, subject to the technological constraint described by the production function. Suppose that the firm uses n inputs (labor, coal, iron ore, etc.) to produce a single output, let $\mathbf{x}$ be the column vector of inputs,

$$\mathbf{x}' = (x_1, x_2, \ldots, x_n)',$$

and let q be the output. The firm's production function is represented by

$$q = f(\mathbf{x}) = f(x_1, x_2, \ldots, x_n), \tag{1.6}$$

giving output as a function of its inputs. Equation (1.6) assumes nothing but the existence of a maximum output corresponding to any combination of inputs.

Let $\mathbf{r}$ be a row vector of (positive) given prices of the inputs,

$$\mathbf{r} = (r_1, r_2, \ldots, r_n),$$

and p the (positive) given price of the output.

A firm is in a competitive situation if it can buy and sell in any quantities at exogenously given prices, which are independent of its production decisions. In other words, the competitive firm is a price taker.

The firm behaves so as to maximize the profit π, given as the difference between revenue, pq, and cost, given as the total expenditure on all inputs,

$$\mathbf{rx} = \sum_{j=1}^{n} r_j x_j.$$

The problem of the (competitive) firm can then be stated as the following mathematical programming problem:

$$\begin{aligned} &\underset{\mathbf{x}}{\text{maximize}} \quad \pi(\mathbf{x}) = pf(\mathbf{x}) - \mathbf{rx} \\ &\text{subject to} \qquad \mathbf{x} \geqq \mathbf{0}. \end{aligned}$$

Another version of this problem often used in production theory is based on the assumption of the given output level q^*. The firm is trying to minimize its cost, M, of the inputs used to produce q^*. The expenditure of the firm is given by

$$M(\mathbf{x}) = r_1 x_1 + r_2 x_2 + \cdots + r_n x_n,$$

and the mathematical programming problem is then

$$\begin{aligned} &\underset{\mathbf{x}}{\text{minimize}} \quad M(\mathbf{x}) \\ &\text{subject to} \quad f(\mathbf{x}) = q^* \quad \text{and} \quad \mathbf{x} \geqq \mathbf{0}. \end{aligned} \tag{1.7}$$

Closely interrelated to this formulation is the problem of the firm with a given level of expenditure or prespecified budget, M^*, and an objective function that maximizes the production q. Thus the firm chooses levels of inputs so as to maximize output, subject to a budget constraint:

$$\begin{aligned} &\underset{\mathbf{x}}{\text{maximize}} \quad f(\mathbf{x}) \\ &\text{subject to} \quad r_1 x_1 + r_2 x_2 + \cdots + r_n x_n \leqq M^* \quad \text{and} \quad \mathbf{x} \geqq \mathbf{0}. \end{aligned} \tag{1.8}$$

For illustration of models (1.7) and (1.8), we consider the technology defined by the Cobb–Douglas production function (for simplicity, but without loss of generality, with just two inputs):

$$q = f(x_1, x_2) = a x_1^{\alpha} x_2^{\beta}$$

with $a > 0$, $0 < \alpha < 1$, and $0 < \beta < 1$.

In this case, problem (1.7) leads to the following mathematical programming model:

$$\begin{aligned} &\underset{x_1, x_2}{\text{minimize}} \quad r_1 x_1 + r_2 x_2 \\ &\text{subject to} \quad a x_1^{\alpha} x_2^{\beta} \geqq q^* \quad \text{and} \quad x_1 > 0,\ x_2 > 0. \end{aligned} \tag{1.9}$$

Problem (1.8) then takes the following form:

$$\begin{aligned} &\underset{x_1, x_2}{\text{maximize}} \quad a x_1^{\alpha} x_2^{\beta} \\ &\text{subject to} \quad r_1 x_1 + r_2 x_2 \leqq M^* \quad \text{and} \quad x_1 > 0,\ x_2 > 0. \end{aligned} \tag{1.10}$$

1.2.4 The Theory of Comparative Advantage

Many years ago, the English economists Robert Torrens[2] (1780–1864) and Ricardo [42] developed independently the classical theory of international trade, referred to as the theory of comparative advantage. Dorfman, Samuelson, and Solow [12, pp. 31–32] slightly modified a traditional numerical example of [42] in the following form: "Portugal can divert resources from food to clothing production and in effect convert one unit of food into one unit of clothing; England, on the other hand, can convert one unit of food into two units of clothing." If there exists an international price ratio $\frac{p_1}{p_2}$, somewhere between 1 and 2, both countries will be better off if they will specialize: Portugal completely in food, England completely in clothing. England will export clothing in exchange for food imports from Portugal. World production will be optimal.

These economic conclusions can be derived using a mathematical programming framework. First, we consider England and denote by x_1 the units of food produced and by x_2 the clothing output. The real value of the national product of England (expressed in clothing units) may be written as

$$Z = \frac{p_1}{p_2}x_1 + x_2, \quad \text{or, say,} \quad Z = 1,5x_1 + x_2. \tag{1.11}$$

The problem is to maximize the national product (1.11) subject to the constraint

$$\begin{aligned} 2x_1 + x_2 &\leqq C, \\ x_1 \geqq 0, \qquad x_2 &\geqq 0. \end{aligned}$$

In this inequality, C is England's maximum output of clothing when no food is produced. According to our assumption, clothing output must be cut back by two units for every unit of food produced. Thus the maximum output of clothing, when x_1 units of food are produced, is $x_2 = C - 2x_1$, and this leads to the inequality stated.

In a similar way, Portugal will maximize its national product:

$$\begin{aligned} Z' = 1,5x_1' + x_2' &= \frac{p_1}{p_2}x_1' + x_2' \\ \text{subject to} \quad x_1' + x_2' &\leqq C', \\ x_1' \geqq 0, \qquad x_2' &\geqq 0, \end{aligned}$$

where x_1' denotes the production of food and x_2' the clothing output for Portugal. C' is Portugal's maximum output of clothing (or food) when no food (or no clothing) is produced.

We will return to this problem in Section 4.6.1.

1.2.5 The Giffen Paradox

In a letter from A. Marshall to F. Y. Edgeworth dated April 22, 1909 [37, p. 441], the following example is described: "I believe that people in Holland travel by canal boat

[2] *An Essay on the External Corn Trade*, 1815.

instead of railway sometimes on account of its cheapness. Suppose a man is in hurry to travel 150 kilos. He had two florins for it and no more. The fare by boat was one cent a kilo, by third-class train two cents. So he decided to go 100 kilos by boat and fifty by train: total costs two florins. Arriving at the boat, he found the charge had been raised to $1\frac{1}{4}$ cents per kilo. 'So then I will travel $133\frac{1}{3}$ (or as near as may be) by boat; I can't afford more than $16\frac{2}{3}$ kilos by train.'"

This decision problem can be formulated as the following optimization problem [3]:

Let $x_1,\ x_2$ denote the distance traveled by train and boat, respectively, and $c_1,\ c_2$ are the times required per kilo covered by the two means of transportation, where $c_1 < c_2$. The objective is to

$$\begin{aligned} \text{minimize} \quad & c_1x_1 + c_2x_2 \\ \text{subject to} \quad & x_1 + x_2 \geqq 150, \\ & 2x_1 + x_2 \leqq 200, \\ & x_1, x_2 \geqq 0. \end{aligned}$$

We write 200 cents for 2 florins.

After the increased boat charge, we are confronted with the following problem:

$$\begin{aligned} \text{minimize} \quad & c_1x_1 + c_2x_2 \\ \text{subject to} \quad & x_1 + x_2 \geqq 150, \\ & 2x_1 + \frac{5}{4}x_2 \leqq 200, \\ & x_1, x_2 \geqq 0. \end{aligned}$$

This problem illustrates the appearance of the Giffen paradox (or the Giffen commodity) if the quantity demanded increases as price increases, and it will be analyzed in Section 4.6.2.

1.2.6 The Transportation Problem

One frequently encountered problem, first introduced by Hitchcock in 1941 and since then in a wide variety of seemingly completely different problems reduced to this model, is the following:

Let a company own m warehouses, denoted by the index i $(i = 1, 2, \ldots, m)$ in each of which is a given amount of a certain commodity, denoted by a_i. Let there also be n retailers, denoted by the index j $(j = 1, 2, \ldots, n)$, with given demand b_j for this commodity. The unit transportation cost between each warehouse–retailer pair, denoted by c_{ij}, is known. The objective of the company is to ship the commodity from the warehouses to the retailers, such that

- no more units of commodity leave a warehouse than there are in stock,
- the demand of each retailer is satisfied, and
- the total transportation cost is minimized.

For simplicity, we assume that the total supply equals the total demand:

$$\sum_{i=1}^{m} a_i = \sum_{j=1}^{n} b_j.$$

If total supply exceeds total demand, exactly one dummy retailer is created to absorb the excess supply; its demand equals

$$\sum_{i=1}^{m} a_i - \sum_{j=1}^{n} b_j.$$

The case of excess demand is handled similarly, using exactly one dummy warehouse. All unit transportation costs from a dummy warehouse or to a dummy retailer are assumed to be equal.

Defining variables x_{ij} to denote the quantity shipped from the ith warehouse to the jth retailer, we can formulate the following model of mathematical programming:

$$\begin{aligned}
\text{minimize} \quad & z(\mathbf{x}) = \sum_{i=1}^{m} \sum_{j=1}^{n} c_{ij} x_{ij} \\
\text{subject to} \quad & \sum_{j=1}^{n} x_{ij} = a_i \quad (i = 1, 2, \ldots, m), \\
& \sum_{i=1}^{m} x_{ij} = b_j \quad (j = 1, 2, \ldots, n), \\
& x_{ij} \geqq 0 \quad (i = 1, 2, \ldots, m), \\
& \qquad\qquad (j = 1, 2, \ldots, n).
\end{aligned}$$

1.2.7 Portfolio Selection Model

Assume that a certain amount of financial funds is going to be invested in n different securities. Let x_j represent the proportion of the available funds that is allocated to the jth security, $j = 1, 2, \ldots, n$. Furthermore, z_j denotes the gain at the end of the planning horizon per dollar invested in security j. The values of z_j are assumed to be a random variable with known expected value:

$$E(z_j) = \mu_j \quad (j = 1, 2, \ldots, n);$$

E stands for mathematical expectation.

Usually the securities with high expected values are securities with high risk. In order to reduce the risk, the brokers recommend to their clients the diversification of financial funds to different securities, which is in contradiction to the profit-maximization hypothesis. As a measure of risk, Markowitz [33, 34] chooses the covariance matrix $V = \{\sigma_{jk}\}$, such that $\sigma_{jk} = E[(z_j - \mu_j)(z_k - \mu_k)]$, $j, k = 1, 2, \ldots, n$. The expected gain of the portfolio selected is expressed by

$$E(P) = \sum_{j=1}^{n} \mu_j x_j$$

and the variance of the total gain by

$$V(P) = \sum_{j=1}^{n} \sum_{k=1}^{n} \sigma_{jk} x_j x_k.$$

The decision problem is now to choose the proportion x_j in order to minimize the variance of the total gain and simultaneously to maximize the expected gain. We have an optimization problem with two objectives that are usually in conflict. One possibility for finding a compromise solution is to minimize the variance of the total gain for the lowest acceptable expected gain, or to maximize the expected gain of the portfolio under the condition that the variance of the total gain does not exceed a certain prescribed level.

Obviously, the solution to this problem depends on the risk aversion of the investors. If we denote the coefficient of the risk aversion by ρ (which is given), the portfolio selection problem can be formulated as the following mathematical programming problem:

$$\text{maximize} \quad Q(\mathbf{x}) = \sum_{j=1}^{n} \mu_j x_j - \rho \sum_{i=1}^{m} \sum_{k=1}^{n} \sigma_{jk} x_j x_k$$

$$\text{subject to} \quad \sum_{j=1}^{n} x_j = 1, \quad x_j \geqq 0 \quad (j = 1, 2, \ldots, n).$$

With respect to his risk aversion, the investor can find a compromise between the maximization of the excepted gain and the minimization of the variance of the total gain. Extensions and modifications of this basic model are now used widely in economic analysis and financial management [14].

1.2.8 Input–Output Analysis and Mathematical Programming

One of the most interesting developments in the field of economics is the Leontief model of industrial interdependence known as input–output analysis. The original idea of developing a detailed accounting of intersectoral activity came from François Quesnay, the founder of the first economic school, the physiocracy. In 1758, Quesnay published a "tableau économique," which was the first model of the economy, describing the relations between three economic sectors:

- the production class, consisting of peasants and tenant farmers;
- the class of the landowners, consisting of the nobility and clergy;
- the sterile class, covering all other occupations, especially commerce and handcraft.

He illustrated how the landowners, who receive a sum of money as rent, spend half this sum on agricultural products and half on products of artisans. In turn, farmers buy industrial products and raw materials, and so on.

More than a century after Quesnay, Leon Walras developed a theory of general equilibrium. In his model, Walras [47] utilized a set of production coefficients defined as the respective quantities of each of the productive services that enter into the production of a unit of each of the products. In the Walrasian general equilibrium system, intermediate goods were expressed as a set of equations with the sales and purchases of the intermediate industries forming the core of the system [28]. At this point, Leontief was retaining Walras's concept of an entirely self-contained, self-determining system of economic interrelationships. The model was completely closed, with all final demand and value added components taken as endogenous. Ten years later, Leontief [29] reformulated the model, with the final demand and value added components treated exogenously. In this way, Leontief's model is the simplest form of the Walrasian general equilibrium allowed to be applied.

Leontief's first paper on input–output analysis appeared in 1936 [27]. Although this was a "novel and important contribution to economic theory," in the paper Leontief emphasized "the numerical description of the American economic structure" [13, p. 434]. His primary concern was linking theory and applications, and he stated that an empirical analysis should be a "descriptive complement of its theoretical analysis" [30, p. 229]. In a 1998 interview, Leontief said, "I am essentially a theorist. But I felt very strongly that theory is just a construction of frameworks to understand how real systems work" [16, p. 123].

The descriptive complement of Leontief's model is the input–output table describing the sales and purchases between the sectors of an economy. Leontief's first input–output tables were constructed for the 1919 and 1929 U.S. economy [27], followed by his first book on the input–output structure of the U.S. economy [28]. The revised and extended version of this book that contained the U.S. input–output table for 1939 appeared in 1951 [29]. The U.S. government continued to construct input–output tables for 1947, 1958, and 1963, and—starting in 1967—for every year ending in a 2 or a 7. Input–output analysis soon spread to Europe and Asia, and today it is one of the most frequently applied techniques in economics. "For the development of the input–output method and its application to important economic problems" [31, p. 147], Wassily Leontief received the Nobel Prize for Economic Sciences in 1973.

In recent years, the input–output framework has been extended to such topics as the interregional flows of products and accounting for energy consumption, environmental pollution and employment associated with industrial production, structural development of an economy, income distribution effects, and others. (For further reading, see, e.g., [35, 25, 26, 11].)

The basic input–output model can be described as follows: Leontief imagined an economy in which goods like iron, coal, plastics, paper, textile products, wood products, etc. are produced in their respective industries by means of a primary factor (which is produced external to this system), such as labor and by means of other inputs such as iron, coal, plastics, paper, textile products, wood products, etc. The economy is thus classified by industries, or sectors. Although sectors may have a variety of commodities as inputs, their outputs are not mixed.[3] Each sector is identified with the commodity that it produces. The observed monetary value of the flow from sector i to sector j (or the monetary value of the good i used in sector j as an input) is denoted by x_{ij}. Sector j's demand for inputs from other sectors during the year will have been related to the value of the good produced by sector j over the same period.

In addition, in any country there are final sales to purchasers, who are external or exogenous to industrial sectors, that constitute the producers and consumers in the economy—for example, households, government, and foreign trade. The demand for these external units, since it has not been used as an input to an industrial production process, is generally referred to as final demand.

Thus if the economy is divided into n sectors, and if we denote x_i as the total output (production) of sector i and y_i as the total final demand for sector i's product, we may write the distribution of sector i's output:

$$\sum_{j=1}^{n} x_{ij} + y_i = x_i \quad (i = 1, 2, \ldots, n). \tag{1.12}$$

Using these data, the technical (or direct) input coefficients, denoted a_{ij}, can be computed:

$$a_{ij} = \frac{x_{ij}}{x_j} \quad (i, j = 1, 2, \ldots, n). \tag{1.13}$$

The coefficients a_{ij} describe the input of good i per unit of good j (produced by sector j). Under the assumption that the demand for input i changes proportionally with the output of sector j, substituting (1.13) for x_{ij} in (1.12) results in

$$\sum_{j=1}^{n} a_{ij} x_j + y_i = x_i \quad (j = 1, 2, \ldots, n), \tag{1.14}$$

or, in matrix notation,

$$A\mathbf{x} + \mathbf{y} = \mathbf{x},$$

where $\mathbf{x}$ is the n-dimensional column vector of total industrial outputs, $\mathbf{y}$ is the n-dimensional column vector of final demand, and A is the $(n \times n)$ matrix of technical input coefficients.

Under the assumption of fixed technical coefficients a_{ij} and for exogenously given levels of final demand $\mathbf{y}$, the levels of total industrial output $\mathbf{x}$ are given by

$$\mathbf{x} = (E - A)^{-1}\mathbf{y}, \tag{1.15}$$

where E is the identity matrix and the matrix $(E - A)^{-1}$ is the so-called Leontief inverse. In contrast to the direct input coefficients a_{ij}, an element of this matrix shows the input of good i per unit of final demand of good j. In this way, the elements of the Leontief inverse describe the direct and indirect interdependencies between the

---

[3] In the new formulation based on so-called make and absorption matrices, multiple outputs of the sectors are included [35, Chapter 5].

sectors of the economy. Equation (1.15) is the fundamental equation of input–output analysis.

One important aspect of the input–output model is that substitutions of inputs are not technologically feasible. However, in more general models, in which substitution is possible, the model can be formulated as a mathematical programming problem (see [12, Chapter 9] and [17, Chapter 9]).

Another formulation of the input–output model as a mathematical programming problem, which can be seen as the reconciliation of input output analysis and neo-classical economics, is given by ten Raa [38, 39]. In contrast to the basic input–output model, the final demand is not exogenously given but is an endogenous variable of the model. Instead of the balance equations (1.14), the sectoral inequalities

$$x_i \geqq \sum_{j=1}^{n} a_{ij} x_j + y_i \quad (i = 1, 2, \ldots, n)$$

are introduced. They express the condition that supply must be sufficient to meet intersectoral and final demand. The objective is to maximize the value of final demand, $\mathbf{p}'\mathbf{y}$, where $\mathbf{p}'$ is a row vector of given prices, say, world prices. For the production of the goods, two primary factors, labor and capital, are used, where K denotes the available stock of capital and L the available labor force. Under the assumption of fixed capital coefficients k_j $(j = 1, 2, \ldots, n)$ and labor coefficients l_j $(j = 1, 2, \ldots, n)$, the total sectoral production $\mathbf{x}$ is bounded by the capital and labor constraints

$$\mathbf{k}'\mathbf{x} \leqq K$$

and

$$\mathbf{l}'\mathbf{x} \leqq L,$$

where $\mathbf{k}'$ is a row vector of capital coefficients and $\mathbf{l}'$ is a row vector of labor coefficients. The following mathematical programming problem is obtained:

$$\begin{aligned}
\text{maximize} \quad & \mathbf{p}'\mathbf{y} && (1.16)\\
\text{subject to} \quad & (E - A)\mathbf{x} - \mathbf{y} \geqq \mathbf{0}, && (1.17)\\
& \mathbf{k}'\mathbf{x} \leqq K, && (1.18)\\
& \mathbf{l}'\mathbf{x} \leqq L, && (1.19)\\
& \mathbf{x} \geqq \mathbf{0}, \quad \mathbf{y} \geqq \mathbf{0}. && (1.20)
\end{aligned}$$

One of the simplifying assumptions of the model (1.16)–(1.20) is that the capital and labor coefficients are fixed. In other words, there is no possibility for continuous substitution between the primary factors labor and capital. In such models "...growth is likely to be impeded by shortages of specific factors rather than by a general scarcity of resources..." [7, p. 29].

Following [18, 45, 7, 44], we postulate substitution possibilities for labor and capital inputs, according to a Cobb–Douglas production function for each sector of the economy. (For the model with CES (constant elasticity of substitution) sectoral production, see [45, pp. 138–164] and [32, pp. 53–61].) The function is written as

$$x_j = \epsilon_j L_j^{\alpha_j} K_j^{\beta_j} \quad (j = 1, 2, \ldots, n), \tag{1.21}$$

where x_j, L_j, and K_j indicate gross output, employment, and capital stock in sector j. We assume an exogenously given final demand $\mathbf{y}$, and the balance equations (1.14) are rewritten—similarly to the model by ten Raa [38]—as inequalities,

$$(E - A)\mathbf{x} \geqq \mathbf{y},$$

or

$$\sum_{j=1}^{n} (\delta_{ij} - a_{ij})x_j - y_i \geqq 0 \quad (i = 1, 2, \ldots, n),$$

where $\delta_{ij} = 0$ for $i \neq j$ and $\delta_{ij} = 1$ for $i = j$ (Kronecker delta). Now we substitute function (1.21) for x_j, and after a simple transformation, we get

$$\sum_{j \neq i} d_{ij} L_j^{\alpha_j} K_j^{\beta_j} L_i^{-\alpha_i} K_i^{-\beta_i} + \frac{y_i}{(1 - a_{ii})\epsilon_i} L_i^{-\alpha_i} K_i^{-\beta_i} \leqq 1, \tag{1.22}$$

where

$$d_{ij} = \frac{a_{ij}\epsilon_j}{(1 - a_{ii})\epsilon_i} \geqq 0 \quad (i, j = 1, 2, \ldots, n).$$

The economic interpretation of condition (1.22) is, in essence, the same as the interpretation of condition (1.17) previously given. Particularly, the form (1.22) implies that the sum of proportions of the deliveries from sector i (into all other sectors and to the final demand) to the net production of sector i must be equal or less than one.

The form of constraints for the capital stock depends on the transferability of capital between the particular sectors of the economy. Under the assumption of perfect transferability, we have

$$\sum_{j=1}^{n} K_j \leqq K, \tag{1.23}$$

where K indicates the total capital stock in the economy. In the case of nontransferability of capital, we have constraints for capital stock in each sector of the economy,

$$K_j \leqq \overline{K_j} \quad (j = 1, 2, \ldots, n), \tag{1.24}$$

where $\overline{K_j}$ indicates the disposable capital stock in sector j.

As an objective function, we consider the minimization of labor input,

$$\min L = \sum_{j=1}^{n} L_j. \tag{1.25}$$

For exogenously given final demand $\mathbf{y}$, the objective function (1.25) implies the maximization of labor productivity.

Summarizing, we have the following mathematical programming problem with the substitution possibilities for labor and capital: Minimize (1.25) subject to the constraints (1.22), (1.23), or (1.24) and the constraints

$$L_j > 0, \qquad K_j > 0 \quad (j = 1, 2, \ldots, n). \tag{1.26}$$

1.2.9 Data Envelopment Analysis

In his classic paper on the measurement of productive efficiency. Farrel [15, p. 11] describes the importance of efficient frontier estimation for modern economics in the following way: "The problem of measuring the productive efficiency of an industry is important to both the economic theorist and the economic policymaker. If the theoretical arguments as to the relative efficiency of different economic systems are to be subjected to empirical testing, it is essential to be able to make some actual measurements of efficiency. Equally, if economic planning is to concern itself with particular industries, it is important to know how far a given industry can be expected to increase its output by simply increasing its efficiency, without absorbing further resources."

In the simplest case where a process or unit has a single input and a single output, efficiency is defined as

$$\text{efficiency} = \frac{\text{output}}{\text{input}}.$$

But in most cases, for the production of the particular good, different inputs are used and the output is the result of all inputs operating in combination. A partial factor ratio, such as output per worker, therefore gives a misleading indication of intrinsic labor productivity.

Moreover, there are processes or nonprofit organizational units, such as local authority departments, schools, universities, hospitals, bank branches, and the like, using different inputs to produce different outputs where some of the outputs are nonmarket goods (goods without market prices). How do we define and to measure the efficiency of such relatively homogeneous units in the multiinput and multioutput situations? Built on the work in [15], Charnes, Cooper, and Rhodes [5] developed a mathematical programming–based technique for measuring the efficiency of a particular organizational unit relative to other units and thus estimating a "best practice" or efficient frontier.

We denote the organizational units by $j = 1, 2, \ldots, n$, referred to as decision making units (DMUs), which are to be evaluated. Each DMU uses m different inputs, denoted by $i = 1, 2, \ldots, m$, to produce s different outputs ($r = 1, 2, \ldots, s$). Specifically, the input vector of DMU$_j$ is denoted by $\mathbf{x}_j$ and the output vector by $\mathbf{y}_j$. We assume (as in [6]) that each DMU has at least one positive input and one positive output value.[4]

[4] In the original study in [5], it was assumed that the input vectors $\mathbf{x}_j$ and the output vector $\mathbf{y}_j$ for $j = 1, 2, \ldots, n$ are all positive.

The main idea in [5] is the reduction of the multiple-output/multiple-input situation (for each DMU) to that of a single "virtual" output and "virtual" input. The efficiency is defined as

$$\text{efficiency} = \frac{\text{"virtual" output}}{\text{"virtual" input}} = \frac{\text{weighted sum of outputs}}{\text{weighted sum of inputs}}.$$

This definition requires a set of weights or multipliers to be estimated that proves to be difficult. The problem can be resolved by arguing that the DMUs may have their own particular value systems and therefore may legitimately define their own peculiar set of multipliers. The observed output and input values, respectively, of DMU_0 (the DMU to be evaluated) are described by vectors $\mathbf{y}_0$ and $\mathbf{x}_0$. The efficiency of DMU_0, denoted by h_0, is defined as follows:

$$h_0(\mathbf{u}, \mathbf{v}) = \frac{\sum_{r=1}^{s} u_r y_{r0}}{\sum_{i=1}^{m} v_i x_{i0}},$$

where u_r is the weight given to output r $(r = 1, 2, \dots, s)$ and v_i is the weight given to input i $(i = 1, 2, \dots, m)$. The key feature of the Charnes–Cooper–Rhodes (CCR) model is that the multipliers u_r and v_i are treated as unknown. They will be chosen so as to maximize the efficiency h_0 of DMU_0 subject to efficiencies of all DMUs in the set having an upper bound of 1. In other words, even with their own particular value system, the DMU_0 cannot be better (more efficient) than the "best" DMUs in the group; all DMUs lie on or below the efficiency frontier.

The mathematical programming problem for the CCR ratio is

$$\begin{aligned}
\underset{\mathbf{u},\mathbf{v}}{\text{maximize}} \quad & h_0(\mathbf{u}, \mathbf{v}) = \frac{\sum_{r=1}^{s} u_r y_{r0}}{\sum_{i=1}^{m} v_i x_{i0}} \\
\text{subject to} \quad & \frac{\sum_{r=1}^{s} u_r y_{rj}}{\sum_{i=1}^{m} v_i x_{ij}} \leqq 1 \quad (j = 1, 2, \dots, n), \\
& u_r \geqq 0 \quad (r = 1, 2, \dots, s), \\
& v_i \geqq 0 \quad (i = 1, 2, \dots, m).
\end{aligned} \tag{1.27}$$

1.3 Classification of Mathematical Programming Problems

With respect to the form of the objective function and of the functions in the constraints in the models of the previous section (and in mathematical programming problems generally), mathematical programming problems can be classified in the following way:

A. ***Linear programming***

The diet problem, the models for the theory of comparative advantage and for the Giffen paradox, the transportation problem, and the input–output model from the previous section are characterized by the linearity of the objective function and the linearity of functions in the constraints. The development in this field of mathematical programming started independently with the works by Kantorovic [20] and Dantzig [8, 9]. The work by the Russian mathematician Leonid V. Kantorovic was motivated by practical problems of production planning: How does one combine the available resources in a factory such that production is maximized? The model and solution method, based on the so-called "resolving multipliers," was published in 1939 in Russian [20] and remained unknown in the West until the late 1950s or early 1960s. The translation, entitled "Mathematical Methods of Organizing and Planning of Production," appeared in 1960. According to [23, p. 240], "the importance of this publication is due to the simultaneous presence of several ideas or elements, some of which had also been present in earlier writing in different parts of economics or mathematics." One of these elements is the description of production in terms of finite number of distinct production processes, each characterized by constant ratios between the inputs and outputs, which has a long history in economics; see, e.g., [47, 4, 27, 28].[5]

The work of Leontief, who proposed a simple matrix structure for a description of interindustrial flows in the economy, called the Interindustry Input–Output Model of the American Economy (described in the previous section), fascinated George B. Dantzig in his effort to formulate organization and production planning problems as linear models. In 1946, Dantzig was the Mathematical Advisor to the U.S. Air Force Comptroller. Challenged by his colleagues D. Hitchcock and M. Wood, he was trying to determine "what could be done to mechanize the planning process" [10, p. 79]. He soon saw the possibilities for the generalization of Leontief's model. In his reminiscences about the early days of linear programming, he writes, "In Leontief's model, there was a one-to-one correspondence between the production processes and the items produced by these processes. What was needed was a model with many *alternative activities*. The application was to be *large-scale*, hundreds of items and activities. Finally, it had to be *computable*. Once the model was formulated, there had to be a practical way to compute what quantities of these activities to engage in that was consistent with their respective input–output characteristics and with given resources" [10, p. 79]. The model he formulated was without the objective function "because practical planners simply had no way to implement such a concept" [10, p. 79].

By mid-1947, Dantzig used the linear form of the objective function to be optimized, and the resulting mathematical problem to be solved was the minimization of a linear form with respect to linear equations and inequalities. It is interesting that at first he turned to the economists, assuming they had worked on this problem. He visited T. C. Koopmans in June 1947 at the Cowles Foundation at the University of Chicago to discuss this problem from a mathematical economist's point of view. The implications for general economic planning were immediately evident, and from that time on there began a very intensive and fruitful cooperation between the economists K. Arrow, P. Samuelson, H. Simon, R. Dorfman, R. Solow, T. C. Koopmans, and L. Hurwicz and the mathematicians A. Tucker, H. Kuhn, D. Gale, and G. Dantzig,

[5] For further discussion, see [23].

to name only a few in the field of mathematical programming. It is worth noting that six of them—Arrow, Samuelson, Simon, Solow, Koopmans, and Hurwicz—later received the Nobel Prize in Economic Sciences.

But the visit of Dantzig with Koopmans was not fully successful, because economists did not have a method of solution. Therefore, Dantzig tried his own luck at finding an algorithm, and in the summer of 1947 he proposed the simplex method for "Maximization of Linear Function of Variables Subject to Linear Inequalities"—this is the title of the fundamental paper, which was circulated privately for several years and published in [22].

The term "linear programming" arose out of a discussion of Dantzig with Koopmans. Concerning the first paper by Dantzig dealing with a system of linear inequalities and called "Programming in a Linear Structure," Koopmans said, "Why not shorten 'Programming in Linear Structure' to 'Linear Programming?' I replied, 'That's it! From now on that will be its name"' [10, p. 85]. Linear programming was born and "has been one of the most important postwar developments in economic theory" [12, p. vii]. "For their contributions to the theory of optimum allocation of resources" [31, p. 213], Leonid Kantorovic and Tjalling C. Koopmans received the Nobel Prize for Economic Sciences in 1975.

B. ***Nonlinear programming***

At the meeting of the Econometric Society in Wisconsin in 1948, which was attended by well-known mathematicians and economists like H. Hotelling, Koopmans, J. von Neumann, and many others, Dantzig presented the concept of linear programming. The first comment in the discussion was Hotelling's: "But we all know the world is nonlinear" [10, p. 82]. Before Dantzig could reply, von Neumann said, "The speaker titled his talk 'Linear Programming' and he carefully stated his axioms. If you have an application that satisfies the axioms, use it. If it does not, then don't" [10, p. 82].

The utility function in the theory of the household (the model in Section 1.2.2) is based on positive but decreasing marginal utility, the neoclassical theory of the firm (the model in Section 1.2.3) with the Cobb–Douglas production function is based on positive but decreasing marginal products. The model (1.5) and the portfolio selection model from the Section 1.2.7 consist of quadratic objective function, and the input–output model with the substitution possibilities for labor and capital, according to a Cobb–Douglas production function, contains nonlinear constraints. When at least one of the functions $f_k(\mathbf{x})$ $(k = 0, 1, 2, \ldots, m)$ in problem (1.1) is nonlinear, we are faced with *nonlinear programming*.

Karush [21] appears to have been first to deal with optimization problems with inequalities as side constraints. The results of Karush did not have any significant impact at that time. Therefore, in our opinion, the starting point of the development of nonlinear programming is the paper by Kuhn and Tucker [24] on necessary and sufficient optimality conditions for solutions of nonlinear programming problems. In this context, the contribution by John [19] with his results on the same topic should be mentioned as well.

In nonlinear programming, two classes of problems should be distinguished.

B.1. ***Convex programming.*** The mathematical programming problem (1.1),

$$\begin{aligned} &\text{minimize} \quad f_0(\mathbf{x}) \\ &\text{subject to} \quad f_i(\mathbf{x}) \leqq 0 \quad (i = 1, 2, \ldots, m), \end{aligned} \tag{1.28}$$

is a convex programming problem if the functions $f_k(\mathbf{x})$ $(k = 0, 1, 2, \ldots, m)$ are convex. Obviously, the linear programming problem is a special case of convex programming, since a linear function is also convex. The convexity of the functions $f_i(\mathbf{x})$ in the constraints of problem (1.28) implies the convexity of the set of feasible solutions K. Furthermore, it is well known that the local minimum of a convex function is a global one. For these reasons, the methods with convergence to a local minimum can be used for solving problems of this class.

B.2. ***Nonconvex programming.*** Problem (1.28) is a nonconvex programming problem if either the function $f_0(\mathbf{x})$ or the set K is nonconvex. When the functions $f_i(\mathbf{x})$ $(i = 1, 2, \ldots, m)$ are nonconvex, but the set K is convex, we speak about *quasiconvex* programming.

A well-known example of nonconvex programming is the *integer programming* problem (some or all variables x_j, $j = 1, 2, \ldots, n_1$, where $n_1 \leq n$, are integer), because of the nonconvexity of the set of feasible solutions K.

With respect to the particular type of objective function $f_0(\mathbf{x})$ and of the constraints $f_i(\mathbf{x})$ (see the previous section), we can distinguish special nonlinear programming problems:

(a) *Quadratic programming*—if some of the functions $f_k(\mathbf{x})$ $(k = 0, 1, 2, \ldots, m)$ in problem (1.1) are quadratic. The problems in (1.28) with quadratic convex function $f_0(\mathbf{x})$ and linear constraints (e.g., the portfolio model in Section 1.2.7) are the simplest nonlinear programming problems.

(b) *Separable programming*—if the functions $f_k(\mathbf{x})$ $(k = 0, 1, 2, \ldots, m)$ are separable:

$$f_k(\mathbf{x}) = \sum_{j=1}^{n} f_{kj}(x_j) \quad (k = 0, 1, 2, \ldots, m).$$

(c) The input–output model with substitution possibilities for labor and capital of the previous section is a *geometric programming* problem, when all functions $f_k(x)$ are polynomials with positive coefficients (so-called *posynomials*):

$$f_k(\mathbf{x}) = \sum_{t=1}^{T_k} c_{kt} \prod_{j=1}^{n} x_j^{a_{ktj}} \quad (k = 0, 1, 2, \ldots, m),$$

where T_k is the number of terms in the polynomial k $(k = 0, 1, 2, \ldots, m)$, the coefficients c_{kt} are positive, the exponents a_{ktj} are any real numbers, and the constraints are written in the form

$$f_i(\mathbf{x}) \leqq 1 \quad (i = 1, 2, \ldots, m).$$

(d) The basic model of the data envelopment analysis is a *fractional programming* problem when the functions $f_k(\mathbf{x})$ $(k = 0, 1, 2, \ldots, m)$ are in the form

$$f_k(\mathbf{x}) = \frac{h_k(\mathbf{x})}{g_k(\mathbf{x})}.$$

With respect to the specification of the function $h_k(\mathbf{x})$ and $g_k(\mathbf{x})$, the different classes of fractional programming can be distinguished (see, e.g., [46]).

The economic implications of different mathematical programming models will be discussed in subsequent chapters.

References and Further Reading

[1] K. J. Arrow, George Dantzig in the development of economic analysis, *Discrete Optim.*, **5** (2008), 159–167.

[2] J. L. Balintfy, Applications of mathematical programming to optimize human diets, *Sympos. Math.*, **XIX** (1976), 313–339.

[3] M. J. Beckmann, Comparative statics in linear programming and the Giffen paradox, *Rev. Econ. Stud.*, **23** (1956), 232–235.

[4] G. Cassel, *Theory of Social Economy*, rev. ed., Harcourt–Brace, New York, 1932; originally published in German, 1918.

[5] A. Charnes, W. W. Cooper, and E. Rhodes, Measuring efficiency of decision making units, *Eur. J. Oper. Res.*, **2** (1978), 429–444.

[6] A. Charnes, W. W. Cooper, and L. M. Thrall, Classifying and characterizing efficiencies and inefficiencies in data envelopment analysis, *Oper. Res. Lett.*, **5**-3 (1986), 105–110

[7] H. Chenery and B. Raduchel, Substitution in planning models, in H. B. Chenery, ed., *Studies in Development Planning*, Harvard University Press, Cambridge, MA, 1971.

[8] G. B. Dantzig, *Programming in a Linear Structure*, Office of the Comptroller, United States Air Force, Washington, DC, 1948.

[9] G. B. Dantzig, Maximization of a linear function of variables subject to linear inequalities, in T. C. Koopmans, ed., *Activity Analysis of Production and Allocation*, Cowles Commission Monographs vol. 13, Wiley, New York, 1951, 330–347.

[10] G. B. Dantzig, Reminiscences about the origins of linear programming, in A. Bachem, M. Grötschel, and B. Korte, eds., *Mathematical Programming: The State of the Art: Bonn* 1982, Springer-Verlag, Berlin, 1983, 78–86.

[11] E. Dietzenbacher and M. L. Lahr, eds., *Wassily Leontief and Input–Output Economics*, Cambridge University Press, New York, 2004.

[12] R. Dorfman, P. A. Samuelson, and R. M. Solow, *Linear Programming and Economic Analysis*, McGraw–Hill, New York, 1958.

[13] R. Dorfman, Wassily Leontief's contribution to economics, *Swedish J. Econ.*, **75** (1973), 430–449.

[14] E. J. Elton and M. J. Gruber, *Modern Portfolio Theory and Investment Analysis*, Wiley, New York, 1981.

[15] M. J. Farrel, The measurement of technical efficiency, *J. R. Statist. Soc. Ser.* A, **120**-3 (1957), 253–281.

[16] D. K. Foley, An Interview with Wassily Leontief, *Macroecon. Dynam.*, **2** (1998), 116–140.

[17] D. Gale, *The Theory of Linear Economic Models*, McGraw–Hill, New York, 1960.

[18] L. Johansen, *A Multisectoral Study of Economic Growth*, North-Holland, Amsterdam, 1960.

[19] F. John, Extremum problems with inequalities as side conditions, in K. O. Fridrichs, O. E. Neugebauer, and J. J. Stoker, eds., *Studies and Essays: Courant Anniversary Volume*, Wiley, New York, 1948.

[20] L. V. Kantorovic, *Matematischeskije Metody Organizatsii i Planirovania Proizvodstva*, Leningrad State University Publishers, Leningrad, USSR, 1939 (in Russian); Mathematical methods in the organization and planning of production, *Manage. Sci.*, **6**-4 (1960), 366–422.

[21] W. Karush, *Minima of Functions of Several Variables with Inequalities as Side Conditions*, M. S. thesis, Department of Mathematics, University of Chicago, Chicago, 1939.

[22] T. C. Koopmans, ed., *Activity Analysis of Production and Allocation*, Cowles Commission Monographs vol. 13, Wiley, New York, 1951.

[23] T. C. Koopmans, Concepts of optimality and their uses, Nobel Memorial Lecture, in A. Lindbeck, ed., *Nobel Lectures in Economic Sciences* 1969–1980, World Scientific, Singapore, 1992, 239–256.

[24] H. W. Kuhn and A. W. Tucker, Nonlinear programming, in J. Neyman, ed., *Proceedings of the Second Berkeley Symposium on Mathematical Statistics and Probability*, University of California Press, Berkeley, CA, 1951, 481–492.

[25] H. D. Kurz, E. Dietzenbacher, and C. Lager, eds., *Input–Output Analysis*, Edward Elgar, Cheltenham, UK, 1998.

[26] M. L. Lahr and E. Dietzenbacher, eds., *Input–Output Analysis: Frontiers and Extensions*, Palgrave–Macmillan, New York, 2001.

[27] W. W. Leontief, Quantitative input and output relations in the economic system of the United States, *Rev. Econ. Statist.*, **18** (1936), 105–125.

[28] W. W. Leontief, *The Structure of American Economy*, 1919–1929, Harvard University Press, Cambridge, MA, 1941.

[29] W. W. Leontief, *The Structure of American Economy,* 1919–1939: *An Empirical Application of Equilibrium Analysis*, Oxford University Press, New York, 1951.

[30] W. W. Leontief, Mathematics in economics, *Bull. Amer. Math. Soc.*, **60** (1954), 215–233.

[31] A. Lindbeck, ed., *Nobel Lectures in Economic Sciences* 1969–1980, World Scientific, Singapore, 1992.

[32] M. Luptáčik, *Geometrische Programmierung und ökonomische Analyse*, Mathematical Systems in Economics vol. 32, Verlag Anton Hain, Meisenheim am Glan, Germany, 1977.

[33] H. M. Markowitz, Portfolio selection, *J. Finance*, **7** (1952), 77–91.

[34] H. M. Markowitz, *Portfolio Selection: Efficient Diversification of Investments*, Cowles Foundation Monographs vol. 16, Yale University Press, New Haven, CT, 1959.

[35] R. E. Miller and P. D. Blair, *Input–Output Analysis: Foundations and Extensions*, Prentice–Hall, Englewood Cliffs, NJ, 1985.

[36] W. Nicholson, *Microeconomic Theory: Basic Principles and Extensions*, 5th ed., Dryden Press/Harcourt–Brace–Jovanovich, New York, 1992.

[37] A. Pigou, ed., *Memorials of Alfred Marshall*, Augustus M. Kelley Publishers, London, 1925.

[38] Th. ten Raa, *Linear Analysis of Competitive Economics*, LSE Handbooks in Economics, Harvester Wheatsheaf, London, 1995.

[39] Th. ten Raa, *The Economics of Input–Output Analysis*, Cambridge University Press, New York, 2005.

[40] R. Richter, U. Schlieper, and W. Friedmann, *Makroökonomik*, 2nd ed., Springer-Verlag, Berlin, 1975.

[41] L. Robbins, *An Essay on the Nature and Significance of Economic Science*, 2nd ed., MacMillan, London, 1935.

[42] D. Ricardo, *Principles of Political Economy and Taxation*, reprint of the 3rd ed., Bell and Sons, London, 1924 (1st ed., 1817).
[43] G. J. Stigler, The cost of substance, *J. Farm Econ.*, **27** (1945), 303–314.
[44] P. Schreiner, The role of input–output in the perspective analysis of the Norwegian economy, in A. Brody and A. P. Carter, eds., *Input–Output Techniques: Proceedings of the* 5*th International Conference on Input–Output Techniques, Geneva, January* 1971, North-Holland, Amsterdam, London, 1972.
[45] J. Schumann, *Input–Output Analyse*, Springer-Verlag, Berlin, 1968.
[46] S. Schaible, A survey of fractional programming, in S. Schaible and N. T. Ziemba, eds., *Generalized Concavity in Optimization and Economics*, Academic Press, New York, 1981, 417–440.
[47] L. Walras, *Elements d'economie politique pure*, W. Jaffe, transl., George Allen/Unwin, London, 1954 (orignally published in French, 1874).

2

Kuhn–Tucker Conditions

In this chapter, necessary conditions for optimality of solution points in mathematical programming problems will be studied. Because of the orientation of this book to present optimization theory as an instrument for qualitative economic analysis, the theory to be described is not immediately concerned with computational aspects of solution techniques, which can be found in many excellent books on mathematical programming, e.g., [11, 12, 27, 23, 3].

The discussion begins with the extension of the Lagrange theory by Kuhn and Tucker [18]—note the contributions by Karush [16] and John [15]—with the derivation of necessary optimality conditions for the optimization problems including inequality constraints.

The rationality of Kuhn–Tucker conditions and their relationship to a saddle point of the Lagrangian function will be explored in Sections 2.2 and 2.3, respectively.

Section 2.4 deals with Kuhn–Tucker conditions for the general mathematical programming problem, including equality and inequality constraints, as well as nonnegative and free variables. Two numerical examples are provided for illustration.

Section 2.5 is devoted to applications of Kuhn–Tucker conditions to a qualitative economic analysis. We will show how to derive general qualitative conclusions, even when the parameters of the involved functions are not numerically specified.

2.1 The Kuhn–Tucker Theorem

The basic mathematical programming problem (1.28), as described in Chapter 1, is that of choosing values of n variables so as to minimize a function of those variables subject to m inequality constraints:

$$\begin{aligned} &\text{minimize} \quad f_0(\mathbf{x}) \\ &\text{subject to} \quad f_i(\mathbf{x}) \leqq 0 \quad (i = 1, 2, \ldots, m). \end{aligned}$$

This problem is a generalization of the classical optimization problem (which uses constraints in equation form), since equality constraints are a special case of inequality

M. Luptáčik, *Mathematical Optimization and Economic Analysis*,
Springer Optimization and Its Applications 36, DOI 10.1007/978-0-387-89552-9_2,

constraints. By introducing m additional variables, called slack variables, $y_i (i = 1, 2, \ldots, m)$, the mathematical programming problem (1.28) can be rewritten as a classical optimization problem:

$$\begin{array}{ll} \text{minimize} & f_0(\mathbf{x}) \\ \text{subject to} & f_i(\mathbf{x}) + y_i^2 = 0 \quad (i = 1, 2, \ldots, m). \end{array}$$

A characterization of the solution to the mathematical programming problem (1.28) is then analogous to the Lagrange theorem for classical optimization problems.

Under the assumption of so-called constraint qualifications (for a detailed discussion, the reader is referred to [1, 26, 37]), which was designed to avoid cusps in the feasible set, the Lagrange theory for a classical optimization problem can be extended to problem (1.28) by the following theorem.

Theorem 2.1 (see [18]). *Assume that $f_k(\mathbf{x})$ $(k = 0, 1, \ldots, m)$ are all differentiable. If the function $f_0(\mathbf{x})$ attains at point $\mathbf{x}^0$ a local minimum subject to the set $K = \{\mathbf{x} | f_i(\mathbf{x}) \leqq 0 \ (i = 1, 2, \ldots, m)\}$, then there exists a vector of Lagrange multipliers $\mathbf{u}^0$ such that the following conditions are satisfied:*

$$\frac{\partial f_0(\mathbf{x}^0)}{\partial x_j} + \sum_{i=1}^{m} u_i^0 \frac{\partial f_i(\mathbf{x}^0)}{\partial x_j} = 0 \quad (j = 1, 2, \ldots, n), \tag{2.1}$$

$$f_i(\mathbf{x}^0) \leqq 0 \quad (i = 1, 2, \ldots, m), \tag{2.2}$$

$$u_i^0 f_i(\mathbf{x}^0) = 0 \quad (i = 1, 2, \ldots, m), \tag{2.3}$$

$$u_i^0 \geqq 0 \quad (i = 1, 2, \ldots, m). \tag{2.4}$$

In other words, the conditions (2.1)–(2.4) *are necessary conditions for a local minimum of problem* (1.28). *For a maximization problem, the nonnegativity condition* (2.4) *is replaced by the nonpositivity condition $\mathbf{u}^0 \leqq \mathbf{0}$. Conditions* (2.1)–(2.4) *are called the Kuhn–Tucker conditions.*

Proof. As in the case of the classical optimization problem, the Lagrange function can be defined as a function of the original variables—in our case the variables $\mathbf{x}$ and $\mathbf{y}$—and of the Lagrange multipliers $\mathbf{u}$:

$$L(\mathbf{x}, \mathbf{y}, \mathbf{u}) = f_0(\mathbf{x}) + \sum_{i=1}^{m} u_i (f_i(\mathbf{x}) + y_i^2).$$

The necessary conditions for its local minimum are

$$\frac{\partial L}{\partial x_j} = \frac{\partial f_0(\mathbf{x}^0)}{\partial x_j} + \sum_{i=1}^{m} u_i^0 \frac{\partial (f_i(\mathbf{x}^0) + (y_i^0)^2)}{\partial x_j} = 0 \quad (j = 1, 2, \ldots, n), \tag{2.5}$$

$$\frac{\partial L}{\partial y_i} = 2u_i^0 y_i^0 = 0 \quad (i = 1, 2, \ldots, m), \tag{2.6}$$

$$\frac{\partial L}{\partial u_i} = f_i(\mathbf{x}^0) + (y_i^0)^2 = 0 \quad (i = 1, 2, \ldots, m). \tag{2.7}$$

Now it can be shown that the conditions in (2.6) correspond to the Kuhn–Tucker conditions (2.3).

Suppose $u_i^0 = 0$. Then $u_i^0 y_i^0 = u_i^0 f_i(\mathbf{x}^0) = 0$, and both conditions (2.6) and (2.3) are satisfied.

If $u_i^0 \neq 0$, then it follows from (2.6) that $y_i^0 = 0$ and therefore $(y_i^0)^2 = -f_i(\mathbf{x}^0) = 0$: Condition (2.3) is satisfied. On the other hand, it follows from (2.3) that $f_i(\mathbf{x}^0) = 0$ and therefore $y_i^0 = 0$: Condition (2.6) is fulfilled.

Since the variables y_i $(i = 1, 2, \ldots, m)$ are auxiliary variables, they can be eliminated from conditions (2.5) and (2.7), and we obtain conditions (2.1) and (2.2).

It remains to show that the Lagrange multipliers must be nonnegative. For this purpose, we consider the classical optimization problem:

$$\begin{aligned} &\text{minimize} \quad f_o(\mathbf{x}) \\ &\text{subject to} \quad f_i(\mathbf{x}) \leqq b_i \quad (i = 1, 2, \ldots, m). \end{aligned} \tag{2.8}$$

For the Lagrange multipliers u_i^0 $(i = 1, 2, \ldots, m)$ of problem (2.8), the following holds (see, e.g., [23, 1st ed., p. 231]):

$$\frac{\partial f_0(\mathbf{x}^0(\mathbf{b}))}{\partial b_i} = -u_i^0 \quad (i = 1, 2, \ldots, m), \tag{2.9}$$

where $\mathbf{x}^0$ denotes the optimal solution of problem (2.8). Hence the Lagrange multipliers u_i^0 $(i = 1, 2, \ldots, m)$ give us the change of the value of the objective function due to a change of the constraint b_i by a small amount. A higher value of the ith component of the vector $\mathbf{b}$ implies an enlargement of the set K. Therefore, the new optimal value of the objective function $f_0(\mathbf{x})$ cannot be worse:

$$\frac{\partial f_0(x^0)}{\partial b_i} \leqq 0 \quad \text{for a minimization problem} \tag{2.10}$$

and

$$\frac{\partial f_0(\mathbf{x}^0)}{\partial b_i} \geqq 0 \quad \text{for a maximization problem.} \tag{2.11}$$

The nonnegativity condition for the Lagrange multipliers (2.4) follows from (2.9) and (2.10). Similarly, conditions in (2.9) and (2.11) imply that the Lagrange multipliers cannot be positive for problem (1.28) with the objective function to be maximized. □

For a geometric interpretation of the Kuhn–Tucker conditions (2.1)–(2.4), we rewrite the conditions in (2.1) as follows:

$$\frac{\partial f_0(\mathbf{x}^0)}{\partial x_j} = -\sum_{i=1}^{m} u_i^0 \frac{\partial f_i(\mathbf{x}^0)}{\partial x_j} \quad (j = 1, 2, \ldots, n),$$

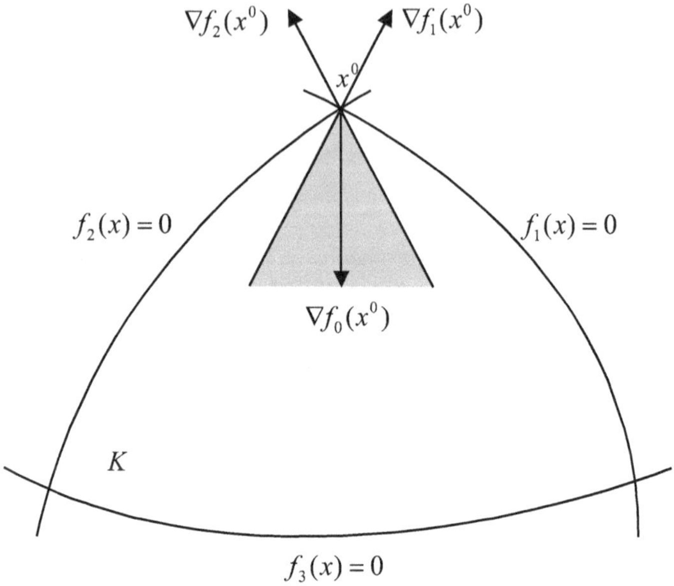

Fig. 2.1. Kuhn–Tucker conditions.

or

$$\nabla f_0(\mathbf{x}^0) = -\sum_{i=1}^{m} u_i^0 \nabla f_i(\mathbf{x}^0),$$

where $\nabla f_0(\mathbf{x})$ denotes the gradient vector (the vector of first-order partial derivatives) of the objective function, and $\nabla f_i(\mathbf{x})$ is the gradient vector of the ith constraint function $(i = 1, 2, \ldots, m)$. Thus the gradient of the objective function must, at the optimal solution, be a nonpositive weighted combination of the gradients of the active constraints (the constraints satisfied at the optimal solution as equalities). The gradient vector of the objective function must therefore lie within the cone spanned by the inward-pointing normals to the opportunity set at $\mathbf{x}^0$. This solution is illustrated in Figure 2.1 for the problem in which $n = 2$, $m = 3$.

Using the Lagrange function (without slack variables) for the mathematical programming problem (1.28),

$$\Phi(\mathbf{x}, \mathbf{u}) = f_0(\mathbf{x}) + \sum_{i=1}^{m} u_i f_i(\mathbf{x}), \tag{2.12}$$

the Kuhn–Tucker conditions (2.1)–(2.4) can be rewritten as follows:

$$\frac{\partial \Phi(\mathbf{x}^0, \mathbf{u}^0)}{\partial x_j} = 0 \quad (j = 1, 2, \ldots, n), \tag{2.1$'$}$$

$$\frac{\partial \Phi(\mathbf{x}^0, \mathbf{u}^0)}{\partial u_i} \leqq 0 \quad (i = 1, 2, \ldots, m), \tag{2.2$'$}$$

$$u_i^0 \frac{\partial \Phi(\mathbf{x}^0, \mathbf{u}^0)}{\partial u_i} = 0 \quad (i = 1, 2, \ldots, m), \tag{2.3$'$}$$

$$u_i^0 \geqq 0 \quad (i = 1, 2, \ldots, m). \tag{2.4$'$}$$

The n conditions in (2.1′) are the same as in the classical programming case, or in other words, as in the traditional Lagrange theory from classical differential calculus.

The m conditions in (2.2′) are the constraints of the mathematical programming problem which permits solution at the boundary of the set of feasible solutions or at an interior point of this set.

The m conditions in (2.3′), which are known as the complementary slackness conditions of mathematical programming, serve essentially to determine which of the two regimes will apply: whether the boundary or the interior minimum point occurs. If the ith constraint is not binding (an interior point), then the corresponding Lagrange multiplier will be zero. If the multiplier u_i is positive, then the corresponding ith constraint is binding (boundary solution). The reader should bear in mind that the converse is not true.

The m conditions in (2.4′), requiring that the Lagrange multipliers be nonnegative, stem from the fact that the constraints in (2.2′) are written as inequalities rather than as equalities; if a constraint is an equality, then the corresponding element of $\mathbf{u}^0$ is unrestricted, as in the classical programming case.

In most of the models of mathematical programming in economics (see Chapter 1), nonnegativity conditions are required. Obviously, it would be possible to include nonnegativity conditions in the set of constraints $f_i(\mathbf{x}) \leqq 0$ $(i = 1, 2, \ldots, m)$. But as we will show now, the Lagrange multipliers corresponding to the nonnegativity conditions can be eliminated. It is therefore useful to consider the nonnegativity conditions separately.

We consider the following mathematical programming problem:

$$\begin{aligned} \text{minimize} \quad & f_0(\mathbf{x}) \\ \text{subject to} \quad & f_i(\mathbf{x}) \leqq 0 \quad (i = 1, 2, \ldots, m), \\ & -x_j \leqq 0 \quad (j = 1, 2, \ldots, n). \end{aligned} \tag{1.28a}$$

First, we write the Lagrange function for problem (1.28a):

$$\Psi(\mathbf{x}, \mathbf{u}, \mathbf{w}) = f_0(\mathbf{x}) + \sum_{i=1}^{m} u_i f_i(\mathbf{x}) + \sum_{j=1}^{n} w_j(-x_j).$$

Then according to Theorem 2.1, the Kuhn–Tucker conditions become

$$\frac{\partial \psi}{\partial x_j} = \frac{\partial f_0(\mathbf{x}^0)}{\partial x_j} + \sum_{i=1}^{m} u_i^0 \frac{\partial f_i(\mathbf{x}^0)}{\partial x_j} - w_j^0 = 0 \qquad (j = 1, 2, \ldots, n),$$

or, equivalently,

$$\frac{\partial f_0(\mathbf{x}^0)}{\partial x_j} + \sum_{i=1}^{m} u_i^0 \frac{\partial f_i(\mathbf{x}^0)}{\partial x_j} = w_j^0 \quad (j = 1, 2, \ldots, n). \tag{2.13}$$

Furthermore,

$$\frac{\partial \psi}{\partial u_i} = f_i(\mathbf{x}^0) \leqq 0 \qquad (i = 1, 2, \ldots, m), \tag{2.14}$$

$$u_i^0 \frac{\partial \psi}{\partial u_i} = u_i^0 f_i(\mathbf{x}^0) = 0 \qquad (i = 1, 2, \ldots, m), \tag{2.15}$$

$$\frac{\partial \psi}{\partial w_j} = -x_j^0 \leqq 0 \qquad (j = 1, 2, \ldots, n), \tag{2.16}$$

$$w_j^0 \frac{\partial \psi}{\partial w_j} = w_j^0(-x_j^0) = 0 \qquad (j = 1, 2, \ldots, n),$$

which, because of (2.13), can be rewritten as

$$x_j^0 \left(\frac{\partial f_0(\mathbf{x}^0)}{\partial x_j} + \sum_{i=1}^{m} u_i^0 \frac{\partial f_i(\mathbf{x}^0)}{\partial x_j} \right) = 0 \qquad (j = 1, 2, \ldots, n), \tag{2.17}$$

$$u_i^0 \geqq 0 \qquad (i = 1, 2, \ldots, m), \tag{2.18}$$

$$w_j^0 \geqq 0 \qquad (j = 1, 2, \ldots, n). \tag{2.19}$$

Using the Lagrange function (2.12), the Kuhn–Tucker conditions (2.13)–(2.19) can be summarized symmetrically with respect to $\mathbf{x}$ and $\mathbf{u}$ as

$$\frac{\partial \Phi(\mathbf{x}^0, \mathbf{u}^0)}{\partial \mathbf{x}} \geqq \mathbf{0}, \tag{2.20}$$

$$\mathbf{x}^0 \frac{\partial \Phi(\mathbf{x}^0, \mathbf{u}^0)}{\partial \mathbf{x}} = 0, \tag{2.21}$$

$$\mathbf{x}^0 \geqq \mathbf{0}, \tag{2.22}$$

$$\frac{\partial \Phi(\mathbf{x}^0, \mathbf{u}^0)}{\partial \mathbf{u}} \leqq \mathbf{0}, \tag{2.23}$$

$$\mathbf{u}^0 \frac{\partial \Phi(\mathbf{x}^0, \mathbf{u}^0)}{\partial \mathbf{u}} = 0, \tag{2.24}$$

$$\mathbf{u}^0 \geqq \mathbf{0}. \tag{2.25}$$

The reader will note that in the case of nonnegativity conditions for the variables $\mathbf{x}$, condition (2.1) of the Kuhn–Tucker theorem has been replaced by two sets of conditions (2.20)–(2.21). An intuitive explanation of this matter will be given in the next section.

2.2 Rationale of the Kuhn–Tucker Conditions

As already mentioned, the Kuhn–Tucker conditions are the natural generalization of the Lagrange multiplier approach, from classical differential calculus replacing equality constraints by inequality constraints, to take account of the possibility that the maximum or minimum in question can occur not only at a boundary point but also at an interior point. The calculus requirements are generally appropriate only if the extremum (i.e., the maximum or minimum) occurs at a point at which all of the variables (including the slack variables) take nonzero values.

Now we consider—for simplicity, but without loss of generality—the minimization of the function $f(x)$ subject to $x \geqq 0$. In this case, the matter can be illustrated graphically. Suppose first that we are at a point at which the value of x can either be increased or decreased (the interior point A in Figure 2.2). By the usual logic of marginal analysis, we must have $\frac{df}{dx} = 0$, for otherwise either a rise or a fall in the value of x could increase the value of f, and f would not be at its minimum.

On the other hand, suppose we are testing for the possibility of a boundary minimum at which $x = 0$. In Figure 2.2, two possibilities for local minimum of the function $f(x)$ subject to $x \geqq 0$ can be observed. If $\frac{df}{dx} = 0$, the point with $x = 0$ (point B in Figure 2.2) may be a minimum for the usual reasons, and if $\frac{df}{dx} > 0$, it may be a minimum point simply because it is impossible to reduce the value of x any further (point C in Figure 2.2).

Direct generalization for the function with n variables leads to the following conclusions. Given a differentiable function $f(x_1, x_2, \ldots, x_n)$,

- for an interior minimum (maximum), it is necessary that $\frac{\partial f}{\partial x_j} = 0\ (j = 1, 2, \ldots, n)$;
- for a boundary minimum, it is necessary that $\frac{\partial f}{\partial x_j} \geqq 0\ (j = 1, 2, \ldots, n)$.

The reader may check that—by the same reasoning—for a boundary maximum it is necessary that $\frac{\partial f}{\partial x_j} \leqq 0$.

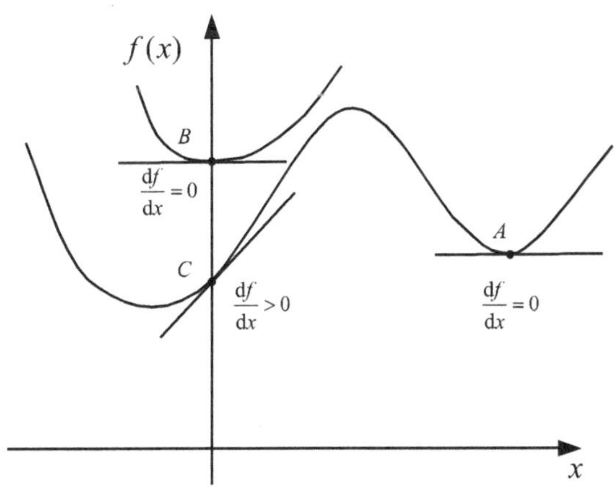

Fig. 2.2. Minimum of the function $f(x)$ subject to $x \geqq 0$.

Similar to the interpretation of the complementary slackness conditions (2.3′) or (2.24), the conditions in (2.21) serve to determine which solution case occurs; if the value of x_j under consideration is positive (interior minimum case), then (2.21) requires $\frac{\partial \Phi}{\partial x_j} = 0$. If $\frac{\partial \Phi}{\partial x_j} > 0$, then we can only have a boundary minimum ($x_j = 0$).

2.3 Kuhn–Tucker Conditions and a Saddle Point of the Lagrange Function

We consider the Lagrange function $\Phi(\mathbf{x}, \mathbf{u})$ as defined in (2.12). The necessary conditions for a local minimum of the Lagrange function (2.12), regarded as a function of $\mathbf{x}$ only, subject only to the nonnegativity conditions $x_j \geqq 0$ ($j = 1, 2, \ldots, n$) are exactly the Kuhn–Tucker conditions (2.20)–(2.22) for problem (1.28a). At the same time, the Kuhn–Tucker conditions (2.23)–(2.25) provide the necessary conditions for a local maximum of the Lagrange function (2.12), regarded as a function of $\mathbf{u}$ only, subject only to the nonnegativity conditions $u_i \geqq 0$ ($i = 1, 2, \ldots, m$). A graphical illustration of this property of the point $(\mathbf{x}^0, \mathbf{u}^0)$ from the Kuhn–Tucker conditions (2.20)–(2.25) is depicted in Figure 2.3. This leads to the following concept of a saddle point.

Definition 2.1. A point $(\mathbf{x}^0, \mathbf{u}^0)$ with $\mathbf{x}^0 \geqq \mathbf{0}$ and $\mathbf{u}^0 \geqq \mathbf{0}$ is said to be a *saddle point* of the Lagrange function $\Phi(\mathbf{x}, \mathbf{u})$ if

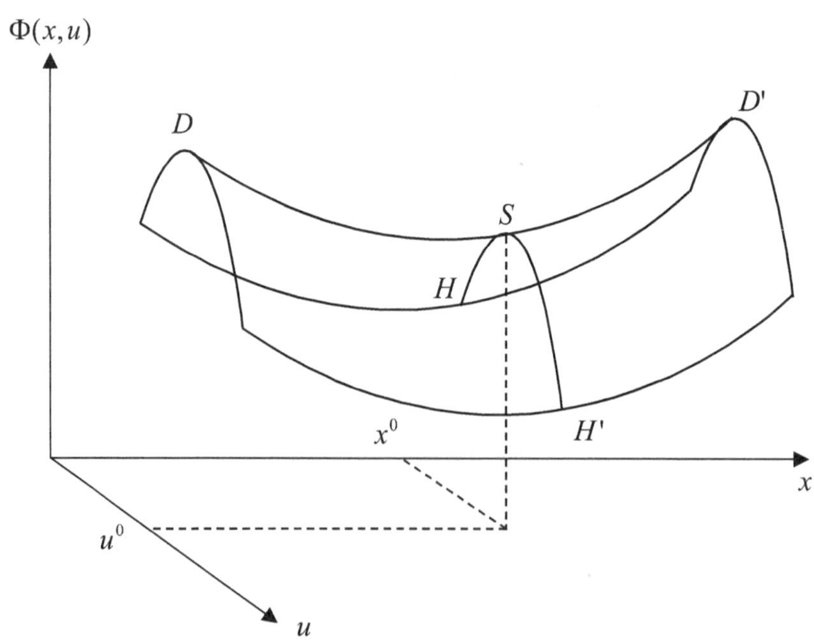

Fig. 2.3. Saddle point of the Lagrange function.

$$\Phi(\mathbf{x}^0, \mathbf{u}) \leqq \Phi(\mathbf{x}^0, \mathbf{u}^0) \leqq \Phi(\mathbf{x}, \mathbf{u}^0)$$

for all $\mathbf{x} \geqq \mathbf{0}, \mathbf{u} \geqq \mathbf{0}$.

In other words, for a fixed $\mathbf{u} = \mathbf{u}^0$, the Lagrange function is minimized at $\mathbf{x}^0$ (due to the second inequality of the relationship in Definition 2.1), whereas for a fixed $\mathbf{x} = \mathbf{x}^0$, the Lagrange function is maximized at $\mathbf{u}^0$ (which follows from the first inequality of the relationship in Definition 2.1).

Now the relationship between the saddle point of the Lagrange function and the optimal solution of problem (1.28a) can be established.

Theorem 2.2. *If there exists a saddle point* $(\mathbf{x}^0, \mathbf{u}^0)$ *of the Lagrange function* $\Phi(\mathbf{x}, \mathbf{u})$, *then* $\mathbf{x}^0$ *is an optimal solution for problem* (1.28a).[1]

In order to obtain the converse of Theorem 2.2, we need convexity properties of the functions $f_k(\mathbf{x})$ $(k = 0, 1, 2, \ldots, m)$, which will be discussed in the next chapter.

2.4 Kuhn–Tucker Conditions for the General Mathematical Programming Problem

The real applications of mathematical programming in economics contain both types of constraints: inequalities as well as equalities. Therefore, we define the general mathematical programming problem as follows:

$$\begin{aligned} \text{minimize} \quad & f_0(\mathbf{x}, \mathbf{y}) \\ \text{subject to} \quad & f_i(\mathbf{x}, \mathbf{y}) \leqq 0 \quad (i = 1, 2, \ldots, m), \\ & g_h(\mathbf{x}, \mathbf{y}) = 0 \quad (h = m+1, \ldots, r), \\ & \mathbf{x} \geqq \mathbf{0}, \\ & \mathbf{y} \in \mathbb{R}^l. \end{aligned} \tag{1.28b}$$

Obviously, problems (1.28) and (1.28a) are special cases of problem (1.28b).

Writing problem (1.28b) in the form (1.28) with $\Phi(\mathbf{x}, \mathbf{y}, \mathbf{u}, \mathbf{v}) = f_0(\mathbf{x}, \mathbf{y}) + \sum_{i=1}^{m} u_i f_i(\mathbf{x}, \mathbf{y}) + \sum_{h=m+1}^{r} v_h g_h(\mathbf{x}, \mathbf{y})$, the reader may verify that the Kuhn–Tucker conditions take the symmetric form

$$\frac{\partial \Phi^0}{\partial \mathbf{x}} \geqq \mathbf{0}, \qquad \frac{\partial \Phi^0}{\partial \mathbf{y}} = \mathbf{0}, \qquad \frac{\partial \Phi^0}{\partial \mathbf{u}} \leqq \mathbf{0}, \qquad \frac{\partial \Phi^0}{\partial \mathbf{v}} = \mathbf{0},$$

$$\mathbf{x}^0 \frac{\partial \Phi^0}{\partial \mathbf{x}} = 0, \qquad \mathbf{u}^0 \frac{\partial \Phi}{\partial \mathbf{u}} = 0, \qquad \mathbf{x}^0 \geqq \mathbf{0}, \qquad \mathbf{u}^0 \geqq \mathbf{0},$$

where $\Phi^0 = \Phi(\mathbf{x}^0, \mathbf{y}^0, \mathbf{u}^0, \mathbf{v}^0)$, $(\mathbf{x}^0, \mathbf{y}^0)$ denotes the local minimum of the function $f_0(\mathbf{x}, \mathbf{y})$ under the constraints of problem (1.28b), and $(\mathbf{u}^0, \mathbf{v}^0)$ are the corresponding

[1] For the proof, see, e.g., [24, pp. 215–217] or [10, p. 539].

Lagrange multipliers. It is worth noting that the Lagrange multipliers **v** related to the equalities are not restricted to the nonnegativity (as in the classical Lagrange theory).

A summary of the rules for the formulation of the Kuhn–Tucker conditions for the general mathematical programming problem is as follows:

Rule 1. For a minimization (maximization) problem write all inequality constraints in the form

$$f_i(x) \leqq 0 \quad (f_i(x) \geqq 0).$$

Rule 2. Write the Lagrange function as the sum of the objective function and the weighted constraints.

Rule 3. The partial derivatives of the Lagrange function

(a) with respect to the nonnegative variables are *nonnegative* (*nonpositive*) for a *minimization* (*maximization*) problem and the complementary slackness condition

$$\mathbf{x}\frac{\partial \Phi}{\partial \mathbf{x}} = 0$$

is fulfilled;

(b) with respect to the free variables are equal to zero;

(c) with respect to the Lagrange multipliers corresponding to the inequality constraints are *nonpositive* (*nonnegative*) for a *minimization* (*maximization*) problem and the complementary slackness condition

$$\mathbf{u}\frac{\partial \Phi}{\partial \mathbf{u}} = 0$$

is fulfilled;

(d) with respect to the Lagrange multipliers corresponding to the equality constraints are equal to zero.

For a numerical illustration, we consider the following example:

$$\begin{aligned} \text{minimize} \qquad & f_0(\mathbf{x}) = x_1^2 - 4x_1 + x_2^2 - 6x_2 \\ \text{subject to} \qquad & x_1 + x_2 \leqq 3, \\ & -2x_1 + x_2 \leqq 2. \end{aligned}$$

The Lagrange function is

$$\Phi(\mathbf{x}, \mathbf{u}) = x_1^2 - 4x_1 + x_2^2 - 6x_2 + u_1(x_1 + x_2 - 3) + u_2(-2x_1 + x_2 - 2).$$

Application of the Kuhn–Tucker conditions (2.1′)–(2.4′) gives

$$\frac{\partial \Phi}{\partial x_1} = 2x_1 - 4 + u_1 - 2u_2 = 0, \tag{2.26}$$

$$\frac{\partial \Phi}{\partial x_2} = 2x_2 - 6 + u_1 + u_2 = 0, \tag{2.27}$$

$$\frac{\partial \Phi}{\partial u_1} = x_1 + x_2 - 3 \leqq 0, \tag{2.28}$$

$$u_1 \frac{\partial \Phi}{\partial u_1} = u_1(x_1 + x_2 - 3) = 0, \tag{2.29}$$

$$\frac{\partial \Phi}{\partial u_2} = -2x_1 + x_2 - 2 \leqq 0, \tag{2.30}$$

$$u_2 \frac{\partial \Phi}{\partial u_2} = u_2(-2x_1 + x_2 - 2) = 0, \tag{2.31}$$

$$u_1 \geqq 0, \qquad u_2 \geqq 0. \tag{2.32}$$

There is in general no simple computational procedure for the solution of these conditions. In order to show how to use the Kuhn–Tucker conditions, it is necessary to explore various cases defined principally by reference to whether each u_i is zero.

For the first case, suppose that $u_1 = 0$ and $u_2 = 0$. From conditions (2.26)–(2.27), we get $x_1 = 2$ and $x_2 = 3$. This vector cannot be a solution of our problem because it violates the first constraint $x_1 + x_2 \leqq 3$.

Second, suppose that $u_1 \neq 0$ and $u_2 = 0$. Then equations (2.26) and (2.27) are reduced to

$$2x_1 + u_1 = 4,$$
$$2x_2 + u_1 = 6.$$

Due to the complementary slackness condition (2.29), inequality (2.28) must be fulfilled as the equality

$$x_1 + x_2 = 3.$$

The above system of equations yields the solution $x_1 = 1$, $x_2 = 2$, and $u_1 = 2$, which also satisfies the remaining conditions (2.30)–(2.31). In other words, all Kuhn–Tucker conditions (2.26)–(2.32) are satisfied.

The third case corresponds to $u_1 = 0$, $u_2 \neq 0$. The resulting system of equations,

$$2x_1 - 2u_2 = 4,$$
$$2x_2 + u_2 = 6,$$
$$-2x_1 + x_2 = 2,$$

yields the solution $x_1 = \frac{4}{5}$, $x_2 = \frac{18}{5}$, $u_2 = -\frac{6}{5}$, which violates conditions (2.28) and (2.32).

The last possibility is $u_1 \neq 0$ and $u_2 \neq 0$. Because of the complementary slackness conditions (2.29) and (2.31), both inequality constraints (2.28) and (2.30) must be satisfied as equalities:

$$x_1 + x_2 = 3,$$
$$-2x_1 + x_2 = 2.$$

The solution is $x_1 = \frac{1}{3}$ and $x_2 = \frac{8}{3}$. Substituting these values in (2.26)–(2.27), we obtain a negative value for the Lagrange multiplier $u_2 = -\frac{8}{9}$, which is a contradiction to condition (2.32).

Only the values $x_1 = 1$, $x_2 = 2$, $u_1 = 2$, and $u_2 = 0$ satisfy all Kuhn–Tucker conditions and a simple inspection of the graph of the feasible solutions illustrates that this is indeed the optimal solution of our example.

Without further assumption about the functions $f_k(\mathbf{x})$ $(k = 0, 1, 2, \ldots, m)$, Theorem 2.1 provides only necessary conditions for a local optimal solution of problem (1.28).

In order to illustrate that the Kuhn–Tucker conditions are not sufficient conditions for a local minimum (maximum) of mathematical programming problems, we consider the following very simple one-variable example:

$$\text{maximize} \quad f_0(x) = (x-1)^3 \tag{2.33}$$
$$\text{subject to} \quad x \leqq 2, \tag{2.34}$$
$$x \geqq 0. \tag{2.35}$$

According to Rule 1 for the formulation of the Kuhn–Tucker condition, we rewrite the constraint as $2 - x \geqq 0$. Then the Lagrange function is

$$\Phi(x, u) = (x-1)^3 + u(2-x).$$

Application of the Kuhn–Tucker conditions (2.20)–(2.25) for the maximization problem gives

$$\frac{\partial \Phi}{\partial x} = 3(x-1)^2 - u \leqq 0, \tag{2.36}$$
$$x\frac{\partial \Phi}{\partial x} = x[3(x-1)^2 - u] = 0, \tag{2.37}$$
$$\frac{\partial \Phi}{\partial u} = 2 - x \geqq 0, \tag{2.38}$$
$$u\frac{\partial \Phi}{\partial u} = u(2-x) = 0, \tag{2.39}$$
$$u \geqq 0 \quad \text{(because of Rule 1)}. \tag{2.40}$$

The reader may verify that $x^0 = 1$ and $u^0 = 0$ satisfy the Kuhn–Tucker conditions (2.36)–(2.40). By simple inspection, it can be shown that the maximum of function (2.33) under the constraints (2.34)–(2.35) is at the point $x = 2$ and not at the point $x^0 = 1$.

The question of sufficiency of the Kuhn–Tucker conditions or the "second-order conditions" for the optimal solution of mathematical programming problems will be explored in the next chapter.

2.5 The Kuhn–Tucker Conditions and Economic Analysis

As illustrated in the previous section, the Kuhn–Tucker conditions can be helpful in the solution of specific numerical problems. Many algorithms of quadratic programming are based on these conditions. For economists, the Kuhn–Tucker conditions can be more useful for derivation of qualitative results without the necessity of specifying numerically the parameters of a mathematical programming problem. The primary aim is to characterize the optimal behavior of an economic agent under consideration. "As a result the Kuhn–Tucker conditions may perhaps constitute the most powerful single weapon provided to economic theory by mathematical programming" [4, p. 165].

A few examples will illustrate how the Kuhn–Tucker conditions can be used as an instrument for qualitative economic analysis.

2.5.1 Peak Load Pricing

Many profit-maximizing firms are confronted with the situation that the demand for a given product varies by the hour of the day so that at some times the capacity of the firm is fully utilized (peak periods), while at other times demand is slow so that some capacity remains underutilized (off-peak periods). As shown by Littlechild [20] with the aid of Kuhn–Tucker analysis and previously formulated by Steiner [35] and Williamson [38], in such situations the differential pricing is—in the sense of profit maximization—optimal. According to [4, p. 167], the following proposition can be formulated.

Proposition 2.1. *The profit-maximizing outputs will be such that prices at off-peak periods will merely cover marginal operating costs (raw materials, labor, etc.), while in peak periods the prices will exceed marginal operating costs. The sum of the excesses of these prices over marginal operating costs for all peak periods will just add up to marginal capital cost, i.e., they will sum to the marginal cost of increasing capacity.*

Proof. We denote the quantity demanded during each of the 24 hours of the day by $x_1, x_2, \ldots, x_{24}$ and the corresponding prices (e.g., telephone rates) by $p_1, p_2, \ldots, p_{24}$. It is assumed that all $x_i > 0$, i.e., that some output is sold during each hour of the day. The hourly output capacity is denoted by y. The function $C(x_1, x_2, \ldots, x_{24})$ describes the daily total operating cost and $g(y)$ the daily cost of capital (capacity). We assume that the marginal operating cost, $\frac{\partial C}{\partial x_i}$, as well as the marginal capacity cost, $\frac{dg}{dy}$, are positive. Furthermore, it is assumed that prices are not affected by the firm's output, i.e., $\frac{\partial p_i}{\partial x_i} = 0$. (Perfect competition prevails. The prices $p_1, \ldots, p_{24}$ can therefore be regarded as given and fixed.)

The firm seeks to maximize the total profit per day,

$$\pi = \sum_{i=1}^{24} p_i x_i - C(x_1, x_2, \ldots, x_{24}) - g(y),$$

subject to the 24 hourly capacity constraints,

$$x_i \leqq y \quad (i = 1, 2, \ldots, 24),$$
$$x_i \geqq 0 \quad (i = 1, 2, \ldots, 24),$$

$$y \geqq 0.$$

The Lagrange function has the form

$$\Phi(\mathbf{x}, y, \mathbf{u}) = \sum_{i=1}^{24} p_i x_i - C(x_1, x_2, \ldots, x_{24}) - g(y) + \sum_{i=1}^{24} u_i (y - x_i).$$

Under the assumption of a perfectly competitive firm, the Kuhn–Tucker conditions are then

$$\frac{\partial \Phi}{\partial x_i} = p_i - \frac{\partial C}{\partial x_i} - u_i \leqq 0 \quad (i = 1, 2, \ldots, 24), \tag{2.41}$$

$$x_i \frac{\partial \Phi}{\partial x_i} = x_i \left(p_i - \frac{\partial C}{\partial x_i} - u_i \right) = 0 \quad (i = 1, 2, \ldots, 24), \tag{2.42}$$

$$\frac{\partial \Phi}{\partial y} = -\frac{dg}{dy} + \sum_{i=1}^{24} u_i \leqq 0, \tag{2.43}$$

$$y \frac{\partial \Phi}{\partial y} = y \left(-\frac{dg}{dy} + \sum_{i=1}^{24} u_i \right) = 0, \tag{2.44}$$

$$\frac{\partial \Phi}{\partial u_i} = y - x_i \geqq 0 \quad (i = 1, 2, \ldots, 24), \tag{2.45}$$

$$u_i \frac{\partial \Phi}{\partial u_i} = u_i (y - x_i) = 0 \quad (i = 1, 2, \ldots, 24), \tag{2.46}$$

$$u_i \geqq 0 \quad (i = 1, 2, \ldots, 24). \tag{2.47}$$

Since we have assumed that $x_i > 0$ $(i = 1, 2, \ldots, 24)$, it follows from (2.45) that $y > 0$ (that is, if capacity, y, were zero, nothing could be produced).

Because we are only interested in solutions in which all x_i and y are positive, (2.41) and (2.43) become the following:

$$p_i - \frac{\partial C}{\partial x_i} - u_i = 0 \quad (i = 1, 2, \ldots, 24), \tag{2.41$'$}$$

$$-\frac{\partial g}{\partial y} + \sum_{i=1}^{24} u_i = 0. \tag{2.43$'$}$$

In any off-peak period t, there is by definition excess capacity $(y > x_t)$. Therefore, by the complementary slackness condition (2.46), we must have $u_t = 0$ for off-peak periods.

Then the first part of Proposition 2.1 follows immediately from (2.41′):

$$p_t = \frac{\partial C}{\partial x_t} \quad \text{for any off-peak period } t;$$

that is, for any off-peak period, it is optimal to set the price equal to the marginal operating cost, $\frac{\partial C}{\partial x_t}$. Since there is excess capacity, demand should be encouraged by charging a price as low as possible without incurring a loss on the marginal unit sold.

For any peak period, s, the capacity of the firm is fully utilized ($x_s = y$). Since we have assumed that $\frac{dg}{dy} > 0$ (increasing output capacity requires additional capital), it follows from (2.43′) that

$$\frac{dg}{dy} = \sum_s u_s > 0,$$

that is, at least for some of the peak periods, the Lagrange multipliers must be positive. Then we obtain from (2.41′) that

$$p_s = \frac{\partial C}{\partial x_s} + u_s \quad \text{for any peak period } s.$$

The price will exceed the marginal operating cost by a supplementary amount equal to the value of the Lagrange multiplier u_s. Moreover, by (2.43′) the sum of these supplements for all peak periods together will be exactly equal to the marginal capacity cost, $\frac{\partial g}{\partial y}$. Since peak period demand presses on capacity, any increase in this demand must require additional capital, and it must therefore cover its marginal capital cost, $\frac{dg}{dy}$. □

This completes the proof of Proposition 2.1 as the basic principles for the setting of daytime and evening telephone rates, for the higher accommodation prices in the peak season, etc. This principle can be applied in the recent discussion about the road pricing system as well.

2.5.2 Revenue Maximization under a Profit Constraint[2]

Suppose that a firm produces a single product whose output is q and that its sales are affected by its advertising expenditure a. The firm will maximize its total revenue $R(q,a)$ subject to a profit constraint,

$$\Pi = R(q,a) - C(q) - a \geqq m,$$

where $C(q)$ indicates the cost of production and where the marginal revenue of advertising and the marginal cost of output are both positive ($\frac{\partial R}{\partial a} > 0$, $\frac{dC}{dq} > 0$). Then the behavior of the firm is described by the following.

Proposition 2.2. *The revenue-maximizing output will be such that the profit is equal to the prescribed level m, the marginal revenue $\frac{\partial R}{\partial q}$ is positive, and the marginal profit $\frac{\partial \Pi}{\partial q}$ is negative.*

Proof. The firm's decision problem is

$$\begin{aligned} \text{maximize} \quad & R(q,a) \\ \text{subject to} \quad & R(q,a) - C(q) - a \geqq m, \\ & q \geqq 0, \qquad a \geqq 0. \end{aligned}$$

[2] Formulation of the problem by Baumol [4, p. 170].

The Lagrange function becomes

$$\Phi(q,a,u) = R(q,a) + u(R(q,a) - C(q) - a - m),$$

and the Kuhn–Tucker conditions are

$$\frac{\partial \Phi}{\partial q} = \frac{\partial R}{\partial q} + u\left(\frac{\partial R}{\partial q} - \frac{dC}{dq}\right) \leqq 0,$$

or

$$(1+u)\frac{\partial R}{\partial q} - u\frac{dC}{dq} \leqq 0, \tag{2.48}$$

$$q\frac{\partial \Phi}{\partial q} = q\left[(1+u)\frac{\partial R}{\partial q} - u\frac{dC}{dq}\right] = 0, \tag{2.49}$$

$$\frac{\partial \Phi}{\partial a} = \frac{\partial R}{\partial a} + u\frac{\partial R}{\partial a} - u \leqq 0,$$

or

$$(1+u)\frac{\partial R}{\partial a} \leqq u,$$

or

$$\frac{\partial R}{\partial a} \leqq \frac{u}{1+u}, \tag{2.50}$$

$$a\frac{\partial \Phi}{\partial a} = a\left[(1+u)\frac{\partial R}{\partial a} - u\right] = 0, \tag{2.51}$$

$$\frac{\partial \Phi}{\partial u} = R(q,a) - C(q) - a - m \geqq 0, \tag{2.52}$$

$$u\frac{\partial \Phi}{\partial u} = u\,[R(q,a) - C(q) - a - m] = 0, \tag{2.53}$$

$$u \geqq 0. \tag{2.54}$$

Assuming $q > 0$ in the solution, condition (2.48) can be written as

$$\frac{\frac{\partial R}{\partial q}}{\frac{dC}{dq}} = \frac{u}{1+u}. \tag{2.48$'$}$$

Since we have assumed $\frac{\partial R}{\partial a} > 0$, it follows from (2.50) and (2.54) that $u > 0$. The complementary slackness condition (2.53) then implies $\Pi = m$. Taking into account our assumption that $\frac{dC}{dq} > 0$, it follows from (2.48′) that the marginal revenue $\frac{\partial R}{\partial q}$ is positive and smaller than the marginal cost $\frac{dC}{dq}$. Therefore, the marginal profit with respect to output, $\frac{\partial \Pi}{\partial q} = \frac{\partial R}{\partial q} - \frac{dC}{dq}$, must be negative. □

From the economic interpretation point of view, it is interesting to compare the obtained results with the results for a profit-maximizing firm. The reader may verify that the necessary condition for profit-maximizing output is that the marginal revenue is equal to the marginal cost. In our model, at the constrained revenue-maximizing output, the marginal revenue is lower than the marginal cost. The implication of this result for the linear revenue function

$$R(q,a) = \alpha_1 q + \alpha_2 a \quad \text{with } \alpha_1 > 0, \quad \alpha_2 > 0$$

and the quadratic cost function

$$C(q) = cq^2 \quad \text{with } c > 0^3$$

is that the constrained revenue-maximizing output from our model is higher than the profit-maximizing output. The optimal solution q_π for the profit-maximizing firm follows directly from the condition

$$\frac{\partial R}{\partial q} = \alpha_1 = \frac{dC}{dq} = 2cq,$$

i.e.,

$$q_\pi = \frac{\alpha_1}{2c}.$$

For the revenue-maximizing firm, the necessary condition for the optimal solution becomes

$$\frac{\partial R}{\partial q} = \alpha_1 = \frac{u}{1+u}\frac{dC}{dq} = \frac{u}{1+u}2cq,$$

and consequently

$$q_R = \frac{\alpha_1}{2c}\left(\frac{1+u}{u}\right) > \frac{\alpha_1}{2c} = q_\pi.$$

2.5.3 Behavior of the Firm under Regulatory Constraint

The regulation of monopolies is an important subject in applied economic analysis. In the sectors with network structure, such as telecommunications, electricity and gas, and railway systems with high fixed and irreversible (sunk) costs, it is cheaper to produce goods by a single firm than by many firms. These situations are called *natural monopolies* and occur whenever the average costs of production for a single firm are declining over a broad range of output levels. The reason lies in the so-called "bundling advantage": With increasing diameter of the pipe, the volume increases

[3] It can be shown that in this case the second-order conditions are fulfilled.

more rapidly than its girth, which is crucial for the costs. The average costs of production fall as the scale of production increases; we say there are economies of scale. A natural monopoly with ireversible costs implies a barrier to market entry and is characterized by sustainable market power. In order to prevent this monopoly power over the customers and to guarantee the reliability and quality of supply at economically or politically desired prices, the regulation of monopolistic firms has been introduced. For the regulation of interstate telephone and telegraph service and of radio and television broadcasting in the United States, the Federal Communications Commission was created in 1934, and the Civil Aeronautics Board, which regulated the prices charged by the interstate scheduled airlines as well as entry into the industry, was established in 1938. The Federal Energy Regulatory Commission was established in the United States in 1977. Independent regulatory agencies operate now in all countries of the European Union.

The monopoly profit-maximizing level of output is that one for which marginal revenue equals marginal cost. At this output level, price will exceed marginal cost. The profitability of the monopolist will depend on the relationship between price and average cost.

One approach to devising monopoly pricing schemes that is followed in many regulatory situations is to permit the monopoly to charge a price above average cost that is sufficient to earn a "fair" rate of return on investment. From an economic point of view, the interesting question concerns the impact of regulation on the firm's input choices. In the most frequently quoted paper in regulatory economics Averch and Johnson [2] showed that under the rate of return constraint the profit-maximizing firm chooses an inefficient input mix in the sense "that (social) cost" is not minimized at the output it selects [2, p. 1052].

Let us start the analysis considering a basic model of the monopoly firm producing a single output using two inputs, capital and labor, where the respective quantities are denoted q, x_1 and x_2; the production function permits inputs to be employed in any proportion. The unit price of the firm's output is denoted p. Suppose that it can buy as much as it wants of the two inputs at constant unit prices of c_1 and c_2, respectively, so that its profit function is

$$\Pi = pq - c_1 x_1 - c_2 x_2. \tag{2.55}$$

Assuming that $x_1 > 0$ and $x_2 > 0$ (in other words, both production factors are essential), the profit maximization requires that

$$\frac{\partial \Pi}{\partial x_1} = \frac{\partial pq}{\partial x_1} - c_1 = 0, \tag{2.56}$$

$$\frac{\partial \Pi}{\partial x_2} = \frac{\partial pq}{\partial x_2} - c_2 = 0, \tag{2.57}$$

and consequently,

$$\frac{\frac{\partial pq}{\partial x_1}}{\frac{\partial pq}{\partial x_2}} = \frac{c_1}{c_2}. \tag{2.58}$$

The ratio of marginal revenue products will equal the ratio of the input prices. The marginal revenue product $\frac{\partial pq}{\partial x_i}$ describes the extra revenue that accrues to a firm when it sells the output that is produced by one more unit of input i $(i = 1, 2)$. The marginal revenue product of factor i (MR_i) is given by the multiplication of marginal revenue (MR) by the marginal physical product (MP_i) of factor i: $\mathrm{MR}_i = \mathrm{MR} \cdot \mathrm{MP}_i$. The marginal revenue is the additional revenue obtained by a firm when it is able to sell one more unit of output. The marginal physical product describes the additional output that can be produced by one more unit of a particular input while holding all other inputs constant. According to (2.58), the firm uses an efficient mix of capital and labor in the sense that cost is minimized at the output it selects. Rewriting (2.58) as

$$\frac{\frac{\partial pq}{\partial x_1}}{c_1} = \frac{\frac{\partial pq}{\partial x_2}}{c_2}, \tag{2.58$'$}$$

every additional Euro given to any input yields the same revenue.

Now, following [2], suppose that the firm is regulated by government, which imposes a constraint on its rate of return. The introduction of such regulatory constraint is motivated by the following argument: "In judging the level of prices charged by firms for services subject to public control, government regulatory agencies commonly employ a 'fair rate of return' criterion: After the firm subtracts its operating expenses from gross revenues, the remaining net revenue should be just sufficient to compensate the firm for its investment in plant and equipment. If the rate of return, computed as the ratio of net revenue to the value of plant and equipment (the rate base), is judged to be excessive, pressure is brought to bear on the firm to reduce prices. If the rate is considered to be too low the firm is permitted to increase prices" [2, p. 1052]. The profit-maximizing behavior of the firm under such a regulatory constraint can then be described by the following.

Proposition 2.3. *The firm does not equate the marginal rate of factor substitution to the ratio of the input prices. The firm has an incentive to increase its investment: The amount of capital used with the regulatory constraint is not less than the amount used without a constraint.*

Proof. We define the firm's production function as

$$q = f(x_1, x_2), \quad \text{where } f_1 = \frac{\partial f}{\partial x_1} > 0, \quad f_2 = \frac{\partial f}{\partial x_2} > 0,$$
$$f(0, x_2) = f(x_1, 0) = 0.$$

That is, marginal products are positive, and production requires both inputs.

The inverse demand function can be written

$$p = p(q), \quad \text{where } p'(q) = \frac{dp}{dq} < 0.$$

The profit Π is defined by (2.55).

Let x_1 denote the physical quantity of plant and equipment in the rate base, b_1 the acquisition cost per unit of plant and equipment in the rate base, β_1 the value of depreciation of plant and equipment during a time period in question, and B_1 the cumulative value of depreciation.

The regulatory constraint of [2] is

$$\frac{pq - c_2x_2 - \beta_1}{b_1x_1 - B_1} \leqq s, \tag{2.59}$$

where the profit net of labor cost and capital depreciation constitutes a percentage of the rate base (net depreciation) no greater than a specified maximum s.

For simplicity, in [2] it was assumed that depreciation (β_1 and B_1) is zero and the acquisition cost b_1 is equal to 1 (i.e., the value of the rate base is equal to the physical quantity of capital). The price, or the "cost of capital," c_1 is the interest cost involved in holding plant and equipment (to be distinguished from the acquisition cost b_1). The regulatory constraint (2.59) can then be rewritten as

$$\frac{pq - c_2x_2}{x_1} \leqq s,$$

or

$$pq - sx_1 - c_2x_2 \leqq 0. \tag{2.60}$$

The "fair rate of return" s is the rate of return allowed by the regulatory agency on plant and equipment in order to compensate the firm for the cost of capital.

If $s < c_1$, the allowable rate of return is less than the actual cost of capital and the firm would withdraw from the market. Therefore, we shall assume that $s \geqq c_1$; the allowable rate of return must at least cover the actual cost of capital.

The problem of the firm is to maximize the profit described by function (2.55) subject to (2.60) and $x_1 \geqq 0$, $x_2 \geqq 0$. The Lagrange function is defined as

$$\Phi(x_1, x_2, u) = p(q)q - c_1x_1 - c_2x_2 - u(p(q)q - sx_1 - c_2x_2),$$

where $q = f(x_1, x_2)$.

The Kuhn–Tucker necessary conditions for a maximum at x_1^0, x_2^0, u^0 are

$$\frac{\partial \Phi}{\partial x_1} = (1-u)[p + p'q]f_1 - c_1 + us \leqq 0, \tag{a}$$

$$x_1 \frac{\partial \Phi}{\partial x_1} = x_1\{(1-u)[p + p'q]f_1 - c_1 + us\} = 0, \tag{b}$$

$$\frac{\partial \Phi}{\partial x_2} = (1-u)[p + p'q]f_2 - (1-u)c_2 \leqq 0, \tag{c}$$

$$x_2 \frac{\partial \Phi}{\partial x_2} = x_2\{(1-u)[p+p'q]f_2 - (1-u)c_2\} = 0, \tag{d}$$

$$\frac{\partial \Phi}{\partial u} = -(p(q)q - sx_1 - c_2x_2) \geqq 0, \tag{e}$$

$$u\frac{\partial \Phi}{\partial u} = u(p(q)q - sx_1 - c_2x_2) = 0, \tag{f}$$

$$u \geqq 0. \tag{g}$$

Because the production requires both inputs $x_1^0 > 0$, $x_2^0 > 0$, and assuming $u^0 > 0$ (i.e., the regulatory constraint (2.60) is binding at (x_1^0, x_2^0)), conditions (a), (c), and (e) can be rewritten as the following equalities:

$$(1-u)[p+p'q]f_1 + us = c_1, \tag{2.61}$$

$$[p+p'q]f_2 = c_2, \tag{2.62}$$

$$pq - sx_1 - c_2x_2 = 0. \tag{2.63}$$

The expression $(p + p'q)$ describes the marginal revenue and f_1, f_2 denote the marginal physical products of capital and labor, respectively. Note that (2.61)–(2.63) will determine the values of x_1^0, x_2^0, and u^0.

If there is no regulatory constraint (2.60) so that the constraint is not active ($u = 0$), (2.61) and (2.62) reduce to (2.56)–(2.57) with the familiar rule that the marginal revenue product of each factor is equal to its price.

It follows from (2.61) that $u^0 > 0$ (the binding regulatory constraint (2.60)) will distort the equality of the marginal revenue product of capital $(p + p'q)f_1$ with its actual cost c_1. Consequently, the relative proportions of capital and labor used by the firm will be changed. The marginal rate of factor substitution $\frac{f_1}{f_2}$ is no longer equal to the ratio of the input prices. The first part of Proposition 2.3 is proved.

Assuming that $u > 0$, it is clear from (2.61) that $u = 1$ implies $c_1 = s$. On the other hand, if $c_1 = s$, (2.61) reduces to (2.56), which corresponds to the behavior of unregulated monopoly. Therefore, we sharpen our assumption $s \geqq c_1$ to $s > c_1$ and get $u^0 \neq 1$.

Let the superscript 0 denote the solution of the optimization problem for the regulated monopoly and asterisk the solution for the unregulated monopoly. Furthermore, denote the expression $[p + p'q]f_1$ for the marginal revenue product of capital and the expression $[p + p'q]f_2$ for the marginal revenue of labor by MR_1 and MR_2, respectively.

Adding c_1u^0 to both sides of (2.61) and rearranging terms yields

$$\mathrm{MR}_1^0 = c_1 - \frac{(s-c_1)}{1-u^0}u^0. \tag{2.64}$$

Under the assumption that $s > c_1$ and $u^0 < 1$ (as claimed in [2]), it follows from (2.64) that $\mathrm{MR}^0 < c_1$.

If the revenue function $G \equiv pf(x_1, x_2)$ is concave (this assumption is not mentioned in [2]; it was introduced by Takayama [36]), then the marginal revenue product of capital MR_1 is a nonincreasing function of capital used, and consequently the amount of capital used under the regulatory constraint (x_1^0) is not less than the amount used without a constraint (x_1^*). If G is assumed to be strictly concave, then $\frac{\partial \mathrm{MR}_1}{\partial x_1} < 0$; hence $x_1^0 > x_1^*$. Furthermore, it follows from (2.61) and (2.62) that

$$\frac{\mathrm{MR}_1}{\mathrm{MR}_2} = \frac{c_1}{c_2} - \frac{(s - c_1)}{c_2} \frac{u^0}{(1 - u^0)} < \frac{c_1}{c_2}.$$

The marginal rate of substitution between inputs ($\frac{\mathrm{MR}_1}{\mathrm{MR}_2}$) is lower than the ratio of input prices. Each output is produced with more capital and less labor as compared to the unregulated optimum. This effect of overcapitalization contained in the second part of Proposition 2.3 is known as the *Averch–Johnson effect*. In their own words, "If the rate of return allowed by the regulatory agency is greater than the cost of capital but is less than the rate of return that would be enjoyed by the firm were it free to maximize profit without regulatory constraint, then the firm will substitute capital for the other factor of production and operate at an output where cost is not minimized" [2, p. 1053]. □

This inefficiency derives from the fact that the net return of the monopolist on every unit of capital is $s - c_1$, and this creates an incentive to substitute capital for labor. Under regulatory constraint $\Pi^0 + c_1 x_1^0 - s x_1^0 = 0$, and consequently $\Pi^0 = (s - c_1) x_1^0$.

An important question in the Averch–Johnson analysis is whether u^0 is indeed less than one. The argument by Averch–Johnson roughly goes as follows: Since $s > c_1$, u^0 cannot be equal to one, as shown before, for the unconstrained rate of return is $u^0 = 0$. Because of the continuity of u^0 with respect to s, u^0 should always be less than one.

But the continuity of u^0 is not intuitively obvious. The value of the Lagrange multiplier may jump from zero to some nonzero value as the constraint moves from a nonactive to an active stage. Takayama [36] showed that the continuity of u^0 in the Averch–Johnson model depends on the continuity of x_1^0 and x_2^0 with respect to s.

Another way to obtain the condition $u^0 < 1$ uses the optimality conditions (2.61)–(2.63). As already mentioned, under the assumptions $x_1^0 > 0$, $x_2^0 > 0$, and $u^0 > 0$, these equations determine the values of x_1^0, x_2^0, and u^0. The value of u^0 can be obtained explicitly from (2.64), assuming that $s - \mathrm{MR}_1^0 > 0$:

$$u^0 = \frac{c_1 - \mathrm{MR}_1^0}{s - \mathrm{MR}_1^0} > 0. \tag{2.65}$$

Under our assumption that $s > c_1$ and the new condition $s > \mathrm{MR}_1^0$, it follows directly from (2.65) that $u^0 < 1$.

If $u^0 = 1$, then due to (2.61), $s = c_1$, which contradicts the assumption of $s > c_1$.

A condition similar to our condition $s > \mathrm{MR}_1^0$ is used by El-Hodiri and Takayama [9]. Assuming that G is concave, the Averch–Johnson effect, $x_1^0 \geqq x_1^*$, occurs if and only if $\mathrm{MR}_1^0 - c_1 \leqq 0$. They assert that $u^0 < 1$ if and only if $\mathrm{MR}_1^0 \leqq c_1$ (i.e., if and only if $x_1^0 \geqq x_1^*$). They can prove the Averch–Johnson effect without assuming anything about u^0 but with the requirement that $\mathrm{MR}_1^0 \leqq c_1$.

Averch and Johnson applied the model to one particular regulated industry—the domestic telephone and telegraph industry. They found that "the model does raise issues relevant to evaluating market behavior" [2, p. 1052]. The scientific discussion as well as the real applications of the rate-of-return regulation has been initiated.

In the 1980s, the discussion—connected with privatization and deregulation policies in several countries—began to concentrate on the question of how a regulating agency could give the best incentives for efficient production in the regulated firm. The way to increase the efficiency is based on the promotion of the competition. Internal subsidization was increasingly considered undesirable and this view has led to reconsideration of the internal organization of firms which claimed to be natural monopolies. Because in the utilities like telecommunications, electricity, and gas, it is only the distributive grid which has the properties of a natural monopoly, vertical desintegration or unbundling has been proposed. The electricity generation must be separated from the transmission and distribution activities. With respect to these grids, economies of scale are still predominant, maybe even increasing in recent decades. With unbundling a market entry in those parts where no natural monopoly properties prevail can ensure and a presupposition for effective competition is created. Competition is effective when each firm cannot appreciably raise the price above that of its rivals for fear of losing its market share, and can only increase profit by cutting costs. Regulation of networks which remain natural monopolies is needed in order to make the entry possible for different providers and in this way to promote a competition. How to design regulation to fulfill the above requirements and so provide incentives for grid companies to reduce the costs and consequently the prices? The main drawback of the rate of return regulation is the lack of incentives for cost reduction and technological innovation. "A profit maximizing firm subject to a fair return on investment regulation will overcapitalize and select those technical changes which will allow to continue to do so—namely labor augmenting innovations" [34, p. 630].[4] We speak about costs based regulation, in which firms' allowed rate of return is based directly on the reported costs of the individual firm.

During the privatization of British Telecom, Littlechild [21], Director of the Office for Electricity Regulation, proposed a new type of regulation, the so-called *price-cap* regulation.[5] The basic idea is that the price index of the monopolistically supplied goods (or services) must not exceed the retail price index minus an exogenously fixed productivity factor. The customer must be able to buy at the prices of given period the same basket of goods (or services) as in the base period without increasing expenditures. The retail price index (RPI) measuring the inflation rate is the consumer price index, which is a Laspeyers index of the usual type:

$$\mathrm{RPI} = \frac{\sum_i p_i q_i^0}{\sum_i p_i^0 q_i^0}.$$

The superscript 0 defines variables of the base period in which the fixed commodity

[4] For further reading, see [13].

[5] For further reading on price-cap regulation, see [7, 19, 17], and for a survey comparing rate of return and price-cap regulation, see [22].

basket of the index was empirically determined. We denote the price of commodity i by p_i and the quantity by q_i $(i = 1, 2, \ldots, n)$.

The profit-maximizing firm is regulated by the following constraint:

$$\sum_{i=1}^{n} p_i q_i^0 \leqq \sum_{i=1}^{n} p_i^0 q_i^0 (1 + \text{RPI} - X), \tag{2.66}$$

where X describes the productivity factor of the sector.

Because both the consumer price index (RPI) as well as the expenditures of the base period ($\sum_{i=1}^{n} p_i^0 q_i^0$) are for the regulator exogenously given, the only control variable for him remains the productivity factor (X). From the regulatory point of view the relevant question therefore concerns the impact of X on the behavior of the regulated firm.

For this purpose, we consider the following optimization problem:

$$\begin{aligned} &\text{maximize} \quad && \Pi(q) = p(q)q - C(q) \\ &\text{subject to} \quad && p(q)q^0 \leqq b^0, \end{aligned} \tag{2.67}$$

where $\Pi(q)$ describes the profit function, $C(q)$ is the cost function, and $p(q)$ is the inverse demand function. b^0 denotes the right side of the regulatory constraint (2.66):

$$b^0 = p^0 q^0 (1 + \text{RPI} - X).$$

The price cap (b^0) can be faced as a function of the expenditures in the base period ($p^0 q^0$), of the retail price index (RPI), and of the productivity factor (X):

$$b^0 = b\left(\underset{(+)}{R^0}, \underset{(+)}{\text{RPI}}, \underset{(-)}{X}\right),$$

where $R^0 = p^0 q^0$. With increasing expenditures (R^0) and increasing consumer price index (RPI), the price cap (b^0) rises. The increasing productivity (X) makes the constraint (2.66) tighter. In other words, the increasing productivity implies lower prices for the consumers.

Furthermore, we postulate positive marginal cost and a declining demand function:

$$\frac{dC}{dq} = \text{MC} > 0, \qquad \frac{dp}{dq} < 0.$$

The Lagrange function for the maximization problem (2.67) is

$$\Phi(q, u) = p(q)q - C(q) + u(b^0 - p(q)q^0).$$

Application of the Kuhn–Tucker conditions (2.20)–(2.21) yields

$$p - \frac{dC}{dq} = -\frac{dp}{dq}(q - uq^0). \tag{2.68}$$

Multiplying both sides of (2.68) by $\frac{1}{p}$, we obtain

$$\frac{p - \mathrm{MC}}{p} = -\frac{dp}{dq}\frac{q}{p} + u\frac{q^0}{p}\frac{dp}{dq}.$$

Using the notion of the price elasticity of demand $\varepsilon = \frac{pdq}{qdp} < 0$, the following optimality condition results:

$$\frac{p - \mathrm{MC}}{p} = -\left(1 - u\frac{q^0}{q}\right)\frac{1}{\varepsilon}. \tag{2.69}$$

For the unregulated monopolistic firm with the Lagrange multiplier $u = 0$, the form (2.69) reduces to the well-known Lerner index (see, e.g., [5, p. 26]):

$$\frac{p - \mathrm{MC}}{p} = -\frac{1}{\varepsilon}.$$

The Lerner index measures the market power of monopoly.

For a perfectly competitive firm, price equals marginal cost, so the Lerner index equals 0. The higher the Lerner index is, the higher is the degree of monopoly power. For the profit-maximizing firm, the Lerner index is equal to the reciprocal value of the price elasticity of demand for the firm's product. The lower the price elasticity of demand for the firm's product is, the higher is the degree of monopoly power.

Under the price-cap regulation the Lerner index is modified by the expression $1 - u \cdot \frac{q^0}{q}$.

From the Kuhn–Tucker condition (2.25) follows the nonnegativity of the Lagrange multiplier u. According to (2.9), this multiplier describes the change of the monopoly profit due to a change of the price cap:

$$u = \frac{\partial \Pi}{\partial b^0} \geqq 0.$$

Moreover, it can be shown [33, pp. 166–168] that under the assumption of concavity of Π and convexity of $p(q)q^0$,

$$\frac{\partial u}{\partial b^0} \leqq 0.$$

The lower price cap—due to the stronger regulation by setting the productivity factor (X) higher—implies higher Lagrange multiplier, and according to (2.69) the lower the degree of monopoly power. In this way, price-cap regulation is an appropriate instrument to reduce the market power of natural monopolies. The form (2.69) reveals the central problem of price-cap regulation for the regulatory agency, the determination of the productivity factor (X). "Too high a price ceiling makes the firm an unregulated monopolist, too low cap conflicts with viability, and in between the 'right' price level is difficult to compute" [19, p. 17]. One possibility of how to calculate the X-factor in order to provide incentives for cost reduction and technological innovation and consequently for reduction of network tariffs in the electricity sector is described by [25].

An application of regulatory constraints for environmental economics and policy will be discussed in the next section.

2.5.4 Environmental Regulation: The Effects of Different Restrictions

The forms of standards used in current environmental regulation vary tremendously. The most frequently discussed forms are systems of permits which determine a fixed amount of emission allowed for each emission source independent of the production level of this source. The deficiencies of such a source-based system of permits are investigated and summarized in [31, Chapter VIII] or [32]. Another form of environmental regulation relates to restrictions on pollution per unit of output or input [8]. In an economic sense, restrictions that are based on a unit of output or input are equivalent to a productivity or intensity regulation, well known from the literature, beginning with the Averch–Johnson model [2], and are discussed in the previous section.

In the paper by Helfand [14], the effects of five different forms of pollution standards on input decisions, the level of production, and firm profits are examined using a graphical approach. In this section, we analyze the effects of different kinds of pollution control standards in a more general way using Kuhn–Tucker conditions.

The model used in [14] involves one firm, facing a horizontal output demand curve and using two inputs, x_1 and x_2, with horizontal supply curves. The assumption that there are only two inputs is only for simplicity but without loss of generality. The assumption of a horizontal output demand curve is more limiting and is realistic only for a good whose price is unaffected by production of the firm. In [14], this assumption makes the problem tractable and permits a graphical presentation.

Assume that the firm produces a single output in the quantity q according to the production function $f(x_1, x_2)$ with the usual properties:

$$\begin{aligned} f_1 &= \frac{\partial f}{\partial x_1} > 0, \qquad & f_2 &= \frac{\partial f}{\partial x_2} > 0, \\ f_{11} &= \frac{\partial^2 f}{\partial x_1^2} < 0, \qquad & f_{22} &= \frac{\partial^2 f}{\partial x_2^2} < 0. \end{aligned} \tag{2.70}$$

In other words, the marginal products of both inputs are positive but declining.

The firm also causes pollution, the level of which depends on the level of production and the technology. In order to reduce the level of pollution, the firm can use an abatement activity or invest in new technology. The resulting level of pollution (or net pollution) can be described as follows:

$$P = G(f(x_1, x_2)) - \mathrm{Ab}(x_3),$$

where $\mathrm{Ab}(x_3)$ denotes the abatement activity as a function of abatement expenditure x_3 (or expenditure for development of a new technology). It is assumed that $\frac{d\mathrm{Ab}}{dx_3} > 0$, that is, more abatement equipment (or higher expenditure for development of a new technology) reduces the level of pollution. More generally, we describe the level of net pollution as follows:

$$P = P(x_1, x_2, x_3)$$

with $P_1 = \frac{\partial P}{\partial x_1} > 0$, $P_2 = \frac{\partial P}{\partial x_2} > 0$, and $P_3 = \frac{\partial P}{\partial x_3} < 0$.[6]

[6] In this formulation of the net pollution function P, we differ slightly from the model in [14].

The firm is assumed to maximize profits while facing an output price p and input prices c_1 and c_2 as well as the price of abatement equipment c_3 as given.

The necessary conditions for a profit-maximizing firm without regulatory constraint (i.e., pollution restrictions) are given by (2.56)–(2.57) in the previous section. In economic terms, the value of the marginal product of input i ($i = 1, 2$) must be equal to its price, i.e., the ratio of marginal revenue products will equal the ratio of the input prices.

Now, similar to the regulatory constraint by [2], pollution restrictions in the form of different kinds of pollution-control standards will be taken into account. What are the effects for the level of production and the firm's profit?

2.5.4.1 Standard as a Set Level of Emissions

Let Z_p be the amount of total pollution permissible in a certain period of time. It can be represented as a constraint in the form $P(\mathbf{x}) \leqq Z_p$. The optimization problem of the profit-maximizing firm is

$$\begin{aligned} \underset{x_1, x_2, x_3}{\text{maximize}} \quad & \Pi = pf(x_1, x_2) - c_1 x_1 - c_2 x_2 - c_3 x_3 \\ \text{subject to} \quad & P(x_1, x_2, x_3) \leqq Z_p, \\ & x_1 \geqq 0, \quad x_2 \geqq 0, \quad x_3 \geqq 0. \end{aligned}$$

We write the Lagrange function

$$\Phi(\mathbf{x}, u) = pf(x_1, x_2) - c_1 x_1 - c_2 x_2 - c_3 x_3 + u(Z_p - P(x_1, x_2, x_3))$$

and the resulting Kuhn–Tucker conditions

$$\frac{\partial \Phi}{\partial x_1} = pf_1 - c_1 - uP_1 \leqq 0, \tag{2.71}$$

or

$$pf_1 \leqq c_1 + uP_1,$$

$$x_1 \frac{\partial \Phi}{\partial x_1} = x_1(pf_1 - c_1 - uP_1) = 0. \tag{2.72}$$

Assuming that $x_1 > 0$, it follows from (2.72) that

$$pf_1 = c_1 + uP_1. \tag{2.73}$$

Furthermore,

$$\frac{\partial \Phi}{\partial x_2} = pf_2 - c_2 - uP_2 \leqq 0, \tag{2.74}$$

or

$$pf_2 \leqq c_2 + uP_2,$$
$$x_2 \frac{\partial \Phi}{\partial x_2} = x_2(pf_2 - c_2 - uP_2) = 0. \tag{2.75}$$

Assuming that $x_2 > 0$, (2.75) implies that

$$pf_2 = c_2 + uP_2. \tag{2.76}$$

We conclude that the value of the marginal product of the input i ($i = 1, 2$) is equal to the marginal input costs, plus the pollution cost, uP_i, where $u = u^0 = \frac{\partial \Pi(\mathbf{x}^0(Z_p))}{\partial Z_p}$ and $P_i = \frac{\partial P}{\partial x_i}$ ($i = 1, 2$).

The Lagrange multiplier u^0 describes the effect of a change of the environmental standards for the profit of the firm and P_i expresses the increase of pollution caused by increasing the ith input by a small unit (i.e., the marginal pollution with respect to the input i):

$$\frac{\partial \Phi}{\partial x_3} = -c_3 - uP_3 \leqq 0,$$
$$x_3 \frac{\partial \Phi}{\partial x_3} = x_3(-c_3 - uP_3) = 0.$$

$x_3 > 0$ implies that $c_3 = -uP_3$, where $P_3 < 0$. Therefore, the value of the pollution reduction caused by one additional unit of abatement equipment is equal to its cost.

Finally, we obtain

$$\frac{\partial \Phi}{\partial u} = Z_p - P(x_1, x_2, x_3) \geqq 0,$$
$$u \frac{\partial \Phi}{\partial u} = u(Z_p - P(x_1, x_2, x_3)) = 0,$$
$$u \geqq 0.$$

We conclude that $P(x_1, x_2, x_3) < Z_p$ implies that $u = 0$. In this case, equalities (2.73) and (2.76) reduce to $pf_1 = c_1$, $pf_2 = c_2$, and $c_3 > 0$ implies that $x_3 = 0$. The economic interpretation of this result is straightforward: If the net pollution is below the given level of emissions, no abatement will be necessary, and we get the same solution as in the unregulated case.

For $u > 0$, we obtain $P(x_1, x_2, x_3) = Z_p$, and in the case of essential production factors ($x_1 > 0$, $x_2 > 0$), conditions (2.73) and (2.76). Because $u > 0$ and $P_1 > 0$, $P_2 > 0$, the value of marginal product of the input i ($i = 1, 2$) under regulation must be higher than in the unregulated case (see conditions (2.56)–(2.57), (2.73), and (2.76)).

Under the assumption (2.70) on the production function $f(x_1, x_2)$ that the marginal products are decreasing, we can conclude

$$pf_1^{\mathrm{I}} > pf_1^0 \text{ implies } x_1^{\mathrm{I}} < x_1^0,$$
$$pf_2^{\mathrm{I}} > pf_2^0 \text{ implies } x_2^{\mathrm{I}} < x_2^0,$$

where the superscript I denotes the model with environmental constraint expressed as a permissible amount of total pollution and 0 denotes the model without regulation. The effect of this type of environmental regulation is obvious: Both inputs are decreasing, and therefore the level of production also decreases. This is the only way the firm can meet the environmental constraint.

2.5.4.2 Standard as Emissions per Unit of Output

Let Z_{PF} be the emission standard expressed as a set level of pollution per unit of output. This amount of emission may be discharged into the environment at a zero price. The regulatory constraint then becomes

$$\frac{P(x_1, x_2, x_3)}{f(x_1, x_2)} \leqq Z_{\mathrm{PF}},$$

and the objective function of the firm is the profit maximization as in the previous model.

The Lagrange function is

$$\Phi = pf(x_1, x_2) - c_1x_1 - c_2x_2 - c_3x_3 + u(Z_{\mathrm{PF}}f(x_1, x_2) - P(x_1, x_2, x_3)),$$

and the Kuhn–Tucker conditions are

$$\frac{\partial \Phi}{\partial x_1} = pf_1 - c_1 + u(Z_{\mathrm{PF}}f_1 - P_1) \leqq 0, \tag{2.77}$$

$$x_1\frac{\partial \Phi}{\partial x_1} = x_1[pf_1 - c_1 + u(Z_{\mathrm{PF}}f_1 - P_1)] = 0, \tag{2.78}$$

$$\frac{\partial \Phi}{\partial x_2} = pf_2 - c_2 + u(Z_{\mathrm{PF}}f_2 - P_2) \leqq 0, \tag{2.79}$$

$$x_2\frac{\partial \Phi}{\partial x_2} = x_2[pf_2 - c_2 + u(Z_{\mathrm{PF}}f_2 - P_2)] = 0, \tag{2.80}$$

$$\frac{\partial \Phi}{\partial x_3} = -c_3 - uP_3 \leqq 0, \tag{2.81}$$

$$x_3\frac{\partial \Phi}{\partial x_3} = x_3(-c_3 - uP_3) = 0, \tag{2.82}$$

$$\frac{\partial \Phi}{\partial u} = Z_{\mathrm{PF}}f(x_1, x_2) - P(x_1, x_2, x_3) \geqq 0, \tag{2.83}$$

$$u\frac{\partial \Phi}{\partial u} = u[Z_{\mathrm{PF}}f(x_1, x_2) - P(x_1, x_2, x_3)] = 0, \tag{2.84}$$

$$u \geqq 0. \tag{2.85}$$

If $Z_{\mathrm{PF}}f(x_1, x_2) > P(x_1, x_2, x_3)$, then $u = 0$, and because $c_3 > 0$, it follows from (2.82) that $x_3 = 0$.

If the price of abatement equipment c_3 is higher than the value of the pollution reduced by one additional unit of abatement equipment $-uP_3$, then the abatement expenditure x_3 will be zero.

Assuming essential production factors ($x_1 > 0,\ x_2 > 0$), the Kuhn–Tucker conditions (2.77) and (2.79) become equalities:

$$pf_1 - c_1 + u(Z_{\mathrm{PF}} f_1 - P_1) = 0,$$
$$pf_2 - c_2 + u(Z_{\mathrm{PF}} f_2 - P_2) = 0,$$

or

$$f_1 = \frac{c_1 + uP_1}{p + uZ_{\mathrm{PF}}}, \tag{2.86}$$
$$f_2 = \frac{c_2 + uP_2}{p + uZ_{\mathrm{PF}}}, \tag{2.87}$$

and therefore

$$\frac{f_1}{f_2} = \frac{c_1 + uP_1}{c_2 + uP_2}. \tag{2.88}$$

We can see that for $u > 0$ (the pollution constraint is binding), the ratio of marginal products cannot equal the ratio of the input prices, as was the case in the absence of the regulatory constraint.

In order to show the effect of the environmental constraint (2.83) for the behavior of the firm, we compare the optimality conditions (2.86)–(2.87) with the optimality conditions without environmental standard (2.56)–(2.57).

Let the superscript II denote the solution of the model with environmental constraint expressed as the maximum amount of emissions per unit of output, i.e., the model in this section.

Recall the first-order optimality conditions for the unregulated firm:

$$f_1^0 = \frac{c_1}{p} \quad \text{and} \quad f_2^0 = \frac{c_2}{p} \tag{2.89}$$

for a given price p of the output.

Comparison of (2.89) with (2.86)–(2.87) reveals that the effect of the environmental regulatory constraint (2.83) on the production of the firm is ambiguous. It depends on the relation between the expressions on the right side of (2.86)–(2.87) and (2.89), respectively. If $P_i^{\mathrm{II}} < \frac{c_i Z_{\mathrm{PF}}}{p}$, then $\frac{c_i + uP_i^{\mathrm{II}}}{p + uZ_{\mathrm{PF}}} < \frac{c_i}{p}$; therefore, $f_i^{\mathrm{II}} < f_i^0$, and consequently, due to the assumption (2.70), $x_i^{\mathrm{II}} > x_i^0$ $(i = 1, 2)$. If the marginal pollution with respect to the input i is lower than the exogenously given constant $k_i = \frac{c_i Z_{\mathrm{PF}}}{p}$, then the amount of input i used in production will—compared with the basic model—increase.

In the opposite case, if the marginal pollution with respect to the input i is relatively high (higher than the parameter k_i), the amount of input i used in production will—in order to fulfill the environmental standard—decrease.

The effects on production, and therefore (taking into account the possible abatement activity) on the level of pollution, remain ambiguous. To summarize, the effect of the standard defined as emissions per unit of output can lead to similar results as in the Averch–Johnson model: Pollution increases with the imposition of an environmental regulatory constraint. If production increases more rapidly than pollution, the environmental standard can be achieved in spite of increasing pollution.

2.5.4.3 Standard as Emissions per Unit of a Specified Input

Another way in which individual stack policy can be effected is to fix an upper bound for the emissions per unit of specified input, such as restricting the amount of sulfur dioxide emissions per ton of coal used for electricity. Such a type of limitation is referred to in [8] as *intensity regulation* and can be formalized as

$$\frac{P(x_1, x_2, x_3)}{x_i} \leqq Z_{Pi}, \qquad \text{for } i = 1, 2.$$

Without loss of generality we suppose that the intensity regulation is imposed for the second production factor. Then the firm will face the following optimization problem:

$$\begin{array}{ll} \text{maximize} & \Pi = pf(x_1, x_2) - c_1, x_1 - c_2 x_2 - c_3 x_3 \\ \text{subject to} & P(x_1, x_2, x_3) \leqq Z_{P_2} x_2, \\ & x_1 \geqq 0, \quad x_2 \geqq 0, \quad x_3 \geqq 0. \end{array}$$

Using the Lagrange function,

$$\Phi(\mathbf{x}, u) = pf(x_1, x_2) - c_1 x_1 - c_2 x_2 - c_3 x_3 + u(Z_{P_2} x_2 - P(x_1, x_2, x_3)),$$

the Kuhn–Tucker conditions are

$$\frac{\partial \Phi}{\partial x_1} = pf_1 - c_1 - uP_1 \leqq 0,$$

$$x_1 \frac{\partial \Phi}{\partial x_1} = x_1(pf_1 - c_1 - uP_1) = 0, \tag{2.90}$$

$$\frac{\partial \Phi}{\partial x_2} = pf_2 - c_2 + u(Z_{P_2} - P_2) \leqq 0,$$

$$x_2 \frac{\partial \Phi}{\partial x_2} = x_2[pf_2 - c_2 + u(Z_{P_2} - P_2)] = 0, \tag{2.91}$$

$$\frac{\partial \Phi}{\partial x_3} = -c_3 - uP_3 \leqq 0,$$

$$x_3 \frac{\partial \Phi}{\partial x_3} = x_3(-c_3 - uP_3) = 0,$$

$$\frac{\partial \Phi}{\partial u} = Z_{P_2} x_2 - P(x_1, x_2, x_3) \geqq 0,$$

$$u \frac{\partial \Phi}{\partial u} = u[Z_{P_2} x_2 - P(x_1, x_2, x_3)] = 0,$$

$$u \geqq 0.$$

Let the superscript III denote the solution of the model with intensity regulation. Again assuming essential production factors ($x_1 > 0$, $x_2 > 0$) and $u > 0$ (the environmental standard is binding), the Kuhn–Tucker condition (2.90) yields

$$f_1^{\mathrm{III}} = \frac{c_1}{p} + \frac{u}{p} P_1 > \frac{c_1}{p} = f_1^0.$$

Due to the assumption (2.70), we have $x_1^{\mathrm{III}} < x_1^0$; the firm decreases the amount of the first input used.

For the reaction with respect to the regulated input, we look at the Kuhn–Tucker condition (2.91). It provides

$$f_2^{\mathrm{III}} = \frac{c_2}{p} + \frac{u}{p}(P_2 - Z_{P2}).$$

If the marginal pollution with respect to the second input (P_2) is higher than the allowable amount of emissions per unit of this input, then the marginal product f_2^{III} is higher than the marginal product f_2^0 $(= \frac{c_2}{p})$ in the absence of the regulatory constraint. Therefore, due to the declining marginal product, the amount of the regulated input used under the intensity regulation x_2^{III} is lower than without such regulation x_2^0. The firm will decrease the level of production. If the marginal pollution P_2 is lower than the tolerated amount of emissions per unit of the second input, we get the opposite result. Because in this case the marginal product f_2^{III} is lower than the marginal product f_2^0, the amount of the regulated input used in the optimal solution x_2^{III} is higher than the amount x_2^0 used without the regulatory constraint. We have the Averch–Johnson effect with respect to the second input; the substitution of the first input by the second one.

More than fifty years after their formulation, the Kuhn–Tucker conditions became a standard instrument of the analysis used in the textbooks of microeconomic theory (e.g., [28, 33]) and in the monographs devoted to various fields of economics like the theory of money [29], public economics [7], or industrial economics [6, 13]. Moreover, they provide the foundation for the development of more complex optimization models dealing with multiple objectives or with dynamical economic systems.

References and Further Reading

[1] K. J. Arrow, L. Hurwicz, and H. Uzawa, Constraint qualifications in maximization problems, *Naval Res. Logist. Q.*, **8** (1961), 175–190.

[2] H. Averch and L. L. Johnson, Behavior of the firm under regulatory constraint, *Amer. Econ. Rev.*, **52** (1962), 1052–1069.

[3] M. S. Bazaraa, H. D. Sherali, and C. M. Shetty, *Nonlinear Programming*, 2nd ed., Wiley, New York, 1993.

[4] W. J. Baumol, *Economic Theory and Operations Analysis*, 3rd ed, Prentice–Hall, Englewood Cliffs, NJ, 1972.

[5] H. Bester, *Theorie der Industrieökonomik*, Springer-Verlag, Berlin, 2000.

[6] J. Bormann and J. Finsinger, *Markt und Regulierung*, Verlag Vahlen, Munich, 1999.

[7] D. Bös, *Pricing and Price Regulation: An Economic Theory for Public Enterprises and Public Utilities*, North-Holland, Amsterdam, 1994.

[8] F. Dudenhöffer, The regulation of intensities and productivities: Concepts in environmental policy: Zeitschrift für die gesamte Staatswirtschaft, *J. Institutional Theoret. Econ.*, **140** (1984), 276–287.

[9] M. A. El-Hodiri and A. Takayama, Behavior of the firm under regulatory constraints: Clasifications, *Amer. Econ. Rev.*, **63** (1973), 235–237.

[10] H. A. Eiselt, G. Pederzoli, and C. L. Sandblom, *Continuous Optimization Models*, de Gruyter, Berlin, 1987.

[11] A. Fiacco and G. P. McCormick, *Nonlinear Programming*, Wiley, New York, 1968.

[12] R. Fletcher, Penalty functions, in A. Bachem, M. Grötschel, and B. Korte, eds., *Mathematical Programming: The State of the Art: Bonn* 1982, Springer-Verlag, Berlin, 1983, 87–114.

[13] M. W. Frank, *The Impact of Rate-of-Return Regulation on Technological Innovation*, Ashgate, Aldershot, UK, 2001.

[14] G. E. Helfand, Standards versus standards: The effects of different pollution restrictions, *Amer. Econ. Rev.*, **81**-3 (1991), 622–634.

[15] F. John, Extremum problems with inequalities as side conditions, in K. O. Fridrichs, O. E. Neugebauer, and J. J. Stoker, eds., *Studies and Essays: Courant Anniversary Volume*, Wiley, New York, 1948.

[16] W. Karush, *Minima of Functions of Several Variables with Inequalities as Side Conditions*, M. S. thesis, Department of Mathematics, University of Chicago, Chicago, 1939.

[17] G. Knieps, *Wettbewerbsökonomie: Regulierungstheorie, Industrieökonomie, Wettbewerbspolitik*, Springer-Verlag, Berlin, 2001.

[18] H. W. Kuhn and A. W. Tucker, Nonlinear programming, in J. Neyman, ed., *Proceedings of the Second Berkeley Symposium on Mathematical Statistics and Probability*, University of California Press, Berkeley, CA, 1951, 481–492.

[19] J. J. Laffont and J. Tirole, *A Theory of Incentives in Procurement and Regulation*, MIT Press, Cambridge, MA, 2000.

[20] S. C. Littlechild, Peak-load pricing of telephone calls, *Bell J. Econ. Manage. Sci.*, **1**-2 (1970), 191–210.

[21] S. C. Littlechild, *Regulation of British Telecommunications' Profitability*, Her Majesty's Stationery Office, London, 1983.

[22] C. Liston, Price-cap versus rate-of-return regulation, *J. Regul. Econ.*, **5** (1993), 25–48.

[23] D. L. Luenberger, *Linear and Nonlinear Programming*, 1st and 2nd eds., Addison–Wesley, Reading, MA, 1973, 1984.

[24] M. Luptáčik, *Nichtlineare Programierung mit ökonomischen Anwendungen*, Athenäum, Königstein, Germany, 1981.

[25] M. Luptáčik, Price-Cap-Regulierung und Benchmarking der österreichischen Netzbetrieber (Price-cap regulation and benchmarking of the Austrian electricity grid companies), in T. Kostal, M. Leibrecht, and A. Pasterniak, eds., *Organisation und Funktionen des Staates im Wandel: Festschrift für Gabriel Obermann*, Facultas.wuv Universitätsverlag, Vienna, 2008, 203–216.

[26] O. L. Mangasarian, *Nonlinear Programming*, McGraw–Hill, New York, 1969.

[27] G. P. McCormick, *Nonlinear Programming: Theory, Algorithms, and Applications*, Wiley, New York, 1983.

[28] W. Nicholson, *Microeconomic Theory: Basic Principles and Extensions*, 5th ed., Dryden Press/Harcourt–Brace–Jovanovich, New York, 1992.

[29] J. Niehans, *The Theory of Money*, Johns Hopkins University Press, Baltimore, 1978.

[30] H. Nikaido, *Convex Structure and Economic Theory*, Academic Press, New York, 1968.

[31] H. Siebert, *Economics of the Environment*, Lexington Books, New York, 1981.

[32] H. Siebert, Some open questions, in H. Siebert, ed., *Umweltallokation im Raum*, Lang, Frankfurt am Main, Germany, 1982, 1–30.
[33] E. Silberberg and W. Suen, *The Structure of Economics: A Mathematical Analysis*, 3rd ed., McGraw–Hill, New York, 2001.
[34] V. K. Smith, The implications of regulation for induced technical change, *Bell J. Econ. Manage. Sci.*, **5**-2 (1974), 623–636.
[35] P. O. Steiner, Peak loads and efficient pricing, *Q. J. Econ.*, **54** (1964), 64–76.
[36] A. Takayama, Behavior of the firm under regulatory constraint, *Amer. Econ. Rev.*, **59** (1969), 255–260.
[37] A. Takayama, *Mathematical Economics*, Cambridge University Press, Cambridge, UK, 1985.
[38] O. E. Williamson, Peak load pricing and optimal capacity under indivisibility constraints, *Amer. Econ. Rev.*, **56** (1966), 810–827.

3

Convex Programming

The notion of convexity plays an important role in economic theory and modeling. The indifference curves generally used in the theory of consumer demand embody the assumption of a diminishing marginal rate of substitution. Denoting by x_i the quantity of the ith good in the consumer bundle ($i = 1, 2$), the diminishing marginal rate of substitution implies that $-\frac{dx_2}{dx_1}$ falls as x_1 increases. The more the consumer has of the first good, the less will be the marginal rate of substitution of this good for the second good. In other words, the more the consumer has of a particular good, the less important to her (relative to other goods) is an extra unit of this good. In mathematical terms, this assumption—very plausible from a consumer behavior point of view—means that the indifference curves are *convex*. The assumption of a diminishing marginal rate of substitution is therefore equivalent to the assumption that all combinations of x_1 and x_2, which are preferred to or indifferent to a particular combination x_1^0, x_2^0, form a convex set. This definition is, on the other hand, related to the concavity property of the utility function. The assumption of a diminishing marginal rate of substitution (and of a convex preference curve) means that the "well-balanced" bundles of commodities are preferred to bundles that are heavily weighted toward one commodity. If the indifference curve is strictly convex (not a straight line), any linear combination of the two indifferent bundles of goods will be preferred to the initial bundles. The same kind of role as indifference curves in the theory of consumer demand plays an isoquant in production theory. An isoquant shows all possible combinations of inputs that result in a certain quantity of output. Under the assumption of diminishing rate of technical substitution, the isoquants must be convex.

In his book on convex structures and economic theory, Nikaido [17] emphasized examples of convexity emerging "from the peculiarity of constraints that bind the behavior of economic magnitudes" [17, p. 4]. Typical examples are the nonnegativity constraints like the nonnegativity of prices and the inequality constraints in the economic models of mathematical programming described in Chapter 1. The convexity assumption is crucial for the workability of mathematical programming models and thereupon for their application in economics. This chapter deals with the basic

M. Luptáčik, *Mathematical Optimization and Economic Analysis*,
Springer Optimization and Its Applications 36, DOI 10.1007/978-0-387-89552-9_3,

properties of mathematical programming under the assumption of convexity for the objective function and for the functions in the constraints.

Section 3.1 furnishes the mathematical background, including the basic definitions and properties. Section 3.2 deals with Kuhn–Tucker conditions as sufficient optimality conditions for convex programming problems. Note that in the previous chapter, in the absence of convexity assumptions, the Kuhn–Tucker conditions provide only *necessary* optimality conditions. In Section 3.3, the important concept of duality will be introduced, and in the last section an economic interpretation of duality in convex programming is discussed.

3.1 Basic Definitions and Properties

The mathematical programming problem

$$\begin{aligned} &\text{minimize} \quad f_0(\mathbf{x}) \\ &\text{subject to} \quad \mathbf{x} \in K \end{aligned} \tag{3.1}$$

is called a *convex programming* problem if the objective function $f_0(\mathbf{x})$ and the set of feasible solutions K are convex, where $K = \{\mathbf{x} | \mathbf{x} \in R^n,\ f_i(\mathbf{x}) \leqq 0 \ (i = 1, 2, \ldots, m)\}$.

The definition of a convex set originates in the intuitive perception that it has no hollows, and it reads as follows: A convex set is a set that contains any segment joining any two points in it. A rectangular playground (Figure 3.1) is convex, whereas a gourd-shaped lake (Figure 3.2) is not. A folded balloon is not convex, but it becomes convex when inflated.

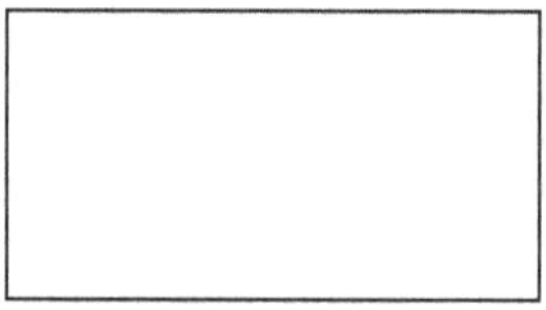

Fig. 3.1. Convex set.

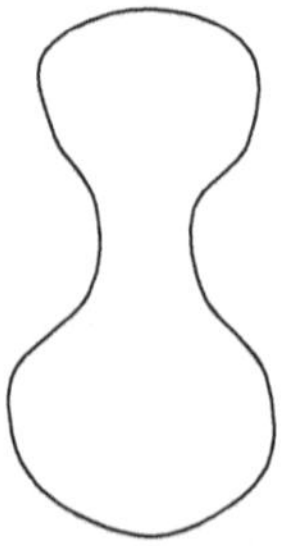

Fig. 3.2. Nonconvex set.

The definition of a convex set involves the operation of joining two points by a segment so that convex sets are conceivable only in a space admitting some sort of linear structure. The simplest kind of such spaces is the Euclidean space R^n.

A formal definition of a convex set is the following.

Definition 3.1. A subset M of R^n is termed convex if $\mathbf{x}, \mathbf{y} \in M$ implies $(\lambda \mathbf{x} + (1 - \lambda)\mathbf{y}) \in M$ for $0 \leqq \lambda \leqq 1$.

This turns out to be a very useful property for the mathematical programming problem (3.1). The convexity of the set K implies that for any two feasible solutions of (3.1), any convex linear combination of the set is a feasible solution.

Let $f(\mathbf{x})$ be a real-valued function defined in R^n. Here $f(\mathbf{x})$ is assumed to have continuous partial derivatives $\frac{\partial f}{\partial x_j}$ $(j = 1, 2, \ldots, n)$.

Definition 3.2. A real-valued function $f(\mathbf{x})$ defined on a convex set D in R^n is called convex on D if

$$f(\lambda \mathbf{x} + (1 - \lambda)\mathbf{y}) \leqq \lambda f(\mathbf{x}) + (1 - \lambda) f(\mathbf{y}) \tag{3.2}$$

for any $\mathbf{x}, \mathbf{y} \in D$ and $0 \leqq \lambda \leqq 1$.

For the single-variable function depicted in Figure 3.3, the graph of the function lies below the line segment joining any two points on the graph.

In the following, the domain of definition D of a convex function is taken to be a convex set in R^n, although this basic assumption will not be stated explicitly each time.

Related to convex sets, the following equivalent definition of a convex function can be given.

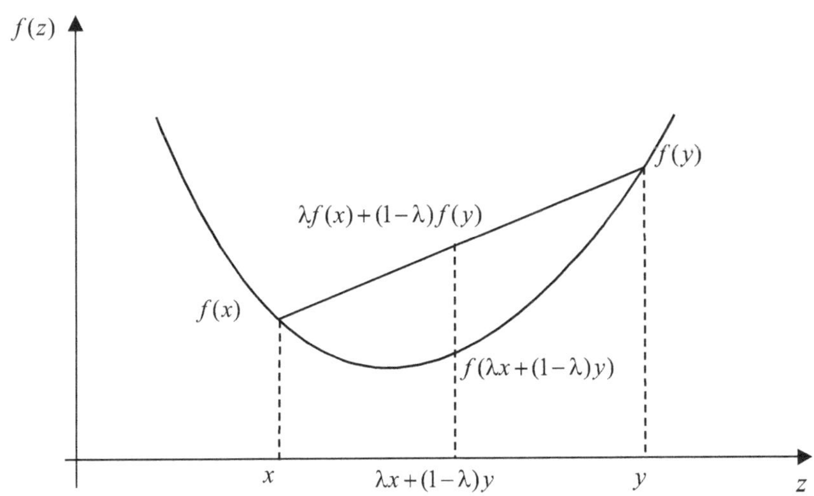

Fig. 3.3. A convex function.

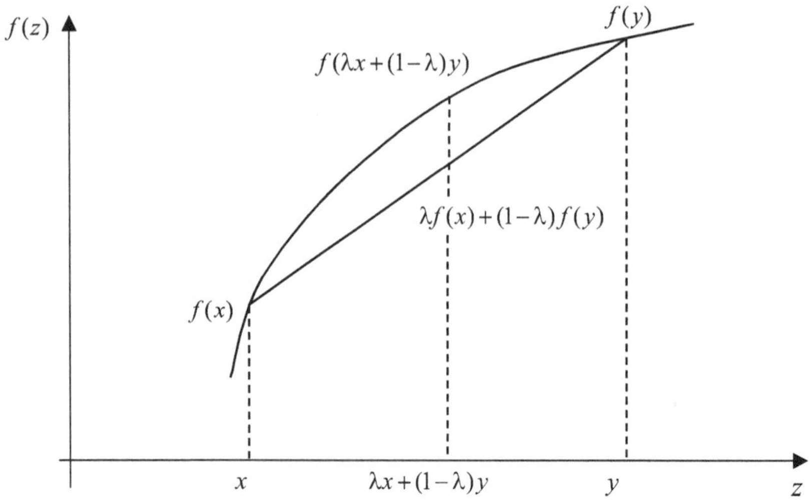

Fig. 3.4. A concave function.

Definition 3.3. A real-valued function $f(\mathbf{x})$ defined on a convex set D in R^n is called convex on D if for any $\alpha \in R$ the set $D_f = \{\frac{(\mathbf{x},\alpha)}{\alpha} \geqq f(\mathbf{x}),\ \mathbf{x} \in D\}$ is a convex set in $R^n \times R$. In other words, the convexity of a function $f(\mathbf{x})$ is equivalent to the convexity of the set of points lying on or above the graph of a function $f(\mathbf{x})$ (the epigraph of f).

Definition 3.4. A real-valued function $f(\mathbf{x})$ defined on a convex set D in R^n is called concave on D if

$$f(\lambda\mathbf{x} + (1-\lambda)\mathbf{y}) \geqq \lambda f(\mathbf{x}) + (1-\lambda) f(\mathbf{y}) \tag{3.3}$$

for any $\mathbf{x}, \mathbf{y} \in D$ and $0 \leqq \lambda \leqq 1$.

The graph of the function depicted in Figure 3.4 lies above the line joining any two points on the graph.

By analogy to Definition 3.3, the concavity of a function $f(\mathbf{x})$ is equivalent to the convexity of the set of points lying on or below the graph of a function $f(\mathbf{x})$. A function $f(\mathbf{x})$ is concave if and only if the function $(-f(\mathbf{x}))$ is convex.

Inequalities (3.2) and (3.3) allow the special case of a straight line, which could be excluded by requiring the inequality to be strict for $0 < \lambda < 1$.

Definition 3.5. A function $f(\mathbf{x})$ is called strictly convex (strictly concave) if in (3.2) (respectively, (3.3)) strict inequality holds.

Obviously, a linear function $f(\mathbf{x})$ defined on a convex set D in R^n is convex and concave (but neither strictly convex nor strictly concave). Only linear functions have this property.

For practical applications and for economic interpretation of convexity, the question arises of how to check a given function for convexity. The direct use of Definition 3.2 can be very cumbersome, and therefore some useful alternative criteria for convexity will be derived in what follows. Let us start with single-variable function $f(x)$ defined in R.

Theorem 3.1. *A function $f(x)$ that has continuous derivatives on D is convex (strictly convex) if and only if the derivative $f'(x)$ is a nondecreasing (increasing) function on D.*

Proof.

(i) Suppose that f is differentiable and convex. Then for $x < y$ and $z = \lambda x + (1 - \lambda)y$, where $0 < \lambda < 1$, the following is true:

$$f(z) \leqq \lambda f(x) + (1 - \lambda) f(y),$$

or

$$f(z) \leqq \frac{y-z}{y-x} f(x) + \frac{z-x}{y-x} f(y), \tag{3.4}$$

where $\lambda = \frac{y-z}{y-x}$. It follows from (3.4) that

$$f(z) - f(x) \leqq \frac{x-z}{y-x} f(x) + \frac{z-x}{y-x} f(y) = \frac{z-x}{y-x}(f(y) - f(x)),$$
$$f(z) - f(y) \leqq \frac{y-z}{y-x} f(x) + \frac{z-y}{y-x} f(y) = \frac{y-z}{y-x}(f(x) - f(y)),$$

and consequently,

$$\frac{f(z) - f(x)}{z-x} \leqq \frac{f(y) - f(x)}{y-x} \leqq \frac{f(y) - f(z)}{y-z}. \tag{3.5}$$

By taking limits in inequality (3.5), $z \to x$ and $z \to y$, we obtain

$$f'(x) \leqq \frac{f(y) - f(x)}{y-x} \leqq f'(y). \tag{3.6}$$

It follows from (3.6) that $f'(x)$ is a nondecreasing function. If the function $f(x)$ is strictly convex, we get a strict inequality in (3.6); in other words, $f'(x)$ is then an increasing function.

(ii) Conversely, suppose that f' is nondecreasing. By the mean value theorem in calculus, there are some numbers θ_1 and θ_2, where $x \leqq \theta_1 \leqq z \leqq \theta_2 \leqq y$, such that

$$f'(\theta_1) = \frac{f(z) - f(x)}{z-x}$$

and

$$f'(\theta_2) = \frac{f(y) - f(z)}{y - z}.$$

Because $f'(\theta_1) \leqq f'(\theta_2)$, we obtain inequality (3.5) and consequently—using inequality (3.4)—the convexity of f.
If f' is increasing, the relation in (3.5) is a strict inequality and, in this case, f is strictly convex. □

Theorem 3.2. *If f is twice differentiable on D, f is convex on D if and only if f'' is nonnegative on D.*

Proof. It is well known in calculus that f' is nondecreasing on D if and only if f'' is nonnegative on D. □

Using the related function

$$\Phi(\mu) = f(\mathbf{x} + \mu\mathbf{s}), \quad \mu \in R, \quad \mathbf{x} \in R^n, \quad \mathbf{s} \in R^n,$$

we can extend the results of Theorems 3.1 and 3.2 to functions of n variables. For this purpose, the following theorem will be proved.

Theorem 3.3. *A function $f(\mathbf{x})$, $\mathbf{x} \in R^n$, is convex if and only if the function $\Phi(\mu)$, $\mu \in R$, is convex for any $\mathbf{x} \in R^n$, $\mathbf{s} \in R^n$.*

Proof.

(i) Suppose that $\Phi(\mu)$ is convex. We can write

$$f(\lambda\mathbf{y} + (1-\lambda)\mathbf{x}) = f(\mathbf{x} + \lambda(\mathbf{y} - \mathbf{x})) = \Phi(\lambda),$$

where $\mathbf{s} = \mathbf{y} - \mathbf{x}$. Then

$$\Phi(\lambda) = \Phi(\lambda \cdot 1 + (1-\lambda) \cdot 0) \leqq \lambda\Phi(1) + (1-\lambda)\Phi(0) = \lambda f(\mathbf{y}) + (1-\lambda) f(\mathbf{x}),$$

which implies the convexity of $f(\mathbf{x})$.

(ii) Conversely, now suppose that $f(\mathbf{x})$ is convex. We can write

$$\begin{aligned} \Phi(\lambda\mu + (1-\lambda)\nu) &= f(\mathbf{x} + \{\lambda\mu + (1-\lambda)\nu\}\mathbf{s}) = f(\mathbf{x} + \lambda\mu\mathbf{s} + (1-\lambda)\nu\mathbf{s}) \\ &= f(\lambda\{\mathbf{x} + \mu\mathbf{s}\} + (1-\lambda)\{\mathbf{x} + \nu\mathbf{s}\}) \\ &\leqq \lambda f(\mathbf{x} + \mu\mathbf{s}) + (1-\lambda) f(\mathbf{x} + \nu\mathbf{s}), \end{aligned}$$

where $f(\mathbf{x} + \mu\mathbf{s}) = \Phi(\mu)$ and $f(\mathbf{x} + \nu\mathbf{s}) = \Phi(\nu)$. Consequently, $\Phi(\lambda\mu + (1-\lambda)\nu) \leqq \lambda\Phi(\mu) + (1-\lambda)\Phi(\nu)$, so that Φ is convex. □

Using the gradient of the function $f(\mathbf{x})$, defined as the vector of partial derivatives, $\nabla f(\mathbf{x}) = (\frac{\partial f}{\partial x_1}, \ldots, \frac{\partial f}{\partial x_n})$, the following theorem turns out to be useful.

Theorem 3.4. *Let $f(\mathbf{x})$ be differentiable at a point $\mathbf{x}$ belonging to $D \in R^n$. f is convex only if*

$$f(\mathbf{y}) - f(\mathbf{x}) \geqq (\mathbf{y} - \mathbf{x})'\nabla f(\mathbf{x}) \quad \textit{and} \quad (\mathbf{y} - \mathbf{x})'\nabla f(\mathbf{y}) \geqq f(\mathbf{y}) - f(\mathbf{x}) \qquad (3.7)$$

for any $\mathbf{x} \in R^n$, $\mathbf{y} \in R^n$, where $\mathbf{x} \neq \mathbf{y}$.

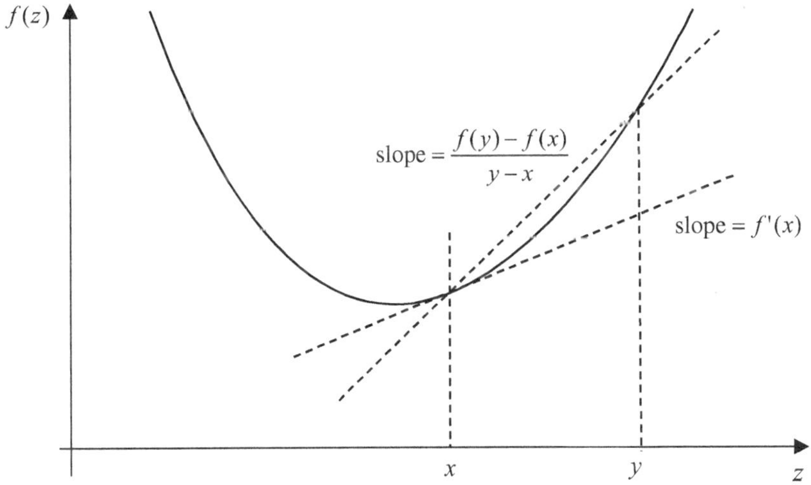

Fig. 3.5. A convex function.

Proof. It follows from (3.6) that a function $\Phi(\mu) = f(\mathbf{x}+\mu\mathbf{s})$ for $\mu \in R$ and $\mathbf{x} \in R^n$, $\mathbf{s} \in R^n$ is convex if and only if

$$\Phi'(\mu)(\nu - \mu) \leqq \Phi(\nu) - \Phi(\mu) \leqq \Phi'(\nu)(\nu - \mu) \tag{3.8}$$

holds for $\mu < \nu$. Now we set $\mu = 0$, $\nu = 1$, and $\mathbf{s} = \mathbf{y} - \mathbf{x}$, and we obtain from (3.8) the relations in (3.7). □

This theorem simply states that a hyperplane tangent to the hypersurface representing f lies below the hypersurface. Figure 3.5 gives a graphical interpretation of this statement when a single-variable function f is defined on a convex subset of R. In this case, the "hypersurface" is the curve $f(x)$, and the "hyperplane" is the straight line tangent $f(x) + f'(x)(y - x)$. The slope of the hyperplane tangent $f'(x)$ is not larger than the slope of the segment joining the points $f(x)$ and $f(y)$.

In terms of the Hessian matrix $[\frac{\partial^2 f(\mathbf{x})}{\partial x_i \partial x_j}]$, Theorem 3.4 can be formulated as follows.

Theorem 3.5. *Let $f(\mathbf{x})$ have continuous partial derivatives of the second order on a convex subset D of R^n. Then f is convex on D if the Hessian matrix $[\frac{\partial^2 f(\mathbf{x})}{\partial x_i \partial x_j}]$ is positive semidefinite for each point $\mathbf{x}$ belonging to D.*

Proof. Let $\mathbf{x} \in D$, $\mathbf{y} \in D$, and $\mathbf{s} = \mathbf{y} - \mathbf{x}$. Suppose that $f(\mathbf{x})$ is convex. Then by Theorem 3.3 the function $\Phi(\mu) = f(\mathbf{x} + \mu\mathbf{s})$ for $\mu \in R$ is convex. We differentiate the function $\Phi(\mu)$ twice:

$$\Phi''(\mu) = \frac{d}{d\mu} \sum_{i=1}^{n} \frac{\partial f(\mathbf{x} + \mu\mathbf{s})}{\partial x_i} s_i = \sum_{i=1}^{n} \sum_{j=1}^{n} \frac{\partial^2 f(\mathbf{x} + \mu\mathbf{s})}{\partial x_i \partial x_j} s_i s_j = \mathbf{s}' \nabla^2 f(\mathbf{x} + \mu\mathbf{s})\mathbf{s}.$$

From Theorem 3.2, we know that $\Phi''(\mu) \geqq 0$ for any $\mu \in R$, and therefore

$$\Phi''(0) = \mathbf{s}'\nabla^2 f(\mathbf{x})\mathbf{s} \geqq 0 \quad \text{for any } \mathbf{x} \in D \text{ and any } \mathbf{s} \in R^n.$$

This means that the Hessian matrix $\nabla^2 f(\mathbf{x})$ is positive semidefinite.

Conversely, suppose that $\nabla^2 f(\mathbf{x})$ is positive semidefinite for all $\mathbf{x} \in R^n$. Then for any $\mathbf{x} \in D$, $\mathbf{y} \in D$, $\mathbf{s} = \mathbf{y} - \mathbf{x}$,

$$\Phi''(\mu) = \mathbf{s}\nabla^2 f(\mathbf{x} + \mu\mathbf{s})\mathbf{s} \geqq 0 \quad \text{for any } \mu \in R.$$

Then it follows from Theorem 3.2 that $\Phi(\mu)$ is convex, and we can write

$$\begin{aligned} f(\lambda\mathbf{y} + (1-\lambda)\mathbf{x}) &= f(\mathbf{x} + \lambda\mathbf{s}) = \Phi(\lambda) = \Phi(\lambda 1 + (1-\lambda)0) \\ &\leqq \lambda\Phi(1) + (1-\lambda)\Phi(0) = \lambda f(\mathbf{y}) + (1-\lambda) f(\mathbf{x}) \end{aligned}$$

for any $\mathbf{x} \in D$ and any $\mathbf{y} \in D$; f is convex on D. □

Remark 3.1. As is well known in matrix theory, a quadratic form $\sum_i \sum_j a_{ij}x_i x_j$ with $a_{ij} = a_{ji}$ is positive semidefinite (definite) if and only if all the principal minors of the matrix $A = (a_{ij})$ are nonnegative (positive).

What are the implications of the convexity property of the functions $f_k(\mathbf{x})$ ($k = 0, 1, 2, \ldots, m$) for the mathematical programming problem (3.1)? The following lemma can be proved.

Lemma 3.1. *Let $f(\mathbf{x})$ be a convex function defined on D, where D is a convex set in R^n. Then the set*

$$S_b = \{\mathbf{x} | f(\mathbf{x}) \leqq b\}$$

is convex (and closed) for any number $b \in R$.

Proof. Let $\mathbf{x}, \mathbf{y} \in S_b$ and $\mathbf{z} = \lambda\mathbf{x} + (1-\lambda)\mathbf{y}$, where $0 \leqq \lambda \leqq 1$. Then

$$f(\mathbf{z}) \leqq \lambda f(\mathbf{x}) + (1-\lambda) f(\mathbf{y}) \leqq \lambda b + (1-\lambda) b = b,$$

and thus $\mathbf{z} \in S_b$. (The set S_b is closed because of the inequality $f(\mathbf{x}) \leqq b$.) □

Then the following property is very useful [17, pp. 17–18].

Theorem 3.6. *The intersection of any given number of convex sets S_b is a convex set.*

An immediate implication of Theorem 3.6 is the following.

Corollary 3.1. *If the functions $f_i(\mathbf{x})$, $\mathbf{x} \in R^n$ ($i = 1, 2, \ldots, m$), are convex, then the set*

$$K = \{\mathbf{x} | f_i(\mathbf{x}) \leqq 0 \ (i = 1, 2, \ldots, m)\}$$

is convex and closed.

In what follows, we restrict the convex programming problem (3.1) to the problem

$$\begin{aligned} &\text{minimize} \quad f_0(\mathbf{x}) \\ &\text{subject to} \quad f_i(\mathbf{x}) \leqq 0 \quad (i = 1, 2, \dots, m), \end{aligned} \tag{3.9}$$

where $f_k(\mathbf{x})$ $(k = 0, 1, \dots, m)$ are convex functions. The assumption of convexity for the objective function $f_0(\mathbf{x})$ in problem (3.1) or (3.9) is important because of the following properties of convex functions.

Theorem 3.7.

(a) *Each local minimum of a convex function $f(\mathbf{x})$ over a convex set D is also a global minimum of $f(\mathbf{x})$ over D.*
(b) *The set of minimizing points for a convex programming problem is convex.*
(c) *A minimum of any strictly convex function is unique.*

Proof.

(a) Assume that a point $\mathbf{x}^0$ in D is a local minimum of a function $f(\mathbf{x})$. Now suppose that there exists another point $\mathbf{y} \in D$ such that $f(\mathbf{y}) < f(\mathbf{x}^0)$. Because of the convexity of the function $f(\mathbf{x})$,

$$\begin{aligned} f(\lambda \mathbf{x}^0 + (1-\lambda)\mathbf{y}) &\leqq \lambda f(\mathbf{x}^0) + (1-\lambda) f(\mathbf{y}) \\ &< \lambda f(\mathbf{x}^0) + (1-\lambda) f(\mathbf{x}^0) = f(\mathbf{x}^0) \end{aligned} \tag{3.10}$$

for $0 < \lambda < 1$. If we let λ be close to one, it follows from (3.10) that in a sufficiently small neighborhood of the point $\mathbf{x}^0$, the value of the function $f(\mathbf{x})$ is smaller than $f(\mathbf{x}^0)$. But this is a contradiction to our assumption that $\mathbf{x}^0$ is a local minimizing point for $f(\mathbf{x})$ over D. Therefore, for an arbitrary point $\mathbf{y} \in D$, we must have $f(\mathbf{y}) \geqq f(\mathbf{x}^0)$.

(b) Let $\mathbf{x}^0 \in D$ and $\mathbf{y} \in D$ be two minimizing points for f over D; that is,

$$f(\mathbf{x}^0) = f(\mathbf{y}) = \min_{\mathbf{x} \in D} f(\mathbf{x}) = f^0.$$

Then we have to show that $\mathbf{z} = \lambda \mathbf{x}^0 + (1-\lambda)\mathbf{y}$, $0 < \lambda < 1$, is also a minimizing point for f over D.
Because of the convexity of the function $f(\mathbf{x})$, we obtain

$$f(\mathbf{z}) = f(\lambda \mathbf{x}^0 + (1-\lambda)\mathbf{y}) \leqq \lambda f(\mathbf{x}^0) + (1-\lambda) f(\mathbf{y}) = f^0,$$

which shows that each point on the segment between $\mathbf{x}^0$ and $\mathbf{y}$ is a minimizing point for f.

(c) Let us assume that a strictly convex function $f(\mathbf{x})$ has two distinct minimizing points $\mathbf{x}^0 \neq \mathbf{y}$. Then

$$f(\lambda \mathbf{x}^0 + (1-\lambda)\mathbf{y}) < \lambda f(\mathbf{x}^0) + (1-\lambda) f(\mathbf{y}) = f(\mathbf{x}^0) = f(\mathbf{y})$$

for $0 < \lambda < 1$, which is a contradiction to the assumption that $\mathbf{x}^0$ is a minimizing point for f. □

For the convex programming problem (3.9), we can summarize as follows:

- Each local minimizing point $\mathbf{x}^0$ is also a global minimizing point.
- The set of optimal solutions is convex.

Now we consider the general mathematical programming problem containing constraints in the form of inequalities and equations:

$$\begin{aligned} \text{minimize} \quad & f_0(\mathbf{x}) \\ \text{subject to} \quad & f_i(\mathbf{x}) \leqq 0 \quad (i = 1, 2, \ldots, m), \\ & g_h(\mathbf{x}) = 0 \quad (h = m+1, \ldots, r), \\ & \mathbf{x} \geqq \mathbf{0}. \end{aligned} \tag{3.11}$$

Substituting each equation $g_h(\mathbf{x})$ by two inequalities $g_h(\mathbf{x}) \leqq 0$ and $-g_h(\mathbf{x}) \leqq 0$, we have a convex programming problem if the functions $f_0(\mathbf{x})$, $f_i(\mathbf{x})$ $(i = 1, 2, \ldots, m)$, $g_h(\mathbf{x})$, and $-g_h(\mathbf{x})$ $(h = m+1, \ldots, r)$ are convex. This is only the case if the functions $g_h(\mathbf{x})$ $(h = m+1, \ldots, r)$ are linear.

Consequently, the mathematical programming problem (3.11) is a convex programming problem if

- the minimized objective function $f_0(\mathbf{x})$ is convex;
- the functions in the inequality constraints $f_i(\mathbf{x}) \leqq 0$ $(i = 1, 2, \ldots, m)$ are convex;
- the functions $g_h(\mathbf{x})$ in the equations $g_h(\mathbf{x}) = 0$ $(h = m+1, \ldots, r)$ are linear.

Remark 3.2. According to Definitions 3.2 and 3.4, the concavity and strict concavity of f correspond to the convexity and strict convexity of $-f$, respectively. Therefore, all the above results can easily be adapted to corresponding results on concave functions.

The implications of the convexity property for the Kuhn–Tucker conditions will be discussed in the next section.

3.2 Kuhn–Tucker Conditions for a Convex Programming Problem

In Chapter 2, we proved the Kuhn–Tucker conditions as necessary conditions for a local minimum of a mathematical programming problem (1.28). In the following theorem, it can be seen that for a convex programming problem (3.9), these conditions are also sufficient.

Theorem 3.8 (see [13]). *If Lagrange multipliers u_i^0 $(i = 1, 2, \ldots, m)$ exist for a convex programming problem* (3.9) *such that*

$$\nabla f_0(\mathbf{x}^0) + \sum_{i=1}^{m} u_i^0 \nabla f_i(\mathbf{x}^0) = 0, \tag{3.12}$$

$$f_i(\mathbf{x}^0) \leqq 0, \tag{3.13}$$

$$u_i^0 f_i(\mathbf{x}^0) = 0, \tag{3.14}$$

$$u_i^0 \geqq 0 \tag{3.15}$$

for $i = 1, 2, \ldots, m$, *then* $\mathbf{x}^0$ *is an optimal solution of problem* (3.9).

Proof. Because of the convexity of the functions $f_k(\mathbf{x})$ $(k = 0, 1, \ldots, m)$, it follows from Theorem 3.4 that

$$f_k(\mathbf{x}) - f_k(\mathbf{x}^0) \geqq (\mathbf{x} - \mathbf{x}^0)' \nabla f_k(\mathbf{x}^0) \quad (k = 0, 1, \ldots, m). \tag{3.16}$$

Using (3.16), (3.12), and then (3.16), (3.13), (3.14), and (3.15), we obtain

$$\begin{aligned} f_0(\mathbf{x}) - f_0(\mathbf{x}^0) &\geqq (\mathbf{x} - \mathbf{x}^0)' \nabla f_0(\mathbf{x}^0) = (\mathbf{x} - \mathbf{x}_0)' \left(-\sum_{i=1}^m u_i^0 \nabla f_i(\mathbf{x}^0) \right) \\ &= -\sum_{i=1}^m u_i^0 \nabla f_i(\mathbf{x}^0)(\mathbf{x} - \mathbf{x}^0) \\ &\geqq -\sum_{i=1}^m u_i^0 (f_i(\mathbf{x}) - f_i(\mathbf{x}^0)) = -\sum_{i=1}^m u_i^0 f_i(\mathbf{x}) + \sum_{i=1}^m u_i^0 f_i(\mathbf{x}^0) \geqq 0, \end{aligned}$$

so that

$$f_0(\mathbf{x}) \geqq f_0(\mathbf{x}^0),$$

and therefore $\mathbf{x}^0$ is an optimal solution to the convex programming problem (3.9). □

Thus for a convex programming problem (3.9) in which a suitable constraint qualification condition is met, the Kuhn–Tucker conditions are both necessary and sufficient for $\mathbf{x}^0$ to solve the mathematical programming problem.

The suitable constraint qualification condition for a convex programming problem (3.9) is formulated in a simple way.

Definition 3.6 (the Slater constraint qualification). The set of feasible solutions for a convex programming problem (3.9) satisfies the Slater constraint qualification if there exists a point $\mathbf{x}^* \in R^n$ for which

$$f_i(\mathbf{x}^*) < 0 \quad (i = 1, 2, \ldots, m),$$

that is, a point at which all inequality constraints are satisfied as strict inequalities.

For an activity analysis model, this condition implies that if the initial endowments of all primary factors are reduced proportionally by a certain amount, a positive amount of each intermediary and each final product can be produced.

As an example of the nonlinear programming problem, we take the *quadratic programming problem* (see the portfolio selection model in Section 1.2.7):

$$\begin{aligned} &\min f_0(\mathbf{x}) = \mathbf{x}'C\mathbf{x} + \mathbf{p}'\mathbf{x} \\ &\text{subject to} \quad A\mathbf{x} \leqq \mathbf{b}\mathbf{x} \geqq \mathbf{0}. \end{aligned} \tag{3.17}$$

Here $\mathbf{p}'$ is a given $1 \times n$ row vector, C is a given $n \times n$ symmetric matrix, A is a given $m \times n$ matrix, and $\mathbf{b}$ is a given $m \times 1$ column vector. Under the assumption that C is positive semidefinite, the objective function $f_0(\mathbf{x})$ is convex (because of Theorem 3.5) and the linear constraints $A\mathbf{x} \leqq \mathbf{b}$ are convex. The problem is therefore one of convex programming.

The Lagrange function is

$$\Phi(\mathbf{x}, \mathbf{u}) = \mathbf{x}'C\mathbf{x} + \mathbf{p}'x + \mathbf{u}'(A\mathbf{x} - \mathbf{b}),$$

and the Kuhn–Tucker conditions are

$$\frac{\partial \Phi}{\partial \mathbf{x}} = 2C\mathbf{x}^0 + \mathbf{p} + A'\mathbf{u}^0 \geqq \mathbf{0}, \tag{3.18}$$

$$\mathbf{x}^{0\prime}\frac{\partial \Phi}{\partial \mathbf{x}} = \mathbf{x}^{0\prime}(2C\mathbf{x}^0 + \mathbf{p} + A'\mathbf{u}^0) = 0, \tag{3.19}$$

$$\mathbf{x}^0 \geqq \mathbf{0}, \tag{3.20}$$

$$\frac{\partial \Phi}{\partial \mathbf{u}} = A\mathbf{x}^0 - \mathbf{b} \leqq \mathbf{0}, \tag{3.21}$$

$$\mathbf{u}^{0\prime}\frac{\partial \Phi}{\partial \mathbf{u}} = \mathbf{u}^{0\prime}(A\mathbf{x}^0 - \mathbf{b}) = 0, \tag{3.22}$$

$$\mathbf{u}^0 \geqq \mathbf{0}. \tag{3.23}$$

The Kuhn–Tucker conditions (3.18)–(3.23) are both necessary and sufficient. The vector $\mathbf{x}^0$ thus solves the quadratic programming problem (3.17) if and only if there is a $\mathbf{u}^0$ such that $\mathbf{x}^0, \mathbf{u}^0$ satisfy the Kuhn–Tucker conditions (3.18)–(3.23). Because of the linearity of conditions (3.18), (3.20), (3.21), and (3.23), many algorithms for quadratic programming are therefore based on the idea of finding solutions $\mathbf{x}^0$ and $\mathbf{u}^0$ to system (3.18)–(3.23). (Readers interested in quadratic programming problems are referred to [14].)

According to Theorem 2.2 in Section 2.3, a sufficient condition for $\mathbf{x}^0$ to solve the nonlinear programming problem (1.28) is that there exists a $\mathbf{u}^0$ such that $(\mathbf{x}^0, \mathbf{u}^0)$ is a saddle point of the Lagrange function $\Phi(\mathbf{x}, \mathbf{u})$. While this theorem does not require any convexity or constraint qualification assumption, the converse of the theorem does require such assumptions.

Theorem 3.9 (see [19]). *Assume that the Slater constraint qualification for a convex programming problem* (3.9) *is met. Then if* $\mathbf{x}^0$ *solves problem* (3.9)*, there exists a nonnegative vector* $\mathbf{u}^0$ *such that* $(\mathbf{x}^0, \mathbf{u}^0)$ *is a saddle point of the Lagrange function*

$$\Phi(\mathbf{x}, \mathbf{u}) = f_0(\mathbf{x}) + \sum_{i=1}^{m} u_i f_i(\mathbf{x})$$

with respect to the set $N \times M$*, where* $N = \{\mathbf{x} | \mathbf{x} \in R^n\}$ *and* $M = \{\mathbf{u} | \mathbf{u} \in R^m, \mathbf{u} \geqq \mathbf{0}\}$.[1]

[1] For the proof, see, e.g., [15, pp. 215–217] or [18, pp. 432–497].

Thus if the Slater constraint qualification is met, then an optimal solution of the convex programming problem (3.9) is equivalent to a saddle point of the Lagrange function.

The Slater theorem is more general than the Kuhn–Tucker theorem because it does not require differentiability of the functions $f_k(\mathbf{x})$ $(k = 0, 1, \ldots, m)$. A disadvantage of this theorem is that the Slater constraint qualification is related to inequality constraints only. Fortunately, it can be shown [2, pp. 35–37] that Theorem 3.9 remains valid if the Slater constraint qualification is met for the nonlinear constraints. (For a convex programming problem (3.11), the equality conditions must be linear.)

Remark 3.3. It is easy to formulate the counterparts of Theorems 3.8 and 3.9 for a constrained maximization problem with the convexity of $f_k(\mathbf{x})$ $(k = 0, 1, \ldots, m)$ replaced by concavity.

More generally, the Kuhn–Tucker conditions as sufficient conditions for $\mathbf{x}^0$ to solve the nonlinear programming problem can also be extended to problems where $f_k(\mathbf{x})$ $(k = 0, 1, 2, \ldots, m)$ are *quasiconvex*.

Definition 3.7. A function $f(\mathbf{x})$ defined on a convex set D in R^n is called quasiconvex if the set $S_b = \{\mathbf{x} | f(\mathbf{x}) \leqq b\}$ is convex for any real number $b \in R$.

Examples of quasiconvex functions are the strictly concave function $f(x) = x^{1/2}$ defined on $D = \{x | x \in R,\ x \geqq 0\}$, the function $f(x) = \ln x$, or the function in Figure 3.6.

Theorem 3.10. *A function $f(\mathbf{x})$ defined on a convex set D is quasiconvex if and only if for any given $\mathbf{x} \in D$, $\mathbf{y} \in D$, and $0 < \lambda < 1$,*

$$f(\lambda\mathbf{x} + (1 - \lambda)\mathbf{y}) \leqq \max[f(x), f(y)] \tag{3.24}$$

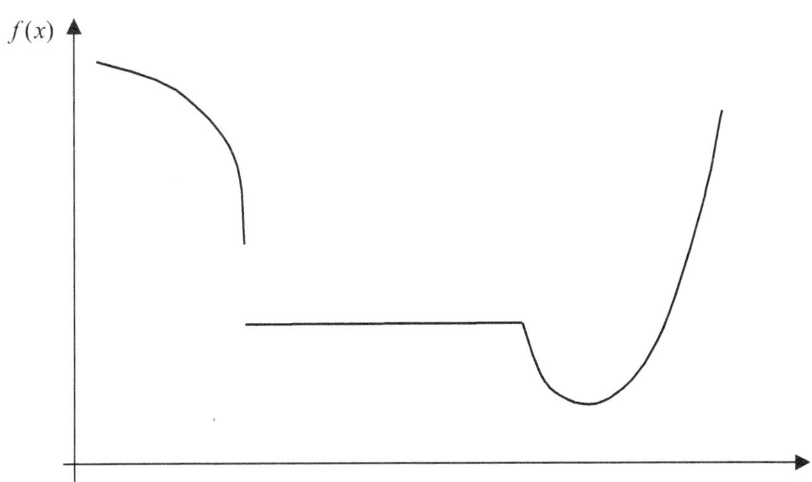

Fig. 3.6. A quasiconvex function.

is strictly quasiconvex if and only if strict inequality holds.[2]

A function $f(\mathbf{x})$ is quasiconcave if and only if $-f(x)$ is quasiconvex. This extension to quasiconvex functions is a generalization since a function that is convex (concave) is also quasiconvex (quasiconcave), but not vice versa. The Cobb–Douglas production function

$$y = ax_1^{\alpha}x_2^{\beta},$$

with $a > 0$, $\alpha > 0$, $\beta > 0$, is quasiconcave but it is not concave if $\alpha + \beta > 1$. Convexity is more restrictive than quasiconvexity.

Quasiconcave functions "are used widely in microeconomics and are the principal ones that will be encountered in this book" [16, p. 67]. The strict quasiconcavity is used to rule out the possibility of indifference curves having linear segments. In the case in which the indifference curve has a linear segment, the utility maximizing consumer will be indifferent among various bundles of commodities. The assumption of a diminishing marginal rate of substitution is equivalent to assuming quasiconcavity of the utility function.

Similarly to the proof of Theorem 3.7(a), the following property for the strictly quasiconvex function can be shown (see [5, p. 89] or [10, p. 86]).

Theorem 3.11. *Each local minimum (maximum) of a strictly quasiconvex (strictly quasiconcave) function $f(\mathbf{x})$ over a convex set D is also a global minimum (maximum) of $f(\mathbf{x})$ over D.*

Thus the strictly quasiconvex functions exhibit the same features as the convex functions. They imply that

- the set of feasible solutions is convex and
- each local minimum is a global one.

These results are very useful because they rest on weaker conditions than those of convexity as supposed in the Kuhn–Tucker theory.

The following theorem, proved by Arrow and Enthoven [1, pp. 783–787], provides an extension of the Kuhn–Tucker conditions as sufficient optimality conditions for a *quasiconvex programming* problem.

Theorem 3.12. *Let $f_k(\mathbf{x})$ $(k = 0, 1, \ldots, m)$ be differentiable quasiconvex functions defined for $\mathbf{x} \geqq \mathbf{0}$. Let $\mathbf{x}^0$ and $\mathbf{u}^0$ satisfy the Kuhn–Tucker conditions* (2.20)–(2.25) *from Chapter* 2*, and let one of the following conditions be satisfied:*

(a) $\frac{\partial f_0(\mathbf{x}^0)}{\partial x_{j_0}} > 0$ *for at least one variable* x_{j_0};
(b) $\frac{\partial f_0(\mathbf{x}^0)}{\partial x_{j_1}} < 0$ *for some relevant variable* x_{j_1};
(c) $\frac{\partial f_0(\mathbf{x}^0)}{\partial \mathbf{x}} \neq \mathbf{0}$ *and $f_0(\mathbf{x})$ is twice differentiable in the neighborhood of* $\mathbf{x}^0$;
(d) $f_0(\mathbf{x})$ *is convex.*

[2] For the proof, see, e.g., [10, p. 82] or [15, p. 82].

Then $\mathbf{x}^0$ *minimizes* $f_0(\mathbf{x})$ *subject to the constraints* $f_i(\mathbf{x}) \leqq 0$ $(i = 1, 2, \ldots, m)$ *and* $\mathbf{x} \geqq \mathbf{0}$.

A *relevant* variable is one that can take on a positive value without necessarily violating the constraints. Or, more formally, x_{j_0} is a *relevant* variable if there is some point in the constraint set, say, $\mathbf{x}^*$, at which $x^*_{j_0} > 0$.

Only one of the above four conditions—and there may be others—need be satisfied for $\mathbf{x}^0$ to minimize $f_0(\mathbf{x})$ subject to the constraints if the Kuhn–Tucker conditions are satisfied at $\mathbf{x}^0$. Condition (a) will be satisfied if any $\frac{\partial f_0(\mathbf{x}^0)}{\partial x_{j_0}} > 0$ and all x_j are relevant (the usual case in economic theory). If no x_{j_0} is relevant, the problem is trivial. From (a) and (b), it follows that $\frac{\partial f_0(\mathbf{x}^0)}{\partial \mathbf{x}} \neq 0$ is sufficient if all x_j are relevant.

Looking at the quasiconvex programming problem (2.33)–(2.35) from Section 2.4 (the objective function $f_0(x) = (x-1)^3$ is quasiconcave), we can see that the Kuhn–Tucker conditions (2.36)–(2.40) alone are not sufficient for $x^0 = 1$ to solve this problem.

None of conditions (a)–(d) is fulfilled, and the Kuhn–Tucker conditions fail to be sufficient in this case. However, condition (c) is fulfilled for the point $x^* = 2$, which is the optimal solution of the quasiconvex problem (2.33)–(2.35).

3.3 Duality Theory

The mathematical programming model (1.1) described in Chapter 1 has been related to the basic economic problem of the allocation of scarce resources among alternative uses. An economist "is familiar with the fact that resource allocation and pricing are two aspects of the same problem. An economist would expect that since linear programming solves the allocation problem, it would solve the pricing problem also, and this, in essence, is what the dualism property consists in" [7, p. 39].

Duality signifies that every mathematical programming problem is closely related to another problem called its "dual." The basic ideas of duality theory were first developed by John von Neumann in his book with Oskar Morgenstern on the theory of games. "What I am doing is conjecturing that the two problems are equivalent. The theory that I am outlining for your problem is an analogue to the one we have developed for games," said von Neumann during his first visit to George Dantzig on October 3, 1974 [6, p. 81]. "Thus I learned about Farkas' lemma, and about duality for the first time" [6, p. 81]. David Gale, Harold Kuhn, and A. W. Tucker are credited as the publishers of the first rigorous proof of the duality theorem.

Far from being confined only to linear programming, the following reasons for the relevance and usefulness of the duality concept for both economists and mathematicians listed by Baumol [4, p. 103] can be affirmed for mathematical programming models:

1. Duality yields a number of powerful theorems which add substantially to our understanding of linear programming.

2. Duality analysis has been very helpful in the solution of programming problems. Indeed, as we shall see, it is frequently easier to find the solution of a programming problem by first solving its associated dual problem.
3. The dual problem turns out to have an extremely illuminating economic interpretation.

Let us consider the general mathematical programming problem (1.28) as defined in Chapter 1:

$$\underset{\mathbf{x}\in R^n}{\text{minimize}} \quad f_0(\mathbf{x}) \tag{3.25a}$$

$$\text{subject to} \quad f_i(\mathbf{x}) \leqq 0 \quad (i = 1, 2, \ldots, m). \tag{3.25b}$$

Using the Lagrange function

$$\Phi(\mathbf{x}, \mathbf{u}) = f_0(\mathbf{x}) + \sum_{i=1}^{m} u_i f_i(\mathbf{x}),$$

problem (3.25) can be written more generally as

$$\min_{\mathbf{x}\in R^n} \sup_{\mathbf{u}\in R^m_+} \Phi(\mathbf{x}, \mathbf{u}). \tag{3.26}$$

That this formulation is valid is shown by the following argument:

$$\begin{aligned} \sup_{\mathbf{u}\in R^m_+} \Phi(\mathbf{x}, \mathbf{u}) &= \sup_{\mathbf{u}\in R^m_+} \Big(f_0(\mathbf{x}) + \sum_{i=1}^{m} u_i f_i(\mathbf{x})\Big) \\ &= \begin{cases} f_0(\mathbf{x}) & \text{for } f_i(\mathbf{x}) \leqq 0, \\ +\infty & \text{otherwise.} \end{cases} \end{aligned}$$

Therefore, $\min\{f_0(\mathbf{x}) | (f_i(\mathbf{x}) \leqq 0)\} = \min_{\mathbf{x}\in R^n} \sup_{\mathbf{u}\in R^m_+} \Phi(\mathbf{x}, \mathbf{u})$. Problem (3.26) is called the *primal problem*.

Definition 3.8. The problem

$$\max_{\mathbf{u}\in R^m_+} \inf_{\mathbf{x}\in R^n} \Phi(\mathbf{x}, \mathbf{u}) \tag{3.27}$$

is called the dual problem associated with the primal problem (3.26).

Then the so-called *weak duality theorem* can be proven.

Theorem 3.13. *If* $\mathbf{x}^0$ *is a feasible solution of the primal problem* (3.26) *and* $\mathbf{u}^0$ *is a feasible solution of the dual problem* (3.27), *then the objective function value of problem* (3.26) *at the point* $\mathbf{x}^0$ *is not less than the objective function value of problem* (3.27) *at the point* $\mathbf{u}^0$.

Proof. We denote the objective function of the dual problem (3.26) by $d(\mathbf{u}) := \inf_{\mathbf{x}\in R^n} \Phi(\mathbf{x}, \mathbf{u})$. Because $\mathbf{u}^0 \geqq \mathbf{0}$, $f_i(\mathbf{x}) \leqq 0$ $(i = 1, 2, \ldots, m)$ and the definition of infimum as a greatest lower bound, we can write

$$\begin{aligned} d(\mathbf{u}^0) &= \inf_{\mathbf{x}\in R^n} \left[f_0(\mathbf{x}) + \sum_{i=1}^{m} u_i^0 f_i(\mathbf{x}) \right] \\ &\leqq f_0(\mathbf{x}^0) + \sum_{i=1}^{m} u_i^0 f_i(\mathbf{x}^0) \leqq f_0(\mathbf{x}^0). \end{aligned} \tag{3.28}$$

The theorem is proved. □

The difference between the optimal objective function values of (3.26) and (3.27) when (3.28) holds as a strict inequality is called the *duality gap*.

Replacing supremum by maximum and infimum by minimum, we can write the primal problem (3.26) as

$$\min_{\mathbf{x}\in R^n} \max_{\mathbf{u}\in R_+^m} \Phi(\mathbf{x}, \mathbf{u}) \tag{3.29}$$

and the dual problem (3.27) as

$$\max_{\mathbf{u}\in R_+^m} \min_{\mathbf{x}\in R^n} \Phi(\mathbf{x}, \mathbf{u}). \tag{3.30}$$

Assuming that the functions $f_0(\mathbf{x})$ and $f_i(\mathbf{x})$ are all real-valued differentiable *convex* functions on R^n, and that regularity conditions in the sense of the Slater constraint qualification are satisfied, problem (3.25) is a convex programming problem and the dual problem (3.30) is (due to Wolfe [21])

$$\text{maximize} \quad \Phi(\mathbf{x}, \mathbf{u}) = f_0(\mathbf{x}) + \sum_{i=1}^{m} u_i f_i(\mathbf{x}) \tag{3.31a}$$

$$\text{subject to} \quad \nabla f_0(\mathbf{x}) + \sum_{i=1}^{m} u_i \nabla f_i(\mathbf{x}) = \mathbf{0}, \tag{3.31b}$$

$$\mathbf{u} \geqq \mathbf{0}. \tag{3.31c}$$

The constraints (3.31b) express the minimization of the Lagrange function $\Phi(\mathbf{x}, \mathbf{u})$ with respect to $\mathbf{x} \in R^n$.

The primal problem (3.29) is then

$$\text{minimize} \quad \Phi(\mathbf{x}, \mathbf{u}) - \sum_{i=1}^{m} u_i \frac{\partial \Phi}{\partial u_i} \tag{3.25a$'$}$$

$$\text{subject to} \quad \frac{\partial \Phi}{\partial u_i} \leqq 0 \quad (i = 1, 2, \ldots, m), \tag{3.25b$'$}$$

which corresponds exactly to problem (3.25). The constraints (3.25b′) of the primal problem are the necessary conditions for the maximization of the Lagrange function $\Phi(\mathbf{x}, \mathbf{u})$ with respect to the nonnegativity condition for the variables $u_1, \ldots, u_m$. In a similar way, the dual problem (3.31) can be written as

$$\text{maximize} \quad \Phi(\mathbf{x}, \mathbf{u}) - \sum_{j=1}^{n} x_j \frac{\partial \Phi}{\partial x_j} \tag{3.31a$'$}$$

$$\text{subject to} \quad \frac{\partial \Phi}{\partial x_j} = 0 \quad (j = 1, 2, \ldots, n), \tag{3.31b$'$}$$

$$u_i \geqq 0 \quad (i = 1, 2, \ldots, m). \tag{3.31c$'$}$$

The first relationship between the primal problem (3.25) and the dual problem (3.31) is given by Theorem 3.13, which states that for any feasible solution $\mathbf{x}$ of problem (3.25) and for any feasible solution $(\mathbf{x}, \mathbf{u})$ of problem (3.31),

$$\Phi(\mathbf{x}, \mathbf{u}) \leqq f_0(\mathbf{x}) \text{ holds.} \tag{3.32}$$

In other words, the value of the objective of the maximization problem over feasible solutions is never greater than the value of the objective of the minimization problem over feasible solutions.

It follows from (3.32) that if the values of the dual objective function (3.31a) and the primal objective function (3.25a) are the same, $(\mathbf{x}, \mathbf{u})$ must be an optimal solution of the dual problem (3.31) and $\mathbf{x}$ an optimal solution of the primal problem (3.25), respectively.

Some relationships between the primal problem (3.25) and the dual problem (3.31) that require the convexity assumption for the functions $f_k(\mathbf{x})$ $(k = 0, 1, \ldots, m)$ are described by the following two theorems.

Theorem 3.14 (strong duality theorem). *If* $\mathbf{x}^0$ *is an optimal solution of problem* (3.25), *then there exists a vector* $\mathbf{u}^0$ *such that* $(\mathbf{x}^0, \mathbf{u}^0)$ *is an optimal solution of the dual problem* (3.31) *and the value of the primal objective function* $f_0(\mathbf{x}^0)$ *is equal to the value of the dual objective function* $\Phi(\mathbf{x}^0, \mathbf{u}^0)$.

Proof. If $\mathbf{x}^0$ is an optimal solution of the primal problem (3.25), where $f_k(\mathbf{x})$ $(k = 0, 1, \ldots, m)$ are convex functions, then there exists—according to the Kuhn–Tucker theorem from Chapter 2—a vector of Lagrange multipliers $\mathbf{u}^0$ such that conditions (2.1)–(2.4) are fulfilled. Then it follows from (2.1) and (2.4) that $(\mathbf{x}^0, \mathbf{u}^0)$ is a feasible solution of the dual problem (3.31), and it follows from (2.2) that $\mathbf{x}^0$ is a feasible solution of problem (3.25). Finally, using condition (2.3), we obtain

$$f_0(\mathbf{x}^0) = \Phi(\mathbf{x}^0, \mathbf{u}^0),$$

and because of the weak duality theorem, the vector $(\mathbf{x}^0, \mathbf{u}^0)$ is an optimal solution of the dual problem (3.31). □

For the converse, we need an additional assumption, for example, that employed in [11] or [10], which is formulated in the following.

Theorem 3.15. *Suppose that the matrix of second partial derivatives*

$$\left(\frac{\partial^2\Phi(\mathbf{x}^0,\mathbf{u}^0)}{\partial x_j\partial x_k}\right)=\frac{\partial^2\Phi(\mathbf{x}^0,\mathbf{u}^0)}{\partial x^2}=\nabla_{\mathbf{x}}^2\Phi(\mathbf{x}^0,\mathbf{u}^0) \tag{3.33}$$

is nonsingular. If $(\mathbf{x}^0,\mathbf{u}^0)$ *solves* (3.31), *then* $\mathbf{x}^0$ *solves* (3.25).

Proof. If $(\mathbf{x}^0,\mathbf{u}^0)$ is an optimal solution to problem (3.31), then it is also an optimal solution to the following problem:

$$\min\left\{-\Phi(\mathbf{x},\mathbf{u})\|\frac{\partial\Phi}{\partial\mathbf{x}}=\mathbf{0},\ \mathbf{u}\geqq\mathbf{0}\right\}. \tag{3.34}$$

The Lagrange function and the Kuhn–Tucker conditions for problem (3.34) yield

$$L(\mathbf{x},\mathbf{u},\boldsymbol{\lambda})=-\Phi(\mathbf{x},\mathbf{u})+\boldsymbol{\lambda}\frac{\partial\Phi}{\partial\mathbf{x}},$$

$$\frac{\partial L^0}{\partial\mathbf{x}}=-\frac{\partial\Phi^0}{\partial\mathbf{x}}+\boldsymbol{\lambda}^0\frac{\partial^2\Phi^0}{\partial\mathbf{x}^2}=\mathbf{0}, \tag{a}$$

$$\frac{\partial L^0}{\partial\mathbf{u}}=-\frac{\partial\Phi^0}{\partial\mathbf{u}}+\boldsymbol{\lambda}^0\frac{\partial^2\Phi^0}{\partial\mathbf{x}\partial\mathbf{u}}\geqq\mathbf{0}, \tag{b}$$

$$\frac{\partial L^0}{\partial u}\mathbf{u}^0=-\frac{\partial\Phi^0}{\partial\mathbf{u}}\mathbf{u}^0+\boldsymbol{\lambda}^0\frac{\partial^2\Phi^0}{\partial\mathbf{x}\partial\mathbf{u}}\mathbf{u}^0=0, \tag{c}$$

$$\mathbf{u}^0\geqq\mathbf{0}, \tag{d}$$

$$\frac{\partial L^0}{\partial\boldsymbol{\lambda}}=\frac{\partial\Phi^0}{\partial\mathbf{x}}=\mathbf{0}, \tag{e}$$

where $L^0=L(\mathbf{x}^0,\mathbf{u}^0,\boldsymbol{\lambda}^0)$.

It follows from (a) and (e) that

$$\boldsymbol{\lambda}^0\frac{\partial^2\Phi^0}{\partial\mathbf{x}^2}=\mathbf{0}.$$

In our assumption, the matrix (3.33) is nonsingular and therefore $\boldsymbol{\lambda}^0=\mathbf{0}$. Substitution of $\boldsymbol{\lambda}^0=\mathbf{0}$ in conditions (a)–(e) provides

$$\frac{\partial\Phi^0}{\partial\mathbf{x}}=\mathbf{0},\qquad\frac{\partial\Phi^0}{\partial\mathbf{u}}\leqq\mathbf{0},\qquad\mathbf{u}^0\frac{\partial\Phi^0}{\partial\mathbf{u}^0}=0,\qquad\mathbf{u}^0\geqq\mathbf{0},$$

which are the Kuhn–Tucker conditions for the primal problem (3.25). Because of the convexity of the functions $f_k(\mathbf{x})$ $(k=0,1,\dots,m)$ and due to Theorem 3.8, the vector $\mathbf{x}^0$ is the optimal solution of problem (3.25). □

Because the nonnegativity conditions very often occur in economic models, we conclude this section with a formulation of the Wolfe dual problem to problem (1.28a):

$$\begin{array}{ll} \text{minimize} & f_0(\mathbf{x}) \\ \text{subject to} & f_i(\mathbf{x}) \leqq 0 \quad (i = 1, 2, \ldots, m), \\ & -x_j \leqq 0 \quad (j = 1, 2, \ldots, n). \end{array}$$

The Lagrange function is

$$\Psi(\mathbf{x}, \mathbf{u}, \mathbf{w}) = f_0(\mathbf{x}) + \sum_{i=1}^{m} u_i f_i(\mathbf{x}) + \sum_{j=1}^{n} w_j(-x_j) = \Phi(\mathbf{x}, \mathbf{u}) - \mathbf{w}'\mathbf{x},$$

and the dual problem becomes

$$\begin{array}{ll} \text{maximize} & \Phi(\mathbf{x}, \mathbf{u}) - \mathbf{w}'\mathbf{x} \\ \text{subject to} & \dfrac{\partial \Phi}{\partial \mathbf{x}} - \mathbf{w} = \mathbf{0}, \\ & \mathbf{u} \geqq \mathbf{0}, \quad \mathbf{w} \geqq \mathbf{0}. \end{array}$$

Elimination of the Lagrange multipliers $\mathbf{w}$ yields (since $\mathbf{x}'\frac{\partial \Phi}{\partial \mathbf{x}} = \mathbf{x}'\mathbf{w} = \mathbf{w}'\mathbf{x} = 0$)

$$\begin{array}{ll} \text{maximize} & \Phi(\mathbf{x}, \mathbf{u}) - \mathbf{x}'\dfrac{\partial \Phi}{\partial \mathbf{x}} \\ \text{subject to} & \dfrac{\partial \Phi}{\partial \mathbf{x}} \geqq \mathbf{0}, \\ & \mathbf{u} \geqq \mathbf{0}. \end{array}$$

3.4 Economic Interpretation of Duality in Convex Programming

Let us consider a primal nonlinear programming problem for determining a product mix that maximizes profit,

$$\Pi(\mathbf{x}) = f_0(x_1, x_2, \ldots, x_n), \tag{3.35}$$

subject to the constraints

$$f_i(x_1, x_2, \ldots, x_n) \leqq b_i \quad (i = 1, 2, \ldots, m), \tag{3.36}$$

which limit the use of resource i for production to the quantity available b_i, and

$$x_j \geqq 0 \quad (j = 1, 2, \ldots, n), \tag{3.37}$$

which ensure nonnegative production of product j $(j = 1, 2, \ldots, n)$.

Assuming that $f_0(\mathbf{x})$ is concave and $f_i(\mathbf{x})$ $(i = 1, 2, \ldots, m)$ are convex functions, the production model (3.35)–(3.37) is a convex programming problem. What are the economic implications of these concavity–convexity assumptions? As shown in Section 3.1, convexity of the functions $f_i(\mathbf{x})$ $(i = 1, 2, \ldots, m)$ implies that the constraint set of the primal problem is convex. Interpreting feasible solutions $\mathbf{x}$ to be possible production levels, this means that given any two feasible levels $\mathbf{x}'$ and $\mathbf{x}''$, all linear convex combinations or "weighted average mixes" $\alpha\mathbf{x}' + (1-\alpha)\mathbf{x}''$, $0 \leqq \alpha \leqq 1$, must also be feasible. In other words, one of the implications of the convexity assumption is that the produced items come in divisible units.

The concavity of the profit function $f_0(\mathbf{x})$ means that marginal profit is nonincreasing. Mathematically, these assumptions imply that if $\mathbf{x}^0$ is a local maximum for f_0 under the constraints (3.36)–(3.37), then it is also a global maximum.

Remark 3.4. The reader can easily verify that a convex cost function (concave production function) implies nondecreasing marginal cost (nonincreasing marginal products) but that the converse implication is, in general, not valid.

Before we turn to the economic interpretation of duality, it will be useful to review the Kuhn–Tucker conditions for the production model (3.35)–(3.37). Using the Lagrange function

$$\Phi(\mathbf{x}, \mathbf{u}) = f_0(\mathbf{x}) + \sum_{i=1}^{m} u_i (b_i - f_i(\mathbf{x})),$$

the necessary and sufficient conditions (due to the concavity–convexity assumptions) for $(\mathbf{x}^0, \mathbf{u}^0)$ to be a solution of the problem (3.35)–(3.37) are

$$\frac{\partial \Phi(\mathbf{x}^0, \mathbf{u}^0)}{\partial x_j} = \frac{\partial f_0(\mathbf{x}^0)}{\partial x_j} - \sum_{i=1}^{m} u_i^0 \frac{\partial f_i(\mathbf{x}^0)}{\partial x_j} \leqq 0,$$

or

$$\frac{\partial f_0(\mathbf{x}^0)}{\partial x_j} \leqq \sum_{i=1}^{m} u_i^0 \frac{\partial f_i(\mathbf{x}^0)}{\partial x_j} \qquad (j = 1, 2, \ldots, n), \qquad (3.38)$$

$$x_j^0 \frac{\partial \Phi(\mathbf{x}^0, \mathbf{u}^0)}{\partial x_j} = x_j^0 \left[\frac{\partial f_0(\mathbf{x}^0)}{\partial x_j} - \sum_{i=1}^{m} u_i^0 \frac{\partial f_i(\mathbf{x}^0)}{\partial x_j} \right] = 0, \qquad (3.39)$$

$$x_j^0 \geqq 0 \qquad (j = 1, 2, \ldots, n), \qquad (3.40)$$

$$\frac{\partial \Phi(\mathbf{x}^0, \mathbf{u}^0)}{\partial u_i} = b_i - f_i(\mathbf{x}^0) \geqq 0 \qquad (i = 1, 2, \ldots, m), \qquad (3.41)$$

$$u_i^0 \frac{\partial \Phi(\mathbf{x}^0, \mathbf{u}^0)}{\partial u_i} = u_i^0 [b_i - f_i(\mathbf{x}^0)] = 0 \qquad (i = 1, 2, \ldots, m), \qquad (3.42)$$

$$u_i^0 \geqq 0 \qquad (i = 1, 2, \ldots, m). \qquad (3.43)$$

Let us, according to (2.9), tentatively interpret the Lagrange multipliers $u_i^0, u_2^0, \ldots, u_m^0$ as the shadow prices or accounting values for the m scarce resources as expressed by the constraints (3.36). Then the Kuhn–Tucker conditions (3.38) have a clear economic interpretation. The left side, $\frac{\partial f_0(\mathbf{x}^0)}{\partial x_j}$, is the marginal profit yield of product j. $\frac{\partial f_i(\mathbf{x}^0)}{\partial x_j}$ is simply the (marginal) quantity of input i needed to produce another unit of product j. Consequently, $\frac{u_i^0 \partial f_i(\mathbf{x}^0)}{\partial x_j}$ is the accounting value of the amount of input i needed to produce an incremental unit of product j. Thus the right side of (3.38) is the value of all (scarce) resources needed to produce an additional unit of product j. Therefore, the inequalities (3.38) imply that the marginal profit of any product j cannot be higher than its marginal accounting input cost.

Now consider conditions (3.39), which assert that if the marginal profit yield of product j is lower than the imputed accounting value of its inputs, then product j will, optimally, not be produced because it incurs an opportunity loss. It is better to use the inputs for the production of other products. If product j is produced, $x_j > 0$, then its marginal profit must equal the accounting value of its marginal input requirements, $\sum_{i=1}^m \frac{u_i^0 \partial f_i(\mathbf{x}^0)}{\partial x_j}$.

The economic interpretation of the Kuhn–Tucker conditions (3.41) is obvious; the amount of input i used by an optimal product mix cannot exceed the available quantity b_i $(i = 1, 2, \ldots, m)$.

In economic terms, the conditions (3.42) are more interesting. They assert that if in an optimal solution, there is an unused amount of input i $(f_i(\mathbf{x}^0) < b_i)$, then it must be a resource without deficit or a resource in excess (i.e., a resource with accounting value zero). An additional amount of this resource does not change the optimal value of the objective function (due to (2.9)). If the accounting value u_i^0 is positive, then all of its available amount b_i will be used by an optimal solution $(f_i(\mathbf{x}^0) = b_i)$.

We now turn to the interpretation of the dual problem corresponding to the primal problem (3.35)–(3.37).

According to the previous section, the dual takes the following form:

$$\text{minimize} \quad \Phi(\mathbf{x}, \mathbf{u}) - \mathbf{x}' \frac{\partial \Phi}{\partial \mathbf{x}} = f_0(\mathbf{x}) + \sum_{i=1}^m u_i [b_i - f_i(\mathbf{x})] - \sum_{j=1}^n x_j \left[\frac{\partial f_0(\mathbf{x})}{\partial x_j} - \sum_{i=1}^m u_i \frac{\partial f_i(\mathbf{x})}{\partial x_j} \right] \tag{3.44}$$

$$\text{subject to} \quad \sum_{i=1}^m u_i \frac{\partial f_i(\mathbf{x})}{\partial x_j} \geqq \frac{\partial f_0(\mathbf{x})}{\partial x_j} \quad (j = 1, 2, \ldots, n), \tag{3.45}$$

$$u_i \geqq 0 \quad (i = 1, 2, \ldots, m). \tag{3.46}$$

The dual constraints (3.45)–(3.46) coincide with the Kuhn–Tucker conditions (3.38) and (3.43); therefore, their economic interpretation is given above.

However, in that discussion, we tentatively took the Lagrange multipliers u_i^0 $(i = 1, 2, \ldots, m)$ to be accounting values or shadow prices. More exactly, one might wish that the value u_i^0 of the ith dual variable at optimum be equal to the marginal profit of the ith input (due to (2.9)). In what sense is this valid?

The first recognized papers dealing with this question are by Gale [9] and Balinski and Baumol [3]. The problem arises because $\frac{\partial \Pi(\mathbf{x}^0(\mathbf{b}))}{\partial b_i}$ is not always defined. For some changes in the values of the b_i, there can be points of discontinuity in the derivatives or points where there exist finite right and left-side partial derivatives, $\frac{\partial \Pi(\mathbf{x}^0(\mathbf{b}))}{\partial b_i^+}$ and $\frac{\partial \Pi(\mathbf{x}^0(\mathbf{b}))}{\partial b_i^-}$, but at which $\frac{\partial \Pi(\mathbf{x}^0(\mathbf{b}))}{\partial b_i}$ itself is not defined.

However, there is an earlier result by Uzawa [20], quoted in [3] and [9], that comes very close to the desired proposition that u_i^0 equal $\frac{\partial \Pi}{\partial b_i}$. While the optimal value of the dual variable u_i^0 cannot always be interpreted as the marginal profit of the ith input, since the latter is not always defined, u_i^0 will invariably lie between the corresponding right- and left-side partial derivatives, which exist and are finite, given that the constraints qualification holds. In symbols, we have

$$\frac{\partial \Pi(\mathbf{x}^0(\mathbf{b}))}{\partial b_i^+} \leqq u_i^0 \leqq \frac{\partial \Pi(\mathbf{x}^0(\mathbf{b}))}{\partial b_i^-} \tag{3.47}$$

for all i. An intuitive geometric as well as a rigorous algebraic proof of this proposition is given in [3, pp. 243–245] and [9, p. 24]. Independently of these papers, a more general proof of the proposition (3.47) was developed by Horst [12].

Theorem 3.16. *Let* (3.35)–(3.37) *be a convex problem, and assume that either Slater's constraint qualification is satisfied or the constraints* (3.36) *are linear. Moreover, let the set* $\mathcal{U}(\mathbf{b})$ *of optimal dual solutions be compact. Then we have*

$$\frac{\partial \Pi^0(\mathbf{b})}{\partial b_i^+} = \min_{\mathbf{u} \in \mathcal{U}(\mathbf{b})} u_i \quad (i = 1, 2, \ldots, m), \tag{3.48}$$

$$\frac{\partial \Pi^0(\mathbf{b})}{\partial b_i^-} = \max_{\mathbf{u} \in \mathcal{U}(\mathbf{b})} u_i \quad (i = 1, 2, \ldots, m), \tag{3.49}$$

where $\frac{\partial \Pi^0(\mathbf{b})}{\partial b_i^+}$ *denotes the right-side partial derivative and* $\frac{\partial \Pi^0(\mathbf{b})}{\partial b_i^-}$ *denotes the left-side partial derivative of the primal objective function* (3.35) *at the optimal point* $\mathbf{x}^0$. *(For the proof, see* [12, *Appendix, pp.* 333–334].)

Using Theorem 3.16, Horst [12] offers a new interpretation of optimal dual variables in convex programming as *equilibrium prices* rather than as shadow prices. For this purpose, we consider a production model (3.35)–(3.37) and the corresponding dual problem (3.44)–(3.46). Suppose we begin with an optimal solution of our primal and dual problem and let b_i vary from its initial value b_i^0. In other words, the firm can buy or sell some quantity of resource i at the given unit price p_i $(i = 1, 2, \ldots, m)$. Denote by $\mathbf{e}_i$ the ith unit vector in R^m and by t a positive real number. The optimal profit obtainable by the firm when the amounts $(b_1, b_2, \ldots, b_m)$ of resources are available is described by the primal objective function $\Pi(\mathbf{x}^0(\mathbf{b}))$. If the firm buys (sells) an extra quantity $t > 0$ of the production factor i, the increase (decrease) in the optimal profit is $\Pi^0(\mathbf{b} + t\mathbf{e}_i) - \Pi^0(\mathbf{b})$, $[\Pi^0(\mathbf{b}) - \Pi^0(\mathbf{b} - t\mathbf{e}_i)]$. The cost to buy the extra quantity t of the ith resource (or the earned amount in the case of selling t units of the ith resource) is described by $p_i t$. Then if for given p_i the inequalities

$$\Pi^0(\mathbf{b} + te_i) - \Pi^0(\mathbf{b}) \leqq p_i t \leqq \Pi^0(\mathbf{b}) - \Pi^0(\mathbf{b} - t\mathbf{e}_i) \tag{3.50}$$

are satisfied, it is neither reasonable to buy nor to sell a quantity $t > 0$ of the production factors i.

The inequalities (3.50) depend on the given unit price p_i and on the quantity t. Under the assumption of Theorem 3.16, dividing (3.50) by $t > 0$ and taking the limits $t \to 0$, we obtain

$$\lim_{t\to 0} \frac{\Pi^0(\mathbf{b} + te_i) - \Pi^0(\mathbf{b})}{t} \leqq p_i \leqq \lim_{t\to 0} \frac{\Pi^0(\mathbf{b}) - \Pi^0(\mathbf{b} - te_i)}{t},$$

or

$$\frac{\partial \Pi^0(\mathbf{b})}{\partial b_i^+} \leqq p_i \leqq \frac{\partial \Pi^0(\mathbf{b})}{\partial b_i^-} \quad (i = 1, 2, \ldots, m). \tag{3.51}$$

For a given unit price p_i within the interval (3.51), "it is neither worth buying nor worth selling a small quantity of the resource i; an equilibrium is attained" [12, p. 331]. Inequality (3.51) is an extension of the classical result that the price of a good is between the appropriate right- and left-side partial derivatives, when they exist [20], which is connected with the Menger–Wieser theory of imputation.

Denote the extreme values from Theorem 3.16 by

$$u_i^- := \min_{\mathbf{u}\in\mathcal{U}(\mathbf{b})} u_i = \frac{\partial \Pi^0(\mathbf{b})}{\partial b_i^+} \quad (i = 1, 2, \ldots, m)$$

and

$$u_i^+ := \max_{\mathbf{u}\in\mathcal{U}(\mathbf{b})} u_i = \frac{\partial \Pi^0(\mathbf{b})}{\partial b_i^-} \quad (i = 1, 2, \ldots, m).$$

Then by the arguments given above, u_i^- approximately describes the gain in the optimal profit when a unit of the ith resource is added; but this should not be done unless $p_i < u_i^-$. Similarly, u_i^+ gives us approximately the loss in the optimal profit when a unit of the ith resource is subtracted; but this should not be done unless $p_i > u_i^+$. Therefore, the components u_i of multiple optimal dual solutions should be interpreted as equilibrium prices in the sense explained above and the extreme values u_i^- and u_i^+ as shadow prices or accounting values; u_i^- as "shadow buying price"; and u_i^+ as "shadow selling price."

The following corollary provides a condition for the uniqueness of the optimal dual solution and hence the validity of the classical shadow price interpretation.

Corollary 3.2. *Let* $\mathbf{x}^0$ *be a solution of the convex problem* (3.35)–(3.37), *and assume that the gradients* $\nabla f_i(\mathbf{x}^0)$, $i \in I(\mathbf{x}^0)$ *are linearly independent.* $I(\mathbf{x}^0)$ *describes the set* *of the active constraints, i.e.,* $I(\mathbf{x}^0) = \{i \mid f_i(\mathbf{x}^0) = 0\}$. *Then the dual problem* (3.44)–(3.46) *has a unique solution* $\mathbf{u}^0$, *and*

$$\frac{\partial \Pi^0(\mathbf{b})}{\partial b_i} = u_i^0 \quad (i = 1, 2, \ldots, m). \tag{3.52}$$

Proof. By Theorem 3.14 applied to the problem (3.35)–(3.37), there exists a vector $\mathbf{u}^0$ such that $(\mathbf{x}^0, \mathbf{u}^0)$ is an optimal solution to the dual problem (3.44)–(3.46). In other words, because of the Kuhn–Tucker condition (3.39), there exists a vector $\mathbf{u}^0$ such that

$$\frac{\partial f_0(\mathbf{x}^0)}{\partial x_j} = \sum_{i=1}^{m} u_i^0 \frac{\partial f_i(\mathbf{x}^0)}{\partial x_j} \qquad \text{for all } x_j^0 > 0. \tag{3.53}$$

Because of the Kuhn–Tucker conditions (3.42), the system of equations (3.53) can be rewritten as

$$\frac{\partial f_o(\mathbf{x}^0)}{\partial x_j} = \sum_{i \in i(\mathbf{x}^0)} u_i^0 \frac{\partial f_i(\mathbf{x}^0)}{\partial x_j} \qquad \text{for all } x_j^0 > 0. \tag{3.54}$$

(3.54), however, has one solution at most, since by assumption the gradients $\nabla f_i(\mathbf{x}^0)$, $i \in I(\mathbf{x}^0)$ are linearly independent. From Theorem 3.16, it then follows that

$$\frac{\partial \Pi^0(\mathbf{b})}{\partial b_i^+} = \frac{\partial \Pi^0(\mathbf{b})}{\partial b_i^-} = u_i^0 \quad (i = 1, 2, \ldots, m);$$

hence we have (3.52). □

Finally, we turn to an economic interpretation of the dual objective function (3.44). One of the two interpretations offered by Balinski and Baumol [3] is based on the assumption that optimal values for the primal variables $x_1^0, \ldots, x_n^0$ are given, and therefore the first term $f_0(\mathbf{x})$ in the objective function (3.44) can be omitted when the objective function as a whole is minimized.

The item in square brackets $[b_i - f_i(\mathbf{x})]$ in the second term of (3.44) gives the unused amount of input i. Interpreting u_i as the accounting value of the ith input implies that the second term $\sum_{i=1}^{m} u_i[b_i - f_i(\mathbf{x})]$ expresses the (accounting) value of all unused inputs.

The expression in square brackets $[\frac{\partial f_0(\mathbf{x})}{\partial x_j} - \sum_{i=1}^{m} u_i \frac{\partial f_i(\mathbf{x})}{\partial x_j}]$ in the third term of the dual objective function (3.44) corresponds to the dual constraints (3.45) and is therefore always nonpositive. According to the economic interpretation of the Kuhn–Tucker condition (3.38) given above, $[\frac{\partial f_0(\mathbf{x})}{\partial x_j} - \sum_{i=1}^{m} u_i \frac{\partial f_i(\mathbf{x})}{\partial x_j}]$ is the marginal opportunity loss incurred by an increment in the output of product j. Multiplying by x_j, the production level of that output, and summing over all products, we obtain the marginal opportunity cost of all outputs taken together. In conclusion, minimization of the dual objective function (3.44) with respect to the accounting values $u_1, \ldots, u_m$

means the minimization of the value of unused inputs together with the marginal opportunity cost of total output.

Another economic interpretation by Balinski and Baumol [3] is based on rewriting the dual objective function (3.44) as follows:

$$\Phi(\mathbf{x},\mathbf{u}) - \mathbf{x}'\frac{\partial \Phi}{\partial x} = \sum_{i=1}^{m} u_i b_i + \left[f_0(\mathbf{x}) - \sum_{i=1}^{m} u_i f_i(\mathbf{x}) \right] - \sum_{j=1}^{n} x_j \left[\frac{\partial f_0(\mathbf{x})}{\partial x_j} - \sum_{i=1}^{m} u_i \frac{\partial f_i(\mathbf{x})}{\partial x_j} \right]. \tag{3.44'}$$

This reformulation allows an interpretation in terms of economic rents. The economic rent is the difference between the firm's total return and the cost of the inputs needed for production, evaluated by their marginal contributions.

The first term $\sum_{i=1}^{m} u_i b_i$ describes the marginal evaluation of the firm's scarce resources. The second term in (3.44′) (i.e., $[f_0(\mathbf{x}) - \sum_{i=1}^{m} u_i f_i(\mathbf{x})]$) is the excess of the total profit $\Pi = f_0(\mathbf{x})$ over the payments given to the inputs actually used $f_i(\mathbf{x})$, if each were paid u_i per unit. In other words, the second term in (3.44′) is the economic rent received by the firm. If in an optimal solution there is excess capacity in every production factor ($f_i(\mathbf{x}^0) < b_i$ for all i) and hence $u_i^0 = 0$ ($i = 1, 2, \ldots, m$), then the entire return becomes economic rent.

Finally, the last term of (3.44′) can be interpreted as a deduction from rent resulting from a product j whose marginal yield is negative, $\frac{\partial f_0(\mathbf{x})}{\partial x_j} < \sum_{i=1}^{m} u_i \frac{\partial f_i(\mathbf{x})}{\partial x_j}$.

As we have seen in Chapter 1, in many economic models of mathematical programming, the concavity property of the objective function (3.35) and the convexity properties of the constraints (3.36) are reduced to their linearity. What the implications are—from a mathematical as well as from an economic interpretation point of view—will be discussed in the next chapter.

References and Further Reading

[1] K. J. Arrow and A. C. Enthoven, Quasi-concave programming, *Econometrica*, **29**-4 (1961), 779–800.

[2] K. J. Arrow, L. Hurwicz, and H. Uzawa, *Studies in Linear and Non-Linear Programming*, Stanford University Press, Stanford, CA, 1958.

[3] M. L. Balinski and W. J. Baumol, The dual in nonlinear programming and its economic interpretation, *Rev. Econ. Stud.*, **35**-3 (1968), 237–256.

[4] W. J. Baumol, *Economic Theory and Operations Analysis*, 3rd ed, Prentice–Hall, Englewood Cliffs, NJ, 1972.

[5] L. Collatz and W. Wetterling, *Optimierungsaufgaben*, 2nd ed., Springer-Verlag, Berlin, 1971.

[6] G. B. Dantzig, Reminiscences about the origins of linear programming, in A. Bachem, M. Grötschel, and B. Korte, eds., *Mathematical Programming: The State of the Art: Bonn* 1982, Springer-Verlag, Berlin, 1983, 78–86.

[7] R. Dorfman, P. A. Samuelson, and R. M. Solow, *Linear Programming and Economic Analysis*, McGraw–Hill, New York, 1958.

[8] H. A. Eiselt, G. Pederzoli, and C. L. Sandblom, *Continuous Optimization Models*, de Gruyter, Berlin, 1987.

[9] D. Gale, A geometric duality theorem with economic applications, *Rev. Econ. Stud.*, **34**-1 (1967), 19–24.

[10] M. Hamala, *Nelineárne programovanie*, Alfa, Bratislava, 1972.

[11] P. Huard, Dual programs, *IBM J. Res. Dev.*, **6** (1962), 137–139.

[12] R. Horst, On the interpretation of optimal dual solutions in convex programming, *J. Oper. Res. Soc.*, **35** (1984), 327–335.

[13] H. W. Kuhn and A. W. Tucker, Nonlinear programming, in J. Neyman, ed., *Proceedings of the Second Berkeley Symposium on Mathematical Statistics and Probability*, University of California Press, Berkeley, CA, 1951, 481–492.

[14] H. P. Künzi, W. Krelle, and R. Randow, *Nichtlineare programierung*, Springer-Verlag, Berlin, 1979.

[15] M. Luptáčik, *Nichtlineare Programierung mit ökonomischen Anwendungen*, Athenäum, Königstein, Germany, 1981.

[16] W. Nicholson, *Microeconomic Theory: Basic Principles and Extensions*, 5th ed., Dryden Press/Harcourt–Brace–Jovanovich, New York, 1992.

[17] H. Nikaido, *Convex Structures and Economic Theory*, Academic Press, New York, 1968.

[18] E. Silberberg and W. Suen, *The Structure of Economics: A Mathematical Analysis*, 3rd ed., McGraw–Hill, Singapore, 2001.

[19] M. Slater, *Lagrange Multipliers Revisited: Contribution to Non-Linear Programming*, Cowles Commission Discussion Paper 403, University of Chicago, Chicago, 1950.

[20] H. Uzawa, A note on the Menger–Wieser theory of imputation, *Z. Nationalökon.*, **18**-3 (1958).

[21] Ph. Wolfe, A duality theorem for non-linear programming, *Q. Appl. Math.*, **XIX** (1961), 239–244.

4 Linear Programming

The simplest and most widely spread models of convex programming are linear programming models; in other words, models with linear objective function and with linear constraints. This might turn out to be a serious restriction on our field of interest. But as shown in Chapter 1, a wide variety of problems can be satisfactorily represented by linear models. In many cases, the problem naturally takes a linear form; in some cases where this is not so, the problem may be approximately represented by a linear model. As mentioned by Vandermeulen [37, p. 4], "At least in the initial stages, linear models yield more economic output from less mathematical input." In the preface to their well-known book, Dorfman, Samuelson, and Solow [12] denote linear programming as "one of the most important postwar developments in economic theory" [12, p. vii]. In two recent books by ten Raa [35, 36] devoted to the analysis of market economies with the profit motive as the driving force, the main tools are linear programming and input–output analysis.

We start this chapter with the general linear programming problem and its basic properties. In Section 4.2, the economic implication of the linearity assumption will be discussed. Section 4.3 deals with the duality theory of linear programming and its economic interpretation. The paradox of "more for less" will be explored in Section 4.4. Then the simplex computational procedure with its intuitive economic explanation will be presented. The last section is devoted to some economic applications of linear programming. The model from Section 1.2.4 dealing with the theory of comparative advantage and the Giffen paradox described by the model in Section 1.2.5 will be analyzed. In the third application, two alternation formulations of Leontief pollution model will be discussed.

4.1 The General Linear Programming Problem

Looking at the structure of the models in Sections 1.2.1, 1.2.4, 1.2.5, and 1.2.6 in Chapter 1, the general linear programming problem can be formulated as follows:

$$\text{minimize} \quad f_0(\mathbf{x}) = \mathbf{c}'\mathbf{x} \tag{4.1}$$

M. Luptáčik, *Mathematical Optimization and Economic Analysis*,
Springer Optimization and Its Applications 36, DOI 10.1007/978-0-387-89552-9_4,

$$\text{subject to} \qquad A\mathbf{x} = \mathbf{b}, \tag{4.2}$$

$$\mathbf{x} \geqq \mathbf{0}, \tag{4.3}$$

where $\mathbf{c}' = (c_1, c_2, \ldots, c_n)$ is a row vector, $\mathbf{x} = (x_1, x_2, \ldots, x_n)$ is a column vector, $A = (a_{ij})$ is an $m \times n$ matrix, and $\mathbf{b} = (b_1, b_2, \ldots, b_m)$ is a column vector.

The formulation of the constraints (4.2) as equalities is without loss of generality taking into account corresponding slack variables in the case of inequalities. We shall always assume that the equations in (4.2) have been multiplied by -1, where necessary, in order to make all $b_i \geqq 0$.

Especially for the purpose of economic interpretation of the computational procedure, the following formulation of the problem (4.1)–(4.3) can be useful:

$$\begin{aligned} &\text{minimize} && f_0(\mathbf{x}) = c_1x_1 + c_2x_2 + \cdots + c_nx_n \\ &\text{subject to} && x_1P_1 + x_2P_2 + \cdots + x_nP_n = P_0, \\ & && \mathbf{x} \geqq \mathbf{0}, \end{aligned}$$

where P_j $(j = 1, 2, \ldots, n)$ is the jth column of the matrix A and $P_0 = \mathbf{b}$. The set of feasible solutions $K = \{\mathbf{x} | A\mathbf{x} = \mathbf{b},\ \mathbf{x} \geqq \mathbf{0}\}$ is determined by the intersection of the finite set of linear constraints (i.e., hyperplanes). The boundary of K (if K is not empty) will consist of sections of some of the corresponding hyperplanes. If K is empty, then our problem does not have any solutions. Otherwise—because of the linearity of the constraints (4.2)–(4.3)—the set K is convex (due to Corollary 3.1 from Chapter 3). Therefore, if a problem has more than one solution, it has, in reality, an infinite number of solutions. Out of all these solutions, it is our task to determine one that minimizes the objective function (4.1). This work is somewhat simplified by the results of Theorem 4.1 below.

Before proceeding with this theorem, we should introduce the notions of an extreme point and a convex hull.

Definition 4.1. A point U in a convex set K is called an extreme point if U cannot be expressed as a convex combination of any other two distinct points in K.

For example, the extreme points of a cube are its vertices.

Definition 4.2. The convex hull $K(S)$ of any given set of points S is the set of all linear convex combinations of points from S.

$K(S)$ is the smallest convex set containing S. If S is just the eight vertices of a cube, then $K(S)$ is the whole cube.

If the set S consists of a finite number of points, the convex hull of S is termed a *convex polyhedron*. $K(S)$ of the eight vertices of a cube is a convex polyhedron. If K is a closed and bounded set with a finite number of extreme points, then any point in the set can be expressed as a convex combination of the extreme points. Thus K is the convex hull of its extreme points.

If the set of feasible solutions K described by (4.2)–(4.3) is not empty, it can either be a convex polyhedron or a convex region that is unbounded in some direction. If it is a convex polyhedron, then our problem has a solution with a finite minimum value for the objective function; if K is unbounded, the problem has a solution, but the minimum might be unbounded.

For computational purposes, we assume that K is a convex polyhedron. As will be shown later, computational devices exist that determine whether K is empty or whether a linear programming problem has an unbounded solution.

Under the assumption that K is a convex polyhedron, we only need to look at its extreme points in order to determine the optimal solution of problem (4.1)–(4.3). This results from the following basic theorem of linear programming [17, Chapter 3].

Theorem 4.1. *The objective function* (4.1) *takes its minimum at an extreme point of the set K of feasible solutions. If it assumes its minimum at more than one extreme point, then it takes on the same value for every convex combination of those particular points.*

Proof. Let us denote the extreme points of the convex polyhedron K by $\mathbf{x}^{(1)}, \mathbf{x}^{(2)}, \ldots, \mathbf{x}^{(r)}$ and the optimal solution by $\mathbf{x}^0$. This means that $f_0(\mathbf{x}^0) \leqq f_0(\mathbf{x})$ for all $\mathbf{x} \in K$.

Every feasible solution $\mathbf{x}$ in K can be represented as a convex combination of the extreme points in K; therefore,

$$\mathbf{x}^0 = \sum_{i=1}^{r} \lambda_i \mathbf{x}^{(i)} \quad \text{with} \quad \lambda_i \geqq 0 \quad \text{and} \quad \sum_{i=1}^{r} \lambda_i = 1.$$

Then since $f_0(\mathbf{x})$ is a linear function, we have

$$\begin{aligned} f_0(\mathbf{x}^0) &= f_0\left(\sum_{i=1}^{r} \lambda_i \mathbf{x}^{(i)}\right) = f_0(\lambda_1 \mathbf{x}^{(1)} + \lambda_2 \mathbf{x}^{(2)} + \cdots + \lambda_r \mathbf{x}^{(r)}) \\ &= \lambda_1 f_0(\mathbf{x}^{(1)}) + \lambda_2 f_0(\mathbf{x}^{(2)}) + \cdots + \lambda_r f_0(\mathbf{x}^{(r)}). \end{aligned}$$

Let $f_0(\mathbf{x}^{(m)}) = \min_i f_0(\mathbf{x}^{(i)})$. Then substituting for each $f_0(\mathbf{x}^{(i)})$ the value $f_0(\mathbf{x}^{(m)})$, we can write

$$f_0(\mathbf{x}^0) \geqq \lambda_1 f_0(\mathbf{x}^{(m)}) + \lambda_2 f_0(\mathbf{x}^{(m)}) + \cdots + \lambda_r f_0(\mathbf{x}^{(m)}) = f_0(\mathbf{x}^{(m)})$$

because of $\sum_{i=1}^{r} \lambda_i = 1$. Since we assumed that $f_0(\mathbf{x}^0) \leqq f_0(\mathbf{x})$ for all $\mathbf{x} \in K$, we must have

$$f_0(\mathbf{x}^0) = f_0(\mathbf{x}^{(m)});$$

there is an extreme point, $\mathbf{x}^{(m)}$, at which the objective function $f_0(\mathbf{x})$ takes its minimum value.

Now let us assume that the objective function (4.1) takes its minimum at more than one extreme point, say, at $\mathbf{x}^{(1)}, \mathbf{x}^{(2)}, \ldots, \mathbf{x}^{(p)}$, where $p \leqq r$. Here we have $f_0(\mathbf{x}^{(1)}) = f_0(\mathbf{x}^{(2)}) = \cdots = f_0(\mathbf{x}^{(p)}) = z_0$. If $\mathbf{x} \in K$ is any convex combination of the extreme points $\mathbf{x}^{(i)}$ $(i = 1, 2, \ldots, p)$, then

$$f_0(\mathbf{x}) = f_0(\lambda_1 \mathbf{x}^{(1)} + \lambda_2 \mathbf{x}^{(2)} + \cdots + \lambda_p \mathbf{x}^{(p)})$$

$$= \lambda_1 f_0(\mathbf{x}^{(1)}) + \lambda_2 f_0(\mathbf{x}^{(2)}) + \cdots + \lambda_p f_0(\mathbf{x}^{(p)}) = \sum_{i=1}^{p} \lambda_i z_0 = z_0$$

because of $\sum_{i=1}^{p} \lambda_i = 1$. □

The question now arises of how to identify the extreme points in the set of feasible solutions K described by (4.2)–(4.3). The answer is provided by the following theorem [17, Chapter 3].

Theorem 4.2. $\mathbf{x} = (x_1, x_2, \ldots, x_n)$ *is an extreme point of* K *if and only if the components* x_j *with positive value are coefficients of linearly independent vectors* P_j *in*

$$\sum_{j=1}^{n} x_j P_j = P_0.$$

This result leads to the notion of basic solution defined as follows.

Definition 4.3. A basic solution to the linear programming problem is a feasible solution $\mathbf{x} = (x_1, x_2, \ldots, x_n)$, provided that the vectors P_j corresponding to the positive coefficients x_j in

$$\sum_{j=1}^{n} x_j P_j = P_0$$

are linearly independent. These positive variables x_j are called basic variables.

Definition 4.4. A basic solution is called nondegenerate if it has exactly m positive x_j; that is, all basic variables are positive.

From the above, we can conclude that the objective function in a linear programming problem takes its optimum at an extreme point of the feasible set K, that is a basic solution to set of constraints (4.2)–(4.3). Since there are at most $\binom{n}{m}$ sets of m linearly independent vectors from the given set of n, the value $\binom{n}{m}$ is the upper bound to the number of possible solutions to the problem. For large n and m, it would be an impossible task to evaluate all basic solutions and select one that optimizes the objective function. Therefore, we need a procedure for finding an extreme point and determining whether it is the optimum. If it is not, it must be possible to find a neighboring extreme point[1] whose corresponding value of the objective function is smaller than or equal to the preceding value. In a finite number of such steps, an optimal solution is found. This is the basic *simplex method*, developed by Dantzig, as a powerful scheme for solving *any* linear programming problem. It makes it possible to discover whether the problem has an unbounded optimal solution or no feasible solution.

Because the focus of this book is economic analysis, before going into computational aspects of the simplex method, in the next section we want to discuss the economic implications following from the linearity of the objective function and the linearity of the constraints.

[1] Two extreme points are called *neighbors* if they are joined by a segment that belongs to a one-dimensional face of the convex polyhedron.

4.2 Implications of Linearity Assumption for Economic Analysis

We consider the product mix problem (3.35)–(3.37) from Chapter 3. Under the linearity assumption of the profit function $f_0(\mathbf{x})$ and of the functions $f_i(\mathbf{x})$ $(i = 1, 2, \ldots, m)$, this production model takes the following form:

$$\text{maximize} \quad f_0(\mathbf{x}) = c_1x_1 + c_2x_2 + \cdots + c_nx_n \tag{4.4}$$

$$\begin{aligned} \text{subject to} \quad & a_{11}x_1 + a_{12}x_2 + \cdots + a_{1n}x_n \leqq b_1, \\ & a_{21}x_1 + a_{22}x_2 + \cdots + a_{2n}x_n \leqq b_2, \\ & \quad \vdots \qquad\qquad\qquad\qquad \vdots \\ & a_{m1}x_1 + a_{m2}x_2 + \cdots + a_{mn}x_n \leqq b_m, \end{aligned} \tag{4.5}$$

$$x_j \geqq 0 \quad (j = 1, 2, \ldots, n), \tag{4.6}$$

where the coefficients c_j $(j = 1, 2, \ldots, n)$ in the objective function and the coefficients a_{ij} in the constraints are given and constant. What does it mean in economic terms, bearing in mind the interpretation of this model as a more general convex programming problem?

The linear profit function (4.4) implies that the marginal profit of product j is constant and independent of the decision made by the producer $(\partial f_0(\mathbf{x})|\partial x_j = c_j, j = 1, 2, \ldots, n)$. This corresponds to the assumption of perfect competition. The profit yield by every (additional) unit of product j is the same. In other words, the demand for product j is completely elastic, and the demand curve for product j is horizontal.

Linear programming replaces the continuous production function by a collection of n independent linear activities or processes representing the production of distinct goods in our model. The jth activity is completely described by its coefficients for the m inputs: $(a_{1j}, a_{2j}, \ldots, a_{mj})$. The coefficient a_{ij} gives the quantity of the ith input required to produce one unit of the commodity j. Activities are linear in the sense that the quantity of the ith input used for production of the jth commodity is a linear function of its level of production. The concept of marginal productivity of an input is meaningless within the linear programming framework. It is not possible to increase an activity level by increasing the quantity of a single input. All inputs must be increased proportionately or in other words the inputs must be used in the given fixed proportion. The behavior of output when all inputs change by the same proportion is in economic theory characterized by the concept of *returns to scale*. If all input quantities are multiplied by the scale factor α, the production function exhibits *constant* (*increasing, decreasing*) returns to scale if output increases by the *same* (*greater, smaller*) proportion. Therefore, the linear programming technology implies constant returns to scale. A doubling of all input quantities leads to doubling of output produced. Because a continuous substitution between the inputs is not possible, in this type of model "growth is likely to be impeded by shortages of specific factors rather than by a general scarcity of resources" [8, p. 29].

The diagrammatic representation of the unit isoquant for two inputs $i = 1, 2$ is given in Figure 4.1 in the case of three processes $j = 1, 2, 3$.

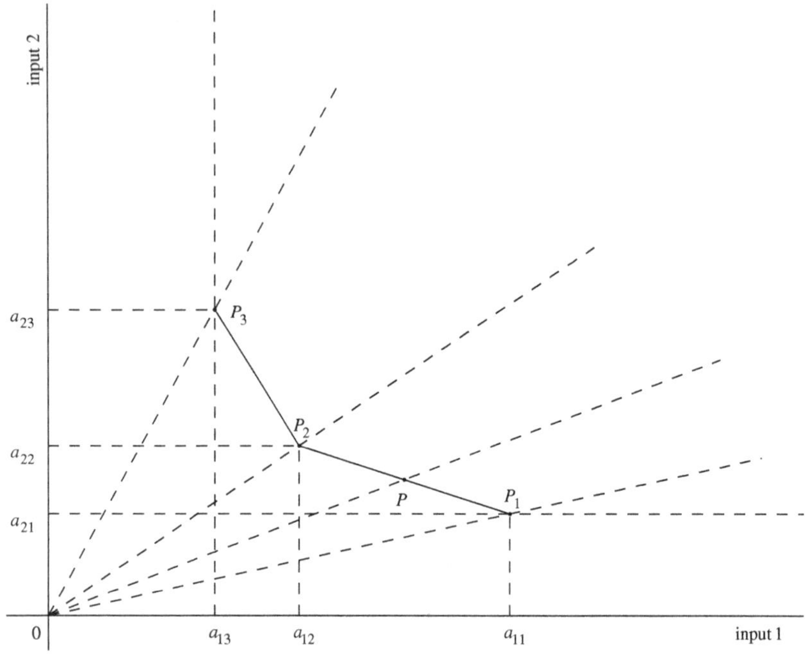

Fig. 4.1. Unit isoquant for a linear programming model with two input factors.

If the first process is used alone, unit output is obtained by combining a_{11} of input 1 and a_{21} of input 2, as shown by the point P_1 with coordinates (a_{11}, a_{21}). Similarly, P_2 and P_3 represent unit output obtained by use of the second and third processes, each taken alone. To combine the first and second processes, assign λa_{11} of the first input to the first process, giving output λ and $(1-\lambda)a_{12}$ of the first input to the second process for output $(1-\lambda)$. Hence a total amount of the first input used for unit output is $\lambda a_{11} + (1-\lambda)a_{12}$. The corresponding amount of the second input used for the first process is λa_{21} and $(1-\lambda)a_{22}$ for the second process. Hence a total usage of the second input is $\lambda a_{21} + (1-\lambda)a_{22}$. We have another point on the unit isoquant with coordinates $(\lambda a_{11} + (1-\lambda)a_{22}; \lambda a_{21} + (1-\lambda)a_{22})$ for $0 < \lambda < 1$, i.e., a point P on the segment $P_1 P_2$. As λ increases from 0 to 1, so P moves from P_2 to P_1 along the segment. In the same way, combinations of the second and third processes are described by points on the segment $P_2 P_3$ and combinations of the first and third processes by points on the segment $P_1 P_3$. It should be noticed that some combinations are efficient and others not. In the case shown in Figure 4.1, combinations of $P_1 P_2$ and of $P_2 P_3$ are efficient, but those of $P_1 P_3$ are not because the same unit output can be produced with lower inputs by using the second process alone or in combination with one of the other two processes. The complete unit isoquant is then described by the piecewise linear curve $P_1 P_2 P_3$, continued horizontally to the right of P_1 and vertically above P_3 to indicate that in these regions the first or second input goes to waste. Because of the fixed input proportions all other isoquants are obtained by blowing up the unit isoquants radially from 0.

The technology is linear in a double sense. The quantities of inputs used for production vary linearly with the level of output and the processes are combined linearly; the activities are additive. Some firms in the economy use one process and some the other, a given proportion of output coming from each process.

The technology described by (4.5)–(4.6) permits any finite number of processes to be used alone or in combination. Figure 4.1 illustrates the production function when there is a small number (here three) of processes. But it is easy to see what happens when the number of efficient processes increases and becomes large. The unit isoquant then consists of a large number of small segments $P_1P_2P_3P_4\ldots$, and the piecewise linear curve tends to become a curve that can be assumed to be continuous and differentiable. The not-so-smooth linear programming technology tends to become a smooth (continuous) technology. That is a reason why linear programming technology with a finite number of processes is more relevant to microeconomic models, and a continuous production function as the limiting case of infinitely many efficient processes is more relevant to macroeconomic models.

The final remark is related to the assumption of independent processes which implies the absence of *external effects*. If such external effects are present, one must take into account the interdependence between the costs of the ith process and the output of the hth. Suppose that the n processes in the production model (4.4)–(4.6) are assigned to different firms. The *external economies* case is one in which an increase in the firm's production produces benefits part of which devolve on others. The standard examples are the training of a labor force or when an expansion of the production of one company makes it cheaper to supply services to all the firms in the industry. An expansion of the scale of a company's operation can also have disadvantageous effects. These are called *external diseconomies* of scale. Increased use of the particular resources can make it harder for others to get these resources.

In conclusion, for the application of mathematical programming models in economics, it is therefore of crucial importance to clarify the implications of the model assumptions. The paper by Baumol and Bushnell [4] provides a qualitative theoretical analysis of the error that might arise when a linear programming model is used to solve a problem involving some nonlinearities.

4.3 Duality in Linear Programming

We turn, again, to the production model (4.4)–(4.6). According to the formulation of the dual problem in Section 3.4, we obtain for $f_0(\mathbf{x}) = \sum_{j=1}^{n} c_j x_j$ and for $f_i(\mathbf{x}) = \sum_{j=1}^{n} a_{ij}x_j$ $(i = 1, 2, \ldots, m)$ the following dual linear programming problem:

$$\text{minimize} \quad \sum_{j=1}^{n} c_j x_j + \sum_{i=1}^{m} u_i \left(b_i - \sum_{j=1}^{n} a_{ij} x_j \right) - \sum_{j=1}^{n} x_j \left(c_j - \sum_{i=1}^{m} a_{ij} u_i \right)$$

$$= \sum_{i=1}^{m} b_i u_i \tag{4.7}$$

$$\text{subject to} \quad \sum_{i=1}^{m} a_{ij} u_i \geqq c_j \quad (j = 1, 2, \ldots, n), \tag{4.8}$$

$$u_i \geqq 0 \quad (i = 1, 2, \ldots, m). \tag{4.9}$$

Problem (4.7)–(4.9) is the *dual* problem corresponding to the *primal* problem (4.4)–(4.6). In matrix notation, we have

PRIMAL PROBLEM	DUAL PROBLEM
maximize $f_0(\mathbf{x}) = \mathbf{c}'\mathbf{x}$	minimize $g(\mathbf{u}) = \mathbf{u}'\mathbf{b}$
subject to $A\mathbf{x} \leqq \mathbf{b}$,	subject to $\mathbf{u}'A \geqq \mathbf{c}'$,
$\mathbf{x} \geqq \mathbf{0}$,	$\mathbf{u}' \geqq \mathbf{0}$,

where $\mathbf{u}' = (u_1, u_2, \ldots, u_m)$ is a row vector of the dual variables.

Comparison of the two problems yields the rules for the construction of the dual problem:

- If the primal problem involves maximization, the dual involves minimization, and vice versa.
- If the primal problem contains inequalities with the $\leqq$ sign, the dual contains inequalities with the $\geqq$ sign, and vice versa.
- The coefficients of the primal objective function are the coefficients of the constraint vector in the dual problem, and vice versa.
- In the constraint inequalities of the primal problem, the matrix A is multiplied by a column vector $\mathbf{x}$, and in those of the dual problem by a row vector $\mathbf{u}'$ (or the matrix A in the dual problem is transposed, i.e., the rows and columns are interchanged and multiplied by a column vector $\mathbf{u}$); and vice versa.
- Apart from the number of nonnegativity conditions, if there are n variables and m inequalities in the primal problem, in the dual problem there will be m variables and n inequalities.

It follows from these characteristics that the dual of the dual problem is the original linear programming problem itself. Therefore, given such a pair of problems, it is entirely arbitrary which of them is referred to as the primal and which as the dual. Each one is the dual of the other.

Before we proceed further, an important question of the economic interpretation of the dual problem (4.7)–(4.9) arises. We are about to describe a simple situation in which one wants to minimize the value of the firm's scarce inputs under the constraints (4.8). The description may seem artificial at first, but it will appear less so as we go on to consider the model (4.7)–(4.9) as a special case of problem (3.44)–(3.46) and its economic interpretation.

A producer just looking for the optimal product mix described by the model (4.4)–(4.6) has a visit from another producer, who wants to buy him out. He makes the following offer to the producer: "If you produce one unit of good 1, you need a_{11} units of the first production factor, a_{21} units of the second factor, and so on until a_{m1} units of the production factor m. I offer to pay you the amount $u_i \geqq 0$ for each unit of the ith production factor in such a way that

$$a_{11}u_1 + a_{21}u_2 + \cdots + a_{m1}u_m \geqq c_1. \tag{4.10}$$

You will get at least the amount corresponding to the profit you will earn by producing one unit of the first commodity. But you can use the production factors for the production of other goods ($j = 2, \ldots, n$), too; therefore, the 'prices' $u_1, u_2, \ldots, u_m$ should satisfy a condition like (4.10) with respect to each commodity:

$$\sum_{i=1}^{m} a_{ij}u_i \geqq c_j \quad (j = 1, 2, \ldots, n).$$

The dual constraints (4.8) come out. Now, if you sell the given production factors to me, your return (and my expenditure) will be $\sum_{i=1}^{m} b_i u_i$. Because of your constraints (4.5) for any product mix $\mathbf{x}$,

$$\begin{aligned} \sum_{i=1}^{m} b_i u_i &\geqq \sum_{i=1}^{m} u_i \sum_{j=1}^{n} a_{ij}x_j = \sum_{j=1}^{n} x_j \sum_{i=1}^{m} a_{ij}u_i \\ &\geqq \sum_{j=1}^{n} c_j x_j \quad \text{(due to (4.8))}, \end{aligned} \tag{4.11}$$

you will be at least as well off." Finally, the competitor minimizes his total expenditures (4.7) under the condition (4.8)–(4.9).[2]

Now consider problem (4.4)–(4.6) as a special case of the convex programming problem (3.35)–(3.37). As shown in Section 4.2, a linear profit function $f_0(\mathbf{x})$ implies a constant (marginal) profit c_j deriving from product j ($j = 1, 2, \ldots, n$), and the linearity of the constraints (4.5) implies the constant (marginal) quantity of input i needed to produce one unit of product j, described by the input coefficients a_{ij} ($i = 1, 2, \ldots, m$; $j = 1, 2, \ldots, n$). Then the dual constraints (4.8) assert that the total (accounting) value of the inputs that are necessary to produce one unit of commodity j ($j = 1, 2, \ldots, n$) cannot be lower than the profit that the firm makes by producing a unit of this commodity. In other words, the sum of the accounting values of the scarce inputs going into the product j is high enough to account for the profit derived by a unit of commodity j.

Returning to the economic interpretation of the dual objective function (4.7) as a special case of the objective function (3.44′), the following question arises: Why does the rent enter the nonlinear dual but not the linear dual? Balinski and Baumol [3] found an explanation in the old "adding-up" discussion of distribution theory: If each input to a production process is paid the value of its marginal product, will the total factor payment exhaust the total profit, or will it leave over some unimputed profit, positive or negative? The answer can be found in *Euler's theorem*, which states that if $f(x_1, x_2, \ldots, x_n)$ is a homogeneous production function of degree k, then

$$\sum_{i=1}^{n} x_i \frac{\partial f}{\partial x_i} = kf(x_1, x_2, \ldots, x_n).$$

[2] In a similar way, the economic interpretation of the dual to the diet problem can be given (see [15, pp. 13–14]).

A production function $f(x_1, x_2, \ldots, x_n)$ is said to be *homogeneous of degree k* if

$$f(\lambda x_1, \lambda x_2, \ldots, \lambda x_n) = \lambda^k f(x_1, x_2, \ldots, x_n) \quad \text{for all } \lambda > 0. \tag{4.12}$$

Homogeneous production functions play an important role in economics because they are intimately connected to the concept of *returns to scale*. Correspondingly, returns to scale are increasing if $k > 1$, constant if $k = 1$, and decreasing if $k < 1$. Now to derive Euler's theorem, we differentiate both sides of (4.12) with respect to λ and find

$$\sum_{i=1}^{n} \frac{\partial f}{\partial(\lambda x_i)} x_i = k\lambda^{k-1} f(x_1, x_2, \ldots, x_n).$$

Setting $\lambda = 1$, we obtain Euler's theorem. As already mentioned, the linear programming model is characterized by constant returns to scale (linear homogeneity), and therefore the profit function $\Pi(\mathbf{x}(b_1, b_2, \ldots, b_m))$, here viewed as a function of the available input quantities $b_1, b_2, \ldots, b_m$, is a differentiable homogeneous function of degree one. Then $\Pi(\mathbf{x}^0(\mathbf{b})) = \sum_{i=1}^{m} b_i \frac{\partial \Pi(\mathbf{x}^0(\mathbf{b}))}{\partial b_i} = \sum_{i=1}^{m} b_i u_i^0$, where $\mathbf{x}^0$, $\mathbf{u}^0$ denote the optimal solution of the primal and the dual problem, respectively.

Before we continue the discussion of the duality in linear programming, it is convenient to rewrite our primal problem (4.4)–(4.6) and dual problem (4.7)–(4.9) by introducing slack variables so that the constraints become equalities. We obtain

PRIMAL PROBLEM: maximize

$$f_0(\mathbf{x}) = \sum_{j=1}^{n} c_j x_j$$

subject to

$$\sum_{j=1}^{n} a_{ij} x_j + l_i = b_i \qquad (i = 1, 2, \ldots, m),$$

$$x_j \geqq 0 \qquad (j = 1, 2, \ldots, n),$$

$$l_i \geqq 0 \qquad (i = 1, 2, \ldots, m),$$

DUAL PROBLEM: minimize

$$g(\mathbf{u}) = \sum_{i=1}^{m} b_i u_i$$

subject to

$$\sum_{i=1}^{m} a_{ij} u_i - q_j = c_j \qquad (j = 1, 2, \ldots, n),$$

$$u_i \geqq 0 \qquad (i = 1, 2, \ldots, m),$$

$$q_j \geqq 0 \qquad (j = 1, 2, \ldots, n).$$

The primal slack variable l_i $(i = 1, 2, \ldots, m)$ describes the unused amount of input i. For the purpose of the economic interpretation of the dual slack variable q_j, we rewrite the jth equation in dual constraints as

$$q_j = \sum_{i=1}^{m} a_{ij} u_i - c_j \quad (j = 1, 2, \ldots, n). \tag{4.13}$$

As already shown, the first expression on the right-hand side of (4.13) is the accounting value of the resources used in producing a unit of output j, and c_j is the unit profit of output j. A positive value of q_j implies that the resources used in producing output j are worth more than the profit yielded by that commodity. Thus we may interpret the dual slack variable q_j as the accounting loss per unit of output j.

Summarizing, our primal and dual problem contain the following four types of variables:

x_j: the quantity of product j (the primal ordinary variables);
l_i: the unused capacity of input i (the primal slack variables);
u_i: the accounting value of input i (the dual ordinary variables);
q_j: the accounting loss per unit of output j (the dual slack variables).

We now turn to the relationship between the primal problem (4.4)–(4.6) and the dual problem (4.7)–(4.9). The first observation, which we already made, is the competitors' assertion (4.11) that the return of the producers by selling the given resources under the primal and dual constraints cannot be lower than their profit for any product mix $\mathbf{x}$. Or, in other words, the firm's total profit will never exceed the accounting value assigned to the firm's scarce inputs.

The reader can verify that this result is a straightforward application of the more general weak duality theorem (Theorem 3.13 from Section 3.3) for the primal problem

$$\max_{\mathbf{x}\in R^n_+} \min_{\mathbf{u}\in R^m_+} \Phi(\mathbf{x}, \mathbf{u})$$

and the dual problem

$$\min_{\mathbf{u}\in R^m_+} \max_{\mathbf{x}\in R^n_+} \Phi(\mathbf{x}, \mathbf{u}),$$

where $f_0(\mathbf{x}) = \sum_{j=1}^n c_j x_j$ and $f_i(\mathbf{x}) = -\sum_{j=1}^n a_{ij} x_j + b_i$ $(i = 1, 2, \ldots, m)$.

We formulate this result in the following lemma.

Lemma 4.1. *Let* $x_1, x_2, \ldots, x_n$ *be a feasible solution of the maximization problem* (4.4)–(4.6), *and let* $u_1, u_2, \ldots, u_m$ *be a feasible solution of the minimization dual problem* (4.7)–(4.9). *Then*

$$\sum_{j=1}^{n} c_j x_j \leq \sum_{i=1}^{m} b_i u_i. \tag{4.11$'$}$$

The producer seeks now to maximize his total profit and the competitor seeks to minimize his expenditure for buying the resources until both values are equal.

Theorem 4.3 (optimality criterion). *If there exists a feasible solution* $x_1^0, x_2^0, \ldots, x_n^0$ *for the maximization problem* (4.4)–(4.6) *and a feasible solution* $u_1^0, u_2^0, \ldots, u_m^0$ *for the dual problem* (4.7)–(4.9) *such that*

$$\sum_{j=1}^{n} c_j x_j^0 = \sum_{i=1}^{m} b_i u_i^0,$$

then both of these solutions are optimal.

Proof. Relation (4.11′) implies that every feasible solution $x_1, x_2, \ldots, x_n$ of problem (4.4)–(4.6) satisfies

$$\sum_{j=1}^{n} c_j x_j \leqq \sum_{i=1}^{m} b_i u_i^0 = \sum_{j=1}^{n} c_j x_j^0$$

and that every feasible solution $u_1, u_2, \ldots, u_m$ satisfies

$$\sum_{i=1}^{m} b_i u_i \geqq \sum_{j=1}^{n} c_j x_j^0 = \sum_{i=1}^{m} b_i u_i^0. \qquad \square$$

In this way, the dual model will impute all of the firm's profit to its scarce resources.

The next important theorem, dealing with the existence of optimal solutions for the pair of linear programming problems (4.4)–(4.6) and (4.7)–(4.9), is the following.

Theorem 4.4 (duality theorem). *If either the primal or the dual problem has a finite optimal solution, then the other problem also has a finite optimal solution and the values of the objective functions are equal; that is,* $\max f_0(\mathbf{x}) = \min g(\mathbf{u})$.

If either problem has an unbounded optimal solution, then the other problem must be infeasible.

Proof. First, let us remember that each one of the problems (4.4)–(4.6) and (4.7)–(4.9) is the dual of the other. Then the first part of this theorem is a straightforward application of the strong duality theorem (Theorem 3.14 from Section 3.3) for convex programming.

To prove the second part, we note that if the primal problem is unbounded, then we have by (4.11) that

$$+\infty \leqq \sum_{i=1}^{m} b_i u_i.$$

Any solution to the dual inequalities (4.8)–(4.9) must have a corresponding value for the dual objective function (4.7) that is an upper bound for the primal objective function (4.4). Since this contradicts the assumption of unboundedness, we must conclude that there are no solutions to the dual problem. By the same argument, if the dual problem is unbounded, then the primal problem must be infeasible. $\square$

However, the primal and dual may both be infeasible at the same time. For example, both the problem

$$\begin{aligned}
\text{maximize} \quad & f_0(\mathbf{x}) = 3x_1 - 2x_2 \\
\text{subject to} \quad & x_1 - x_2 \leqq 2, \\
& -x_1 + x_2 \leqq -3, \\
& x_1 x_2 \geqq 0
\end{aligned}$$

and its dual

$$\begin{aligned}
\text{minimize} \quad & g(\mathbf{u}) = 2u_1 - 3u_2 \\
\text{subject to} \quad & u_1 - u_2 \geqq 3, \\
& -u_1 + u_2 \geqq -2, \\
& u_1, u_2 \geqq 0
\end{aligned}$$

are infeasible. In conclusion, we can summarize that if the primal problem has a feasible solution *and* if the dual problem has a feasible solution, then both problems have optimal solutions.

As shown in Chapter 3, the Kuhn–Tucker conditions are both necessary and sufficient for $\mathbf{x}^0$ to solve the convex programming problem and therefore the linear programming problem as well. The Lagrange function for problem (4.4)–(4.6) is

$$\Phi(\mathbf{x}, \mathbf{u}) = \sum_{j=1}^{n} c_j x_j + \sum_{i=1}^{m} u_i \left(b_i - \sum_{j=1}^{n} a_{ij} x_j \right),$$

and the Kuhn–Tucker conditions yield

$$\frac{\partial \phi}{\partial x_j} = c_j - \sum_{i=1}^{m} a_{ij} u_i^0 \leqq 0 \quad (j = 1, 2, \ldots, n), \tag{4.14}$$

$$x_j^0 \frac{\partial \phi}{\partial x_j} = x_j^0 \left(c_j - \sum_{i=1}^{m} a_{ij} u_i^0 \right) = 0 \quad (j = 1, 2, \ldots, n), \tag{4.15}$$

$$x_j^0 \geqq 0 \quad (j = 1, 2, \ldots, n), \tag{4.16}$$

$$\frac{\partial \phi}{\partial u_i} = b_i - \sum_{j=1}^{n} a_{ij} x_j^0 \geqq 0 \quad (i = 1, 2, \ldots, m), \tag{4.17}$$

$$u_i^0 \frac{\partial \phi}{\partial u_i} = u_i^0 \left(b_i - \sum_{j=1}^{n} a_{ij} x_j^0 \right) = 0 \quad (i = 1, 2, \ldots, m), \tag{4.18}$$

$$u_i^0 \geqq 0 \quad (i = 1, 2, \ldots, m). \tag{4.19}$$

We can see that conditions (4.14) and (4.19) are exactly the dual constraints (4.8)–(4.9), and the Lagrange multipliers $u_i^0, u_2^0, \ldots, u_m^0$ are the dual variables. Conditions (4.16)–(4.17) coincide with the primal constraints (4.6) and (4.5). Therefore, the feasible solutions $x_1^0, x_2^0, \ldots, x_n^0$ and $u_1^0, u_2^0, \ldots, u_m^0$ satisfying conditions (4.15) and (4.18) are optimal, and vice versa.

In this way, the next major theorem of linear programming, the *complementary slackness theorem* (see, e.g., [10, pp. 62–63]), or *equilibrium theorem* [15, pp. 19–20], can be derived.

Theorem 4.5. *The feasible solutions* $x_1^0, x_2^0, \ldots, x_n^0$ *and* $u_1^0, u_2^0, \ldots, u_m^0$ *of* (4.4)–(4.6) *and* (4.7)–(4.9), *respectively, are optimal solutions if and only if*

$$\sum_{i=1}^{m} a_{ij} u_i^0 > c_j \quad \textit{implies} \quad x_j^0 = 0, \tag{4.20a}$$

$$x_j^0 > 0 \quad \textit{implies} \quad \sum_{i=1}^{m} a_{ij} u_i^0 = c_j, \tag{4.20b}$$

and

$$\sum_{j=1}^{m} a_{ij} x_j^0 < b_i \quad \textit{implies} \quad u_i^0 = 0, \tag{4.21a}$$

$$u_i^0 > 0 \quad \textit{implies} \quad \sum_{j=1}^{n} a_{ij} x_j^0 = b_i. \tag{4.21b}$$

These conditions are usually called the *complementary slackness* conditions. Taking into account the primal slack variables $l_1, l_2, \ldots, l_m$ and the dual slack variables $q_1, q_2, \ldots, q_n$, conditions (4.15) and (4.18) can be rewritten as

$$x_j^0 q_j^0 = 0 \quad (j = 1, 2, \ldots, n), \tag{4.20$'$}$$

$$u_i^0 l_i^0 = 0 \quad (i = 1, 2, \ldots, m). \tag{4.21$'$}$$

The primal slack variables l_i $(i = 1, 2, \ldots, m)$ are naturally matched up with the dual variables u_i $(i = 1, 2, \ldots, m)$, and each primal variable x_j is matched with the dual slack q_j. Conditions (4.15) and (4.18) or (4.20′)–(4.21′) require that in each of the $m + n$ matching pairs, at least one variable must have value zero.

The economic interpretation of the above result follows immediately from the interpretation of the Kuhn–Tucker conditions (3.39) and (3.42) from the previous chapter. However, this is only if the system of equations (4.20b) has a unique solution. According to Corollary 3.1, the sufficient condition for the uniqueness of the optimal dual solution is that the gradients $\nabla f_i(\mathbf{x}^0)$, $i \in I(\mathbf{x}^0)$, are linearly independent. For the linear programming problem (4.4)–(4.6), this condition implies that the optimal solution $x_1^0, x_2^0, \ldots, x_n^0$ is a nondegenerate basic solution. Then the dual problem (4.7)–(4.9) has a unique solution, and the classical *shadow price* interpretation of the dual variables according to (3.52) is valid. Hence u_i^0 is equal to the marginal profit contribution of input i. That is, u_i^0 give us the increase of the firm's profits if the company were to increase its amount of the production factor i by one unit. In the interpretation of Chvátal [10, p. 66], u_i^0 "specifies the maximum amount that the firm should be willing to pay, over and above the present trading price, for each extra unit of resource i."

With this interpretation of the dual variables $u_1^0, u_2^0, \ldots, u_m^0$ as shadow prices (or accounting values), condition (4.20a) says that if the cost of an activity (described as the accounting value of the inputs) exceeds the profit derived from it, then it will not be used; i.e., it will be operated at level zero. Because of an opportunity loss, the firm has to use the inputs for the production of other goods. Summarizing, the conditions in (4.20) assert that in the optimal product mix an opportunity loss cannot occur.

Condition (4.21a) states that if there are production factors of which there is a *surplus*, that is, those whose supply is not exhausted, then the shadow price of those production factors must be zero. An extra unit of those resources does not change the firm's profit. According to condition (4.21b), if the shadow price u_i^0 is positive, then there must be a resource with deficit; that is, all of its available amount b_i will be used by an optimal product mix. In this way, the shadow prices reveal the bottleneck in the production and indicate the possibilities of the most profitable expansion of the production.

Thus the dual solution provides a deeper insight into the behavior of the given decision-making units, and thereupon it can be very useful for supporting of the decision process.

An interesting property arising in some linear programming economic models will illustrate this point.

4.4 The More-for-Less Paradox

Let us consider the production problem (1.7) (described in Section 1.2.3) based on the assumption of the minimization of the cost for the given output level. Suppose the firm using n different technologies $T_1, T_2, \ldots, T_n$ has to produce the given amounts $b_1, b_2, \ldots, b_m$ of m different products $P_1, P_2, \ldots, P_m$. The technology T_j is described by the output coefficients a_{ij}, giving the amount of good P_i produced by the technology T_j with unit intensity. The unit cost of technology T_j is described by the coefficient c_j $(j = 1, 2, \ldots, n)$. The decision problem of the firm is how to combine the different technologies, and at which intensities, in order to produce the prescribed amounts of the products $P_1, P_2, \ldots, P_m$ at the minimal cost.

For simplicity and without loss of generality, suppose only two products $(m = 2)$ and six technologies $(n = 6)$ with the matrix of output coefficients

$$A = \begin{pmatrix} 3 & 2 & 3 & 2 & 2 & 4 \\ 1 & 1 & 2 & 2 & 3 & 5 \end{pmatrix},$$

with the output requirements $\mathbf{b}' = (30, 40)$ and with the unit cost $\mathbf{c}' = (2, 1, 4, 2, 6, 8)$, expressed in Euros (€). This problem can be formulated as

$$\begin{aligned} \text{minimize} \quad & f_0(\mathbf{x}) = 2x_1 + x_2 + 4x_3 + 2x_4 + 6x_5 + 8x_6 \\ \text{subject to} \quad & 3x_1 + 2x_2 + 3x_3 + 2x_4 + 2x_5 + 4x_6 = 30, \\ & x_1 + x_2 + 2x_3 + 2x_4 + 3x_5 + 5x_6 = 40, \\ & x_1, x_2, x_3, x_4, x_5, x_6 \geqq 0. \end{aligned} \tag{4.22}$$

The equalities in (4.22) imply that the produced goods are perishable and cannot be put into store. The optimal solution to problem (4.22) is $\mathbf{x}^0 = (0, 0, 0, 0, 5, 5)$ with the objective function value $f(\mathbf{x}^0) = €70$. Although the method of solution will be presented only in the next section, for problem (4.22) it can be easily shown—by means of duality theory from the previous section—that the above solution is optimal.

The dual problem corresponding to problem (4.22) has the following form:

$$\begin{aligned} \text{maximize} \quad & g(\mathbf{u}) = 30u_1 + 40u_2 \\ \text{subject to} \quad & 3u_1 + u_2 \leqq 2, \\ & 2u_1 + u_2 \leqq 1, \\ & 3u_1 + 2u_2 \leqq 4, \\ & 2u_1 + 2u_2 \leqq 2, \\ & 2u_1 + 3u_2 \leqq 6, \\ & 4u_1 + 5u_2 \leqq 8. \end{aligned} \tag{4.23}$$

Due to the Kuhn–Tucker conditions for the general mathematical programming problem (1.28b) from Section 2.4, the dual variables related to the equalities in primal problem are not restricted to the nonnegativity. Therefore, the nonnegativity constraints for u_1 and u_2 in (4.23) are omitted. Applying Theorem 4.5, which provides the necessary and sufficient conditions for the optimal solution of a linear programming problem, we write the system of equations

$$\begin{aligned} 2u_1^0 + 3u_2^0 &= 6 \quad (\text{because of } x_5^0 > 0), \\ 4u_1^0 + 5u_2^0 &= 8 \quad (\text{because of } x_6^0 > 0), \end{aligned}$$

which yields the solution $\mathbf{u}^0 = (-3, 4)$. Because all dual constraints and all complementary slackness conditions are fulfilled, this is the optimal solution for the dual problem (4.23). Substituting $u_1^0 = -3$ and $u_2^0 = 4$ into the dual objective function, we get the same value of €70 as for the primal objective function $f(\mathbf{x}^0)$. The reader can easily verify that the graphic method for solving problem (4.23) provides the same solution.

Suppose we increase the output requirements in the production problem (4.22) from $(30, 40)$ to $(60, 50)$; i.e., we change the right-hand side of problem (4.22) to $(60, 50)$. It can be shown in the same way as before that the optimal solution for the new problem is $x_2^* = 10$ and $x_4^* = 20$ with the objective function value €50. The optimal dual solution for the new problem is $\mathbf{u}^* = (0, 1)$. We can see the firm produces more units of both products (100% more of product P_1 and 25% more of product P_2) for less total cost (71, 4% of the previous cost). Hence we have the more-for-less paradox in this situation.

After the numerical illustration, we define the more-for-less paradox generally. Let us consider the following linear programming problem, with all c_j $(j = 1, 2, \ldots, n)$ and b_i $(i = 1, 2, \ldots, m)$ assumed positive:

$$\begin{aligned} \text{minimize} \quad & f_0(\mathbf{x}) = \mathbf{c}'\mathbf{x} \\ \text{subject to} \quad & A\mathbf{x} = \mathbf{b}, \\ & \mathbf{x} \geqq \mathbf{0}, \end{aligned} \tag{4.24}$$

where A is an $m \times n$ matrix ($m < n$) and of full rank, $\mathbf{b}$ is an $m \times 1$ vector, $\mathbf{x}$ is an $n \times 1$ vector, and $\mathbf{c}'$ is a $1 \times n$ vector.

Problem (4.24) has the more-for-less property if we can increase all or some components of $\mathbf{b}$ and reduce the value of the objective function $f_0(\mathbf{x})$ without reducing other components of $\mathbf{b}$, and keeping all c_j fixed. The following theorem provides the necessary and sufficient condition for the more-for-less paradox.

Theorem 4.6. *The linear programming problem* (4.24) *has the more-for-less property if and only if every optimal solution of the corresponding dual problem*

$$\begin{aligned} \textit{maximize} \quad & g(\mathbf{u}) = \mathbf{b}'\mathbf{u} \\ \textit{subject to} \quad & A'\mathbf{u} \leqq \mathbf{c} \end{aligned}$$

has at least one negative dual variable u_i^0.

The proof of this theorem can be found in two independent papers, the first by Chobot and Turnovec [9, p. 379] and the second by Charnes, Duffuaa, and Ryan [6, p. 195]. The more-for-less paradox in the case of the distribution model was first studied by Charnes and Klingman [7], and Theorem 4.6 is an extension of the main theorem of this paper.

To illustrate, we return to our numerical example (4.22). Because the optimal solution $\mathbf{x}^0 = (0, 0, 0, 0, 5, 5)$ is nondegenerate, the corresponding dual solution $u_1^0 = -3$ and $u_2^0 = 4$ is unique and with one negative component. Due to Theorem 4.6, there exists a vector $\Delta\mathbf{b} > 0$ (in our example, $\Delta b_1 = 30$, $\Delta b_2 = 10$) such that the optimal solution $\mathbf{x}^* = (0, 10, 0, 20, 0, 0)$ for the problem with these new right-hand sides yields a lower objective function value ($\Delta f_0(\mathbf{x}) = -20$). The optimal solution for the corresponding dual problem is degenerate and nonnegative $\mathbf{u}^* = (0, 1)$. The new problem does not have the more-for-less property.

From the economic interpretation point of view, this result is remarkable. For a given linear technology, how do we explain that the increasing production of all goods can lead to a decreasing total production cost? Or, in other words, the reduced level of production is accompanied by increasing total cost.

For the given unit cost c_j ($j = 1, 2, \ldots, n$) and the technological coefficients a_{ij}, there exist a "technological" optimal structure of production, described by $\mathbf{s}^0 = (s_1^0, s_2^0, \ldots, s_m^0)$, which can be found by solving the following linear programming problem [9, p. 348]:

$$\text{minimize} \qquad f_0(\mathbf{x}) = \sum_{j=1}^{n} c_j x_j \tag{4.25}$$

$$\text{subject to} \quad \sum_{j=1}^{n} a_{ij} x_j - s_i = 0 \qquad (i = 1, 2, \ldots, m),$$

$$\sum_{i=1}^{m} s_i = 1, \tag{4.26}$$

$$x_j \geqq 0 \qquad (j = 1, 2, \dots, n),$$
$$s_i \geqq 0 \qquad (i = 1, 2, \dots, m).$$

The "technological" optimal structure $\mathbf{s}^0$ can deviate considerably from the given structure, described by the coefficients b_i:

$$\mathbf{s} = (s_1, s_2, \dots, s_n), \quad \text{where } s_i = \frac{b_i}{\sum_{i=1}^{n} b_i}.$$

Because of the equality constraints in problem (4.24), it is not always possible to realize the "technological" optimal structure $\mathbf{s}^0$. If the value of the structure $\mathbf{s}^0$, described by the optimal value of the objective function (4.25), deviates considerably from the value of the required production structure $\mathbf{s}$, the increasing production in problem (4.24) accompanied by changing the production structure toward the "technological" optimal structure $\mathbf{s}^0$ will lead to decreasing production costs.

To illustrate, let us return to our numerical example (4.22). In order to fulfill the output requirements (with higher output of the product P_2), the firm must use the technologies T_5 and T_6 with higher output of product P_2, but with the highest unit cost ($c_5 = 6$, $c_6 = 8$). Changing the structure of the required outputs (with the higher proportion of the product P_1), the firm will switch to the technologies with higher amount of the product P_1 and with lower unit cost ($c_2 = 1$, $c_4 = 2$). The values of the dual variables u_1 and u_2 show the cost reducing change in the output structure. According to the negative value of the dual variable $u_1 = -3$, an increase of the amount of the product P_1 by one unit reduces the total cost by three units. An increase of the amount of the product P_2 by one unit increases the total cost by four units. Therefore, the increase of the output P_1 from 30 to 60 units overcompensates (because of the possibility to switch to cheaper technologies) for the increasing cost caused by increasing output of the product P_2, and the total costs are lower than before.

In order to characterize the "technological" optimal structure analytically, we consider the dual problem corresponding to problem (4.25)–(4.26):

$$\text{maximize} \qquad g(\mathbf{u}) = u_{m+1} \tag{4.27}$$

$$\text{subject to} \qquad \sum_{i=1}^{m} a_{ij} u_i \leqq c_j \quad (j = 1, 2, \dots, n), \tag{4.28}$$

$$-u_i + u_{m+1} \leqq 0 \qquad (i = 1, 2, \dots, m). \tag{4.29}$$

Assuming $s_i > 0$, the complementary slackness theorem yields

$$u_{m+1} = u_i \quad (i = 1, 2, \dots, m).$$

The dual constraints (4.28) can be then rewritten as

$$\sum_{i=1}^{m} a_{ij} u_{m+1} \leqq c_j \qquad (j = 1, 2, \ldots, n),$$

or

$$u_{m+1} \leqq \frac{c_j}{\sum_{i=1}^{m} a_{ij}} \qquad (j = 1, 2, \ldots, n).$$

The optimal solution of the dual problem (4.27)–(4.29) is given by

$$u_{m+1}^{0} = \min \frac{c_j}{\sum_i a_{ij}} \qquad (j = 1, 2, \ldots, n). \tag{4.30}$$

Using the complementary slackness theorem, we can easily obtain the "technological" optimal structure according to relation (4.30).

To illustrate, we consider our example (4.22) again. According to (4.25)–(4.26), the linear programming problem for the estimation of the "technological optimal" structure is the following:

$$\begin{aligned}
\text{minimize} \quad & f_0(\mathbf{x}) = 2x_1 + x_2 + 4x_3 + 2x_4 + 6x_5 + 8x_6 \\
\text{subject to} \quad & 3x_1 + 2x_2 + 3x_3 + 2x_4 + 2x_5 + 4x_6 - s_1 = 0, \\
& x_1 + x_2 + 2x_3 + 2x_4 + 3x_5 + 5x_6 - s_2 = 0, \\
& s_1 + s_2 = 1, \\
& s_1 \geqq 0, \qquad s_2 \geqq 0, \qquad x_j \geqq 0 \quad (j = 1, \ldots, n).
\end{aligned}$$

The corresponding dual problem

$$\begin{aligned}
\text{maximize} \quad & g(\mathbf{u}) = u_3 \\
\text{subject to} \quad & 3u_1 + u_2 \leqq 2, \\
& 2u_1 + u_2 \leqq 1, \\
& 3u_1 + 2u_2 \leqq 4, \\
& 2u_1 + 2u_2 \leqq 2, \\
& 2u_1 + 3u_2 \leqq 6, \\
& 4u_1 + 5u_2 \leqq 8, \\
& -u_1 + u_3 \leqq 0, \\
& -u_2 + u_3 \leqq 0
\end{aligned}$$

is easy to solve. Assuming $s_1 > 0$, $s_2 > 0$, we obtain $u_3 = u_2 = u_1$. Then the remaining dual constraints reduce to

$$\begin{aligned}
4u_3 &\leqq 2, \\
3u_3 &\leqq 1, \\
5u_3 &\leqq 4,
\end{aligned}$$

$$4u_3 \leqq 2,$$
$$5u_3 \leqq 6,$$
$$9u_3 \leqq 8.$$

Under the maximization of u_3, the optimal solution is $u_3^0 = \frac{1}{3}$, which corresponds exactly to relation (4.30). Due to the complementary slackness theorem the optimal value for the primal variables x_1, x_3, x_4, x_5, and x_6 must be equal to zero. Consequently, $x_2^0 = \frac{1}{3}$, $s_1^0 = \frac{2}{3}$, and $s_2^0 = \frac{1}{3}$. The "technological optimal" is the technology $\mathcal{T}_2$ with the structure of production $s_1^0 : s_2^0 = 2 : 1$. In our numerical example (4.22), we changed the initial output requirements $\mathbf{b} = (30, 40)$ to $\mathbf{b}^* = (60, 50)$, which are closer to the "technological optimal" structure $\mathbf{s}^0$, and the optimal solution was switched from technologies T_5 and T_6 to technologies T_2 and T_4. The reader can easily verify that if we change the output requirements according to the structure $\mathbf{s}^0$ (e.g., $b_1^0 = 100$ and $b_2^0 = 50$), the firm will use technology T_2 only ($x_2^0 = 50$ and the optimal value of the objective function $f(\mathbf{x}^0) = 50$). This is a degenerate optimal solution, which implies that the optimal solution of the dual problems is not unique. One of the optimal solutions is $u_1^0 = \frac{1}{2}$ and $u_2^0 = 0$ (the solution is feasible and the value of the dual objective function $g(\mathbf{u}^0) = f(\mathbf{x}^0) = 50$). Because of the nonnegative dual solution, the modified problem (with the right-hand side vector $\mathbf{b}^*$) has no more the more-for-less property. According to Charnes, Duffuaa, and Ryan [6], we obtained an optimal solution that resolves the paradox. By resolving the paradox, they mean increasing the right-hand side until the total cost starts to increase from the minimum obtained over all possible increases.

The numerical illustration of the more-for-less property given by example (4.22) and the numerical examples in [9] and [6] may lead to the interpretation that the cost reduction can be achieved by switching to the technologies with lower unit cost c_j only. The following example [24, p. 107], with the matrix of output coefficients

$$A = \begin{pmatrix} 3 & 2 & 1 & 2 & 4 \\ 1 & 1 & 2 & 2 & 5 \end{pmatrix},$$

output requirements $\mathbf{b}' = (15, 5)$, and unit cost $\mathbf{c}' = (5, 3, 3, 4, 6)$, shows that the total cost can decrease even if the firm switches to the most expensive (with $\max_j c_j$) technology.

It is obvious that in the above example the firm will choose the first technology only ($x_1^0 = 5$) in order to meet the output requirements with minimal cost ($f_0(\mathbf{x}^0) = 25$). The optimal dual solution is not unique, but for every optimal solution the dual variable y_2^0 is negative. One of the optimal solutions is, e.g., $\mathbf{y}^0 = (2, -1)$.

Changing the output requirements from (15, 5) to (16, 20), it is optimal for the firm to switch from the technology T_1 (with $c_1 = 5$) to the most expensive technology $T_5(c_5 = \max_j c_j = 6)$. Despite the increasing production of both products (the production of the good P_2 increases four times), the total cost decreases from 25 to 24 Euro. The "technological" optimal structure of production provides the explanation for this result.

The minimum in (4.30) is unique in our example, and using the complementary slackness theorem, the "technological" optimal structure $\mathbf{s}_0 = (\frac{4}{9}, \frac{5}{9})$ will be obtained.

It is described by the structure of technology T_5, which coincides with the structure of the new output requirements $\mathbf{b}'$. Only technology T_5 will be chosen ($x_5^0 = 4$), and the total cost decreases to €24.

4.5 Computational Procedure: The Simplex Method

In the preceding sections, we have discussed the basic properties of linear programming models, the economic implications of the linearity assumption, and (relatively extensively) the economic interpretation of the duality. We have not been concerned thus far with the question of how to obtain the solution of the general linear programming problem (4.1)–(4.3). The examples we analyzed in the previous section consisted of two variables (or two constraints) only and could be solved geometrically.

The best-known method for solving the linear programming problems is the simplex method developed (as mentioned in Section 1.3) by G. Dantzig in 1947.

The simplex method can be treated from various points of view. It can be described algebraically, not only as the method for solving linear programs, but as the general method for solving linear inequalities or finding nonnegative solutions to linear equations. A very nice feature of this method is its economic interpretation, and therefore we shall consider this aspect in more detail. Finally, it can be made plausible by geometric interpretation.

Let us consider the general linear programming problem (4.1)–(4.3). As shown in Section 4.1, an optimal solution can always be found among the extreme points of the feasible set or among the basic solutions.

In order to carry on the economic and algebraic discussion simultaneously, let us to interpret problem (4.1)–(4.3) as a diet problem (described in Section 1.2.1) in which it is required to meet nutritional requirements with minimal cost. The following economic interpretation of the simplex method is based on [15, pp. 105–108].

Suppose that we found some diet, not necessarily the cheapest one, that is a basic solution of (4.2). The question of how to find a starting basic solution is a serious problem in itself and will be taken up later.

Let the initial basic solution be $\mathbf{x}_0 = (x_{10}, x_{20}, \dots, x_{m0})$. It is a feasible diet that uses—without loss of generality—the first m foods in amounts $x_{10}, x_{20}, \dots, x_{m0}$ According to our notation from Section 4.1, we have

$$x_{10}P_1 + x_{20}P_2 + \cdots + x_{m0}P_m = P_0, \tag{4.31}$$

where the associated vectors $P_1, P_2, \dots, P_m$ are linearly independent. The corresponding value of the objective function (4.1), the cost of the given diet, is

$$x_{10}c_1 + x_{20}c_2 + \cdots + x_{m0}c_m = f_0^0. \tag{4.32}$$

Since $P_1, P_2, \dots, P_m$ are linearly independent, we can express any vector from the set $P_1, P_2, \dots, P_n$ as a linear combination of the basic vectors $P_1, P_2, \dots, P_m$. Therefore, P_j can be expressed by

$$x_{1j}P_1 + x_{2j}P_2 + \cdots + x_{mj}P_m = P_j \quad (j = 1, 2, \ldots, n). \tag{4.33}$$

Let us consider food $m + 1$, say, carrots, which is not in the current diet. Then (4.33) is an expression for carrots as a linear combination of the foods in the current diet. The dietary interpretation of this relation means that one unit of carrot has the same nutritive content as $x_{1,m+1}$ units of food 1 plus $x_{2,m+1}$ units of food 2, and so on. Gale [15, p. 107] denotes the menu vector $(x_{1,m+1}, x_{2,m+1}, \ldots, x_{m,m+1})$ as *synthetic carrots*. Because of the minimization of the total cost of the diet, the following question arises: What is cheaper, a real carrot or a synthetic carrot (without taking into account that a nice red carrot probably tastes better)? The cost of a real carrot is c_{m+1}. The cost of a synthetic carrot, denoted by z_{m+1}, is obviously given by the expression

$$x_{1,m+1}c_1 + x_{2,m+1}c_2 + \cdots + x_{m,m+1}c_m = z_{m+1},$$

or for any food j, which is not in the current diet,

$$x_{1j}c_1 + x_{2j}c_2 + \cdots + x_{mj}c_m = z_j. \tag{4.34}$$

If real carrots are cheaper than synthetic carrots (that is, if $c_{m+1} < z_{m+1}$), then real carrots should be brought into the new diet. The total cost of the new diet (with the carrot instead of one food from the previous diet) must be lower than the cost of the diet without nice red carrots. This leads to the following rule of the simplex method.

Theorem 4.7. *If, for any fixed j, the condition $z_j - c_j > 0$ holds, then a new feasible solution can be found with the objective function value f_0 smaller than f_0^0. The lower bound of f_0 is either finite or infinite.*

Proof. Multiplying (4.33) by some positive number θ and subtracting from (4.31), for $j = 1, 2, \ldots, n$, we get

$$(x_{10} - \theta x_{1j})P_1 + (x_{20} - \theta x_{2j})P_2 + \cdots + (x_{m0} - \theta x_{mj})P_m + \theta P_j = P_0. \tag{4.35}$$

Similarly, multiplication of (4.34) by the same θ and subtraction from (4.32) yields

$$(x_{10} - \theta x_{1j})c_1 + (x_{20} - \theta x_{2j})c_2 + \cdots + (x_{m0} - \theta x_{mj})c_m + \theta c_j = f_0^0 - \theta(z_j - c_j), \tag{4.36}$$

where θc_j has been added to both sides of (4.36). If all the coefficients of the vectors $P_1, P_2, \ldots, P_m, P_j$ in (4.35) are nonnegative, then we have a new feasible solution whose value of the objective function is, by (4.36), $f_0 = f_0^0 - \theta(z_j - c_j)$. Since the variables $x_{10}, x_{20}, \ldots, x_{m0}$ in (4.35) are all positive, there is a value of $\theta > 0$ for which the coefficients of the vectors $P_1, P_2, \ldots, P_m$ in (4.35) remain positive.

If, for a fixed j, $z_j - c_j > 0$ holds, then we have

$$f_0 = f_0^0 - \theta(z_j - c_j) < f_0^0$$

for $\theta > 0$. The value of the objective function f_0 corresponding to a new feasible solution is lower than the value for the preceding solution.

The two different cases—f_0 is either finite or infinite—can be proved in the following way.

If for fixed j at least one x_{ij} in (4.33) for $i = 1, 2, \ldots, m$ is positive, the largest value of θ for which all coefficients in the brackets of (4.35) remain nonnegative is given by

$$\theta_0 = \min_{x_{ij}>0} \frac{x_{i0}}{x_{ij}} > 0. \tag{4.37}$$

Assuming that the problem is nondegenerate, i.e., that all basic solutions have m positive components, the minimum in (4.37) will be reached for a unique i, e.g., for $i = 1$. Then the coefficient of the vectors P_1 in (4.35) will vanish and we have a new basic feasible solution consisting of P_j and the vectors $P_2, P_3, \ldots, P_m$ of the original basis. This new basis can be used for the next iteration as for the previous one. If for at least one $z_j - c_j > 0$ holds and at least one of the coefficients x_{ij} is positive, another solution with a smaller value of the objective function can be obtained. This process will continue either until all $z_j - c_j \leqq 0$ or until, for some $z_j - c_j > 0$, all $x_{ij} \leqq 0$. In the last case all coefficients of the vectors $P_1, P_2, \ldots, P_m$ in (4.35) remain positive, independently of the value θ. There is no upper bound to θ and the objective function value can be decreased infinitely. In other words, the objective function has a lower bound of $-\infty$, the linear programming problem has an unbounded solution. This completes the proof. □

After the economic interpretation of the simplex method steps, the following theorem (see [16, pp. 66–67]) confirms that these rules will indeed lead us to an optimal solution of the problem.

Theorem 4.8 (optimality criterion). *If for any basic solution* $\mathbf{x}_0 = (x_{10}, x_{20}, \ldots, x_{m0})$ *the conditions* $z_j - c_j \leqq 0$ *hold for all* $j = 1, 2, \ldots, n$, *then* $\mathbf{x}^0$ *is an optimal solution of problem* (4.1)–(4.3).

If the objective function in problem (4.1)–(4.3) is to be maximized, we would use the following criterion instead of changing to a minimization problem: Compute the $z_j - c_j$ and select a new variable with $z_j - c_j < 0$; an optimum solution has been found when all $z_j - c_j \geqq 0$.

The results of Theorems 4.7 and 4.8 enable us to start with a basic solution and generate a set of the new basic solutions that converge to the optimal solution or determine that a finite optimal solution does not exist. The question arising now is that of how to generate a new basic solution.

Let us start with the m linearly independent vectors $P_1, P_2, \ldots, P_m$. We denote the $m \times m$ matrix $(P_1, P_2, \ldots, P_m)$ by B and call it a *basis*. Now we decided to replace the vector P_1 of the old basis—for which the minimum in (4.37) will be obtained by the vector P_{m+1}. According to (4.33), we have

$$P_{m+1} = x_{1,m+1}P_1 + x_{2,m+1}P_2 + \cdots + x_{m,m+1}P_m. \tag{4.38}$$

From (4.38) we obtain

$$P_1 = \frac{1}{x_{1,m+1}}(P_{m+1} - x_{2,m+1}P_2 - \cdots - x_{m,m+1}P_m). \tag{4.39}$$

Substituting (4.38) for P_1 into (4.33) yields

$$\begin{aligned} P_j &= \frac{x_{1j}}{x_{1,m+1}}(P_{m+1} - x_{2,m+1}P_2 - \cdots - x_{m,m+1}P_m) \\ &+ x_{2j}P_2 + \cdots + x_{mj}P_m = (x_{2j} - \frac{x_{1j}}{x_{1,m+1}}x_{2,m+1})P_2 \\ &+ (x_{3j} - \frac{x_{1j}}{x_{1,m+1}}x_{3,m+1})P_3 + \cdots + (x_{mj} - \frac{x_{1j}}{x_{1,m+1}}x_{m,m+1})P_m \\ &+ \frac{x_{1j}}{x_{1,m+1}}P_{m+1}. \end{aligned} \tag{4.40}$$

The reader may verify that the expressions in the brackets in (4.40) are equivalent to the complete elimination formulas of Jordan and Gauss when the pivot element is $x_{1,m+1}$. In other words, to generate a new basic solution, we can use the complete elimination method of Jordan and Gauss, which is well known from linear algebra.

Furthermore, it can be shown [16, p. 60] that the vectors $P_2, P_3, \ldots, P_m, P_{m+1}$ are linearly independent, and therefore the new solution $\mathbf{x}' = (x'_2, x'_3, \ldots, x'_m, x'_{m+1})$ is a basic solution or an extreme point of the feasible set.

The last open question in our description of the simplex procedure is that of how to find the initial basic solution. If the linear programming problem is of the form

$$\begin{aligned} \text{maximize} \quad & f_0(\mathbf{x}) = \mathbf{c}'\mathbf{x} \\ \text{subject to} \quad & A\mathbf{x} \leqq \mathbf{b}, \\ & \mathbf{x} \geqq \mathbf{0}, \end{aligned} \tag{4.41}$$

where $\mathbf{b} \geqq \mathbf{0}$, we write each inequality as an equality by adding a nonnegative slack variable. The corresponding coefficients in the objective function for these slack variables are usually set equal to zero. In this way, we obtain m unit vectors that are linearly independent and build a starting basis.

If the problem was originally of the form $A\mathbf{x} \geqq \mathbf{b}$, the equivalent set of equations is obtained by the subtraction of a nonnegative slack variable from each inequality. Here the equations contain a negative unit matrix which cannot be used for the initial basic solution (because of the nonnegativity constraints). For this case and for the general linear programming problem (4.1)–(4.3), the method of the *artificial basis*, or the *two-phase simplex method* (see, e.g., [16], [10], or any textbook on linear programming) can be used to start the simplex process. This procedure also determines whether or not the problem has any feasible solution.

Once an initial basic solution has been found, the simplex procedure calls for the successive application of the following.

1. The testing of the $z_j - c_j$ elements to determine whether a minimum solution has been found, i.e., whether $z_j - c_j \leqq 0$ for all j.
2. The selection of the vector to be introduced into the basis if some $z_j - c_j > 0$.

3. The selection of the vector to be eliminated from the basis, i.e., the vector with $\min_{x_{ik}>0}(\frac{x_{i0}}{x_{ik}})$, where k corresponds to the vector selected in Step 2. If all $x_{ik} \leqq 0$, then the solution is unbounded.
4. The new solution will be obtained by the complete elimination procedure of Jordan and Gauss.

Each such iteration produces a new basic feasible solution, and according to Theorems 4.7 and 4.8, we shall obtain an optimal solution or find an unbounded solution.

For the illustration of the simplex method, let us solve the following hypothetical production problem. A farmer has 100 acres of land at his disposal, a fixed commitment of 160 man-days of labor, and the amount of €1100 for covering the input cost (for seeding, cultivating, etc.). He wishes to apportion his production factors—land, labor, and capital—between two crops, wheat and potatoes, to yield a maximum return on the market. Assuming good weather and given the prices of both crops, the expected return is €120/acre for wheat and €40/acre for potatoes. The input cost is €20/acre for wheat and €10/acre for potatoes, and the amount of required labor is four man-days per acre for wheat and one man-day per acre for potatoes. How many acres of wheat and how many acres of potatoes should the farmer plant in order to maximize his return revenue?

Denoting the number of acres of planted potatoes by x_1 and the number of acres of planted wheat by x_2, we formulate his decision problem as the following optimization problem:

$$\begin{array}{llr} \text{maximize} & f_0(\mathbf{x}) = 40x_1 + 120x_2 & \\ \text{subject to} & x_1 + x_2 \leqq 100, & \text{(land constraint)} \\ & 10x_1 + 20x_2 \leqq 1100, & \text{(capital constraint)} \\ & x_1 + 4x_2 \leqq 160, & \text{(labor constraint)} \\ & x_1 \geqq 0, & \\ & x_2 \geqq 0. & \end{array}$$

Transforming to equations by adding a nonnegative slack variable to all inequalities except the nonnegativity constraints, we obtain the equivalent linear programming problem:

$$\begin{array}{llr} \text{maximize} & f_0(\mathbf{x}) = 40x_1 + 120x_2 & \\ \text{subject to} & x_1 + x_2 + x_3 = 100, & \text{(land constraint)} \\ & 10x_1 + 20x_2 + x_4 = 1100, & \text{(capital constraint)} \\ & x_1 + 4x_2 + x_5 = 160, & \text{(labor constraint)} \\ & x_j \geqq 0 \quad (j = 1, 2, \ldots, 5). & \end{array}$$

The interpretation of the slack variable is straightforward: x_3 denotes the number of unplanted acres, x_4 is the amount of money, and x_5 is the number of man-days, respectively, not used.

Table 4.1.

i	Basis	$\mathbf{c}_B$	40 P_1	120 P_2	0 P_3	0 P_4	0 P_5	P_0	
1	P_3	0	1	1	1	0	0	100	$R1$
2	P_4	0	10	20	0	1	0	1100	$R2$
3	P_5	0	1	4	0	0	1	160	$R3$
4	$z_j - c_j$		−40	−120	0	0	0	0	$R4$

According to our notation in the general linear programming problem (4.1)–(4.3), the given vectors are then

$$P_1 = \begin{pmatrix} 1 \\ 10 \\ 1 \end{pmatrix}, \quad P_2 = \begin{pmatrix} 1 \\ 20 \\ 4 \end{pmatrix}, \quad P_3 = \begin{pmatrix} 1 \\ 0 \\ 0 \end{pmatrix}, \quad P_4 = \begin{pmatrix} 0 \\ 1 \\ 0 \end{pmatrix}, \quad P_5 = \begin{pmatrix} 0 \\ 0 \\ 1 \end{pmatrix},$$

and

$$P_0 = \begin{pmatrix} 100 \\ 1100 \\ 160 \end{pmatrix}.$$

We see that these equations contain a starting basis of the slack vectors (P_3, P_4, P_5) with the associated first feasible solution of $x_1 = 0$, $x_2 = 0$, $x_3 = 100$, $x_4 = 1100$, and $x_5 = 160$. The first feasible basic solution is that the farmer does nothing. He produces neither potatoes nor wheat and all production factors (land, capital, and labor) are unused. The corresponding value of the objective function is zero, as all coefficients for the slack variables in the objective function are here assumed to be zero. The slack variables do not yield any return for the farmer. In some applications, these coefficients can be other than zero, e.g., penalty costs for not using certain raw materials, storage costs, etc.

For didactic purposes the steps of the simplex method will be described by tableaus. The initial tableau in our example has the form shown in Table 4.1, where $\mathbf{c}_B$ denotes the vector of the objective function coefficients for the basic variables.

Because our initial basis $B = (P_3, P_4, P_5)$ is the identity matrix E_3, the description of all vectors P_j in terms of this basis is very simple. According to (4.33),

$$P_j = B\mathbf{x}_j, \quad \text{or} \quad \mathbf{x}_j = B^{-1}P_j, \tag{4.42}$$

where $\mathbf{x}_j = (x_{1j}, x_{2j}, \ldots, x_{mj})$. For the initial tableau, $B^{-1} = E$ and therefore

$$P_1 = 1P_3 + 10P_4 + 1P_5$$

and

$$P_2 = 1P_3 + 20P_4 + 4P_5.$$

The elements z_j can easily be obtained using (4.34). They are zero because the coefficients c_i for the basic variables are—as indicated in the third column of Table 4.1—all zero. Since we are now maximizing instead of minimizing, we look for negative rather than positive entries in the last row of Table 4.1. All coefficients for the nonbasic variables are negative (the coefficients $z_j - c_j$ for the basic variables are always equal to zero), and therefore we may bring any of the vectors P_1, P_2 into next basis. The production of both products (potatoes and wheat) will yield higher profit for the farmer than to leave the production factors unused. The greatest immediate increase in the value of the objective function is given by

$$\min_j \theta_0(z_j - c_j) \quad \text{for } j = 1, 2,$$

where for each j, θ_0 is determined by (4.37). If there are a number of j for which $z_j - c_j < 0$, the above rule is rather complicated to apply. A much simpler criterion for selecting of a new basic vector is to select the one which corresponds to

$$\min_j (z_j - c_j) \quad \text{or to} \quad \max_j |z_j - c_j|,$$

where $|z_j - c_j|$ denotes the absolute value of $(z_j - c_j)$. In our example,

$$\min_j (z_j - c_j) = z_2 - c_2 = -120.$$

One acre of planted wheat yields higher return than one acre of planted potatoes. The farmer will increase his return revenue if he plants wheat. Mathematically speaking, the vector P_2 will be introduced into the basis.

How many acres of wheat will the farmer plant? He has 100 acres of land, but the input cost for wheat is €20 per acre and he needs four man-day per acre of wheat. Taking into account the availability of €1100 for covering the input cost and a fixed amount of 160 man-days of labor, the number of acres of wheat he can plant is given by

$$\min \left\{ \frac{100}{1}, \frac{1100}{20}, \frac{160}{4} \right\} = 40.$$

This rule corresponds exactly to the form (4.37), which determines the vector leaving the basis:

$$\theta_0 = \min_{x_{i2} > 0} \frac{x_{i0}}{x_{i2}}.$$

The farmer will plant 40 acres of wheat, he needs 160 man-days, therefore the slack variable x_5 will be zero and the vector P_5 will be the one eliminated from the basis. Our new feasible solution will have a new basis consisting of P_3, P_4, and P_2. In economic terms, the farmer will plant wheat (40 acres because of the scarcity of labor), and he will not completely use the production factors land and capital (due to the limitational type of production function). We next wish to compute the new solution explicitly and to express each vector not in the basis in terms of the new basis.

According to (4.40), we obtain the new basic solution using Jordan–Gauss elimination procedure (including the P_0 column and the row with the elements $z_j - c_j$)

Table 4.2.

i	Basis	c_B	40 P_1	120 P_2	0 P_3	0 P_4	0 P_5	P_0	
1	P_3	0	$\frac{3}{4}$	0	1	0	$-\frac{1}{4}$	60	$R5 = R1 - R7$
2	P_4	0	5	0	0	1	-5	300	$R6 = R2 - 20R7$
3	P_2	120	$\frac{1}{4}$	1	0	0	$\frac{1}{4}$	40	$R7 = \frac{1}{4}R3$
4	$z_j - c_j$		-10	0	0	0	30	4800	$R8 = R4 + 120R7$

on Table 4.1 with the pivot element $x_{32} = 4$ (with row 3 being the pivot row or the row for the variable eliminated from the basis and column 2 being the pivot column or the column for the variable entering the basis). The resulting solution is described in Table 4.2. (The reader may verify the computations by continuous numeration of the rows and using the formulas to the right of the tables.)

With 40 acres of wheat ($x_2 = 40$), the farmer leaves 60 acres of land uncultivated ($x_3 = 60$), and he does not invest €300 ($x_5 = 300$). His return revenue is €4800 (obviously higher than at the initial solution). According to (4.40), the elements in the P_1 (P_5) column are the coefficients for the expression of the nonbasic vector P_1 (P_5) in terms of the basic vectors P_3, P_4, and P_2. The reader may verify that

$$P_1 = \frac{3}{4}P_3 + 5P_4 + \frac{1}{4}P_2$$

and

$$P_5 = -\frac{1}{4}P_3 - 5P_4 + \frac{1}{4}P_2.$$

Therefore, as a check on each complete elimination transformation, one should explicitly compute the individual f_0 and $z_j - c_j$ (using the form (4.34)) and compare them with the transformed values of f_0 and $z_j - c_j$.

The solution in Table 4.2 is not optimal because of $z_1 - c_1 < 0$. The return from one acre of planted potatoes ($c_1 = 40$) is higher than the revenue from "synthetic" potatoes ($z_1 = 0 \cdot \frac{3}{4} + 0 \cdot 5 + 120 \cdot \frac{1}{4} = 30$), and therefore the farmer will plant potatoes in order to increase his return revenue. The number of acres of planted potatoes is given by the form (4.37) again:

$$\min\left\{\frac{60}{\frac{3}{4}}, \frac{300}{5}, \frac{40}{\frac{1}{4}}\right\} = 60 \quad \text{(see Table 4.2)}.$$

The farmer will use the total amount of capital; the slack variable x_4 equals zero and the vector P_4 will be eliminated from the basis. Instead of this vector, the vector P_1 will enter the basis. The new basis consists of vectors P_3, P_1, and P_2. The selection of the vector to be eliminated from the basis according to (4.37) ensures the feasibility of the new solution (none of the variables will violate the nonnegativity restrictions).

Table 4.3.

i	Basis	c_B	40 P_1	120 P_2	0 P_3	0 P_4	0 P_5	P_0	
1	P_3	0	0	0	1	$-\frac{3}{20}$	$\frac{1}{2}$	15	$R9 = R5 - \frac{3}{4}R10$
2	P_1	40	1	0	0	$\frac{1}{5}$	-1	60	$R10 = \frac{1}{5}R6$
3	P_2	120	0	1	0	$-\frac{1}{20}$	$\frac{1}{2}$	25	$R11 = R7 - \frac{1}{4}R10$
4	$z_j - c_j$		0	0	0	2	20	5400	$R12 = R8 + 10R10$

This will be obtained by application of Jordan–Gauss elimination formulas (4.40) with the pivot element $x_{21} = 5$ (see Table 4.3).

The farmer should plant 60 acres of potatoes and 25 acres of wheat. Fifteen acres of land remains uncultivated because of the scarcity of capital and labor. The return revenue of the farmer has been raised to €5400.

Since all elements $z_j - c_j$ are nonnegative due to Theorem 4.8, the solution in Table 4.3 is optimal. If for the optimal solution some $z_j - c_j = 0$ for a vector P_j not in the final basis, then this vector can be introduced into the basis without changing the final value of the objective function. The resulting solution will also be a maximum feasible solution, and hence we have determined multiple optimal solutions. Any convex combination of these optimal solutions will also be an optimal solution.

In solving a linear programming problem, the decision maker is interested not only in the solution of the primal problem but in dual solution as well. A very important and useful feature of the simplex method is that we obtain the solution of the dual problem simultaneously with the solution of the primal problem. It is contained in the final simplex tableau without any additional effort. Where can it be found?

The scheme of the initial simplex tableau for problem (4.41) can be written as in Table 4.4.

Table 4.4.

$\mathbf{c}$	$\mathbf{0}$	
A	E	$\mathbf{b}$
$-\mathbf{c}$	$\mathbf{0}$	0

The transformation of the tableau by the complete elimination procedure is nothing more than the multiplication of their columns by the inverse basis B^{-1} (due to (4.42)). Therefore, the final simplex tableau has the form given in Table 4.5.

Table 4.5.

$B^{-1}A$	B^{-1}	$B^{-1}\mathbf{b}$
$\mathbf{c}'_B B^{-1}A - \mathbf{c}$	$\mathbf{c}'_B B^{-1}$	$\mathbf{c}'_B B^{-1}\mathbf{b}$

The interpretation of the elements in Table 4.5 is now straightforward. The elements $B^{-1}A$ are the coefficients of the linear combination of the vectors P_j in terms of the final basis. In the columns that correspond to the unit matrix in the initial tableau (in other words, to the basic vectors of the initial solution), the inverse of the final basis is developed. In the last column, the primal solution $\mathbf{x}^0 = B^{-1}\mathbf{b}$ is obtained.

According to duality theory, the optimal solution to the dual problem is given by

$$\mathbf{u}^0 = \mathbf{c}'_B B^{-1} \quad \text{(due to (4.20b): } \mathbf{u}^{0'} B = \mathbf{c}'_B).$$

Therefore, in the last row of the final simplex tableau and in the columns corresponding to the basic vectors of the initial tableau, we will find the dual solution.

The expression $\mathbf{c}'_B B^{-1}\mathbf{b}$ shows the optimal value of the primal *and* the dual objective function. To see this, let us decompose this expression in two ways:

- $c'_B B^{-1}\mathbf{b} = \mathbf{c}'_B \mathbf{x}^0$, which is the optimal value of the primal objective function, and
- $\mathbf{c}'_B B^{-1}\mathbf{b} = \mathbf{u}^{0'}\mathbf{b}$, which is the optimal value of the dual objective function.

Finally, it is easy to show that the optimality criterion from Theorem 4.8 is equivalent to the dual feasibility. We rewrite the elements z_j as follows:

$$z_j = \sum_{i \in B} x_{ij} \cdot c_i = \mathbf{c}'_B B^{-1} P_j = \mathbf{u}' P_j.$$

According to the optimality criterion from Theorem 4.8,

$$z_j - c_j = \mathbf{u}' P_j - c_j \geqq 0 \quad \text{for all } j \text{ (for a maximization problem)}$$

or

$$\mathbf{u}' A - \mathbf{c}' \geqq \mathbf{0}.$$

Therefore, if in the final tableau $\mathbf{c}'_B B^{-1} A - \mathbf{c}' \geqq \mathbf{0}$ (for the maximization primal problem), rewritten as $\mathbf{u}' A - \mathbf{c}' \geqq \mathbf{0}$, holds, the solution is dual feasible too and according to the duality theory both problems have optimal solutions.

To illustrate, we return to our numerical example in Table 4.3. The basis corresponding to the final solution is given by the vectors P_3, P_1, P_2, that is,

$$B = (P_3, P_1, P_2) = \begin{pmatrix} 1 & 1 & 1 \\ 0 & 10 & 20 \\ 0 & 1 & 4 \end{pmatrix}.$$

We see that in the columns corresponding to the basic vector of the initial table, Table 4.1, the inverse of the final basis B,

$$B^{-1} = \begin{pmatrix} 1 & -\frac{3}{20} & \frac{1}{2} \\ 0 & \frac{1}{5} & -1 \\ 0 & -\frac{1}{20} & \frac{1}{2} \end{pmatrix},$$

is found. The corresponding optimal solution $\mathbf{x}^0$ for the primal problem is

$$\mathbf{x}^0 = B^{-1}\mathbf{b} = (x_3^0, x_1^0, x_2^0) = (15, 60, 25),$$

and the optimal value of the objective function is

$$\mathbf{c}_B'\mathbf{x}^0 = (0, 40, 120)\begin{pmatrix}15\\60\\25\end{pmatrix} = 5400.$$

The optimal solution to the dual problem is given by

$$\mathbf{u}^0 = \mathbf{c}_B' B^{-1} = (u_1^0, u_2^0, u_3^0) = (0, 2, 20).$$

It may be found in the fourth row and in the column of the inverse basis in the final table, Table 4.3. The optimal value of the dual objective function is

$$\mathbf{u}^{0'}\mathbf{b} = (0, 2, 20)\begin{pmatrix}100\\1100\\160\end{pmatrix} = 5400.$$

If a vector that formed the unit matrix (the initial basis) had a $c_j \neq 0$, then the value of this c_j would have to be added back to the corresponding $z_j - c_j$ in the final tableau in order to obtain the correct value for the u_i^0. We note that u_i^0 is equal to the z_j that has, for its corresponding unit vector in the initial simplex tableau, the vector whose unit element is in position i. In our example, $u_1^0 = z_3$, since P_3 is a unit vector with its unit element in position $i = 1$.

Since our problem involves only two variables, it is amenable to graphical analysis. In Figure 4.2, we plot x_1 and x_2 on the two axes. Because of the nonnegativity constraints, we need to consider the nonnegative quadrant only. Let us first pretend that the three constraints are given as equations and plot them as three straight lines. Each of these lines—labeled as land constraint, capital constraint, and labor constraint, respectively—divides the quadrant into two nonoverlapping regions.

Since each constraint is of the $\leqq$ type, only the points (order pairs) lying on or below the border line will satisfy the particular constraint involved. All the points located in the shaded area in Figure 4.2 satisfy all three constraints simultaneously. This shaded area describes the set of feasible solutions. To maximize return revenue f_0, it is necessary to take the objective function into account. Writing the revenue function in the form

$$x_2 = f_0 - \frac{1}{3}x_1$$

and taking f_0 to be a parameter, we can plot this equation as a family of parallel straight lines with the common slope of $-\frac{1}{3}$. For every objective function $f_0 = \mathbf{c}'\mathbf{x}$, we can specify the gradient that indicates in which direction the equation $\mathbf{c}'\mathbf{x} = f_0$ has to be shifted in order to achieve a better (i.e., for maximization problem, higher) value of the objective function. The gradient of the objective function is orthogonal to the straight lines, having the same value of the objective function (see Figure 4.2). These lines (or hyperplanes in a more general treatment) are frequently referred to

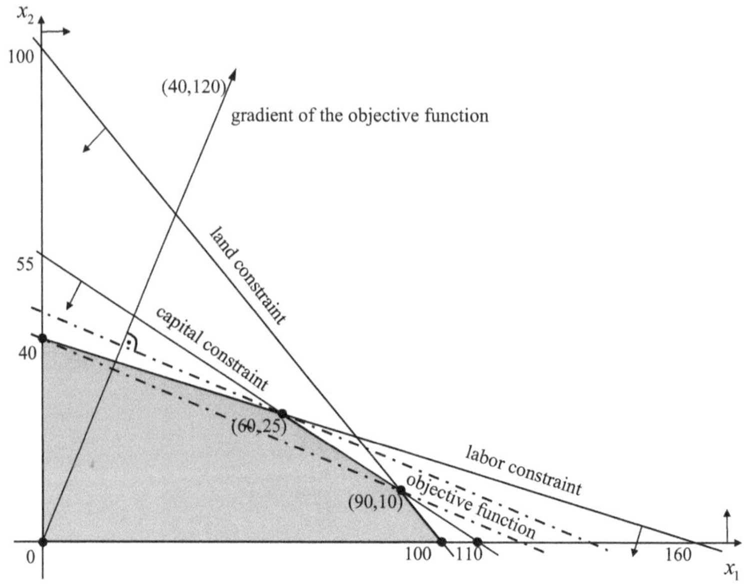

Fig. 4.2. Graphical solution of the farmer problem.

as isoprofit lines, isocost lines (depending on the meaning of the objective function), or sometimes as contour lines. To maximize return revenue, we must select the highest possible isorevenue while still staying in the feasible set. In Figure 4.2, such a selection leads us to the extreme point (60, 25), which is the optimal solution found by the simplex method. In terms of Figure 4.2, the simplex method has led us systematically from the initial extreme point (0, 0) in Table 4.1 to the next extreme point (0, 40) in Table 4.2, followed by a move to the optimal extreme point (60, 25) described by Table 4.3. Note that we have arrived at the optimal solution without having to compare all five extreme points.

Because of the primary aim of this book oriented to qualitative analysis, we do not discuss further the numerical and efficiency aspects of the simplex method: the more efficient revised simplex method with the popular "product form of the inverse" developed by Dantzig and Orchard-Hays [11], the problem of degeneracy and cycling, and problems requiring an unusually large number of iterations. There are examples of linear programming problems [20] in which the simplex algorithm can take an exponential number of computational steps in relation to the size of a model. Fortunately, Khachian [19] showed that for linear programming problems, the number of computational steps increases as a polynomial function of the size of the model. With his "ellipsoid method," new research in solving linear programming problems started. In 1984, the method of project transformation, commonly referred to as Karmarkar's algorithm [18], was discovered. The so-called interior point approach is now the subject of very intensive research and of considerable progress. For more information on this subject, the reader is referred to [29].

4.6 Some Applications of Linear Programming in Economics

In Section 1.2, we formulated several mathematical programming models used in economics. Now we want to analyze some of these models more deeply in order to show how linear programming can be used as an instrument of qualitative analysis.

4.6.1 The Theory of Comparative Advantage

One field of economics in which linear programming is very often applied is international trade (see, e.g., [14]). A well-known example of Ricardo [28], slightly modified by Dorfman, Samuelson, and Solow [12, pp. 31–32], leads—for England—to the following linear programming model (as formulated in Section 1.2.4):

$$\begin{aligned} &\text{maximize} && \mathcal{Z} = \frac{p_1}{p_2} x_1 + x_2 \\ &\text{subject to} && 2x_1 + x_2 \leqq C, \\ &&& x_1 \geqq 0, \quad x_2 \geqq 0, \end{aligned}$$

where $\mathcal{Z}$ denotes the national product (NP) of England.

The optimization problem for Portugal has the same structure:

$$\begin{aligned} &\text{maximize} && \mathcal{Z}' = \frac{p_1}{p_2} x_1' + x_2' \\ &\text{subject to} && x_1' + x_2' \leqq C', \\ &&& x_1' \geqq 0, \quad x_2' \geqq 0, \end{aligned}$$

where $\mathcal{Z}'$ denotes the national product of Portugal.

The graphical representation of the feasible set, or "production possibility" curve, for England is given in Figure 4.3, and for Portugal in Figure 4.4.

We can see that the decision about the production of food and clothing in England and in Portugal depends on the slope of objective function, in other words, on the international price ratio $\frac{p_1}{p_2}$. If there exists a price ratio $\frac{p_1}{p_2}$ somewhere between 1 and 2, it is optimal for England to produce only clothing and for Portugal to produce only food. Although Portugal needs less (or no more) input for both products, the best production pattern for this country involves zero clothing production and complete specialization on food. Portugal will export food in exchange for clothing imports from England, which will specialize completely in clothing. Portugal has comparative advantage in food production (it can convert one unit of food into one unit of clothing, but the price for food is higher than that of clothing), and England in clothing production (it can convert one unit of food into two units of clothing, but it gets for one unit of food less than for two units of clothing). Both countries will be better off than if they do not specialize. The world will, in fact, be at the Ricardo point, where $1 \leqq$ MC (marginal cost) $\leqq 2$. The reader may easily verify that for $\frac{p_1}{p_2} = 2$, the optimal solution for England is not unique (the contour line for NP, the isoincome line, is parallel with the "production possibility" curve). The best production pattern

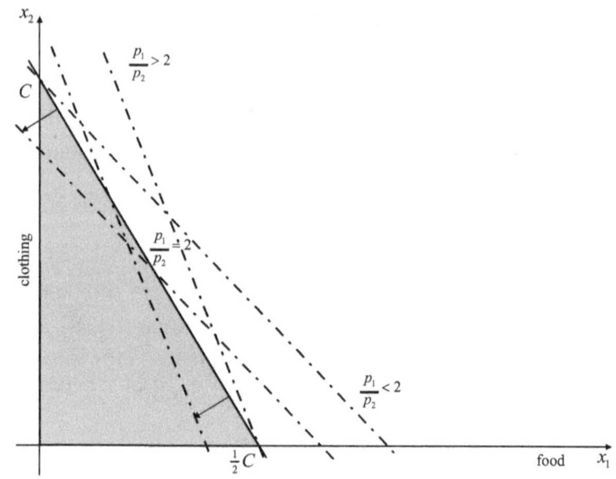

Fig. 4.3. Production possibility set for England.

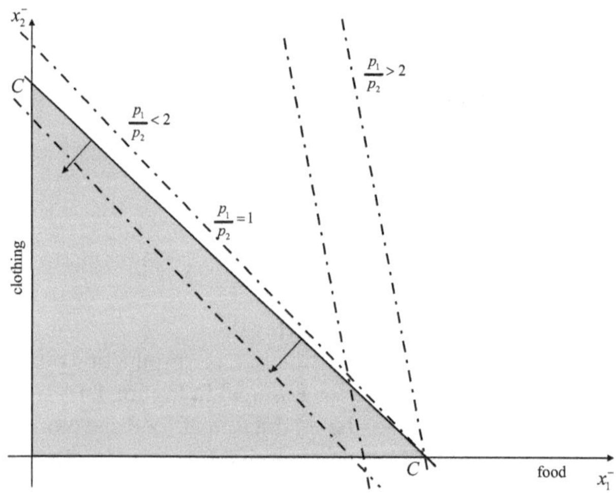

Fig. 4.4. Production possibility set for Portugal.

for Portugal in this situation is a complete specialization on food. When $\frac{p_1}{p_2} = 1$, the optimal solution for Portugal is not unique. When $\frac{p_1}{p_2} < 1$ (or > 2), both countries will specialize completely in clothing (in food).

In the next step, we want to generalize this model for m commodities ($i = 1, 2, \ldots, m$) and n countries ($j = 1, 2, \ldots, n$) (see [14, Chapter 2 and appendix to Chapter 2] and [34, Chapter 6]). The following notation is introduced:

- x_{ij} = quantity of good i produced in country j,
- l_{ij} = constant labor input coefficient in the production of good i in country j,
- L_j = total quantity of labor available in country j,
- p_i = given international price of good i.

In the previous simple model with two countries and two commodities, we have formulated the problem in terms of maximization of the (value of) national product of each country separately considered. Now we will formulate the problem directly in terms of maximization of world output. It can be shown [34, pp. 172–173] that world output will be maximized if and only if each country maximizes its own national output.

The problem of maximizing the value of world output under the constraints that the amount of labor employed in each country cannot exceed the amount disposable and under the nonnegativity constraints for outputs leads to the following linear programming model:

$$\begin{aligned} &\text{maximize} \quad p_1\left(\sum_{j=1}^{n} x_{1j}\right) + p_2\left(\sum_{j=1}^{n} x_{2j}\right) + \cdots + p_m\left(\sum_{j=1}^{n} x_{mj}\right) \\ &\text{subject to} \quad \sum_{i=1}^{m} l_{ij}x_{ij} \leqq L_j \quad (j = 1, 2, \ldots, n), \\ &\qquad x_{ij} \geqq 0 \quad \begin{pmatrix} i = 1, 2, \ldots, m, \\ j = 1, 2, \ldots, n \end{pmatrix}. \end{aligned} \tag{4.43}$$

We are assuming that the resources within each country are completely substitutable so that there is only a single resource limitation (labor) in each country. Further, we postulate constant returns to scale in each country.

The solution to the primal problem (4.43) yields the allocation of m products between n countries. In order to find it, we now ask a different question: What will be the value of labor in each country?

For this purpose, we consider the dual problem to problem (4.43):

$$\begin{aligned} &\text{minimize} \quad \sum_{j=1}^{n} w_j L_j \\ &\text{subject to} \quad l_{ij} w_j \geqq p_i \quad (i = 1, 2, \ldots, m), \\ &\qquad w_j \geqq 0 \quad (j = 1, 2, \ldots, n), \end{aligned} \tag{4.44}$$

where the shadow price of labor in country j, w_j, is interpreted as the money wage rate. Therefore, the dual problem (4.44) consists in minimizing the world total labor reward (world production cost) subject to both the constraint that the value of the resources used will be at least as great as the value of the goods produced and the nonnegativity constraint on the wage rate.

Given $p_i > 0$ $(i = 1, 2, \ldots, m)$ and $l_{ij} > 0$ $(i = 1, 2, \ldots, m;\ j = 1, 2, \ldots, n)$, it follows from the constraints of the dual problem (4.44) that the optimal wage rate w_j^0 must be positive in every country. Due to the complementary slackness theorem,

$$\text{if } w_j^0 > 0, \quad \text{then } \sum_{i=1}^{m} l_{ij} x_{ij}^0 = L_j \quad (j = 1, 2, \ldots, n),$$

where x_{ij}^0 denotes the optimal solution to the primal problem (4.43). Because the optimal money wage rate is positive in the jth country, all of the labor available in that country will be fully utilized. Consequently—due to the nonnegativity of outputs and positivity of labor input coefficients l_{ij}—at least one good must be produced in each country. Because the optimal solution to problem (4.43) must be a basic solution, it consists of n positive components. Assuming that

$$\frac{p_i}{l_{ij}} \neq \frac{p_k}{l_{kj}} \quad \begin{pmatrix} i, k = 1, 2, \ldots, m, \\ j = 1, 2, \ldots, n \end{pmatrix}, \tag{4.45}$$

each country will specialize on its best product. To find it, we rewrite the constraint in the dual problem (4.44) as

$$w_j \geqq \frac{p_i}{l_{ij}} \quad \begin{pmatrix} i = 1, 2, \ldots, m, \\ j = 1, 2, \ldots, n \end{pmatrix}.$$

Then

$$w_j^0 = \max \frac{p_i}{l_{ij}} \quad (i = 1, 2, \ldots, m).$$

Because of the assumption (4.45), the optimal wage rate in country j is unique and determined by the maximal ratio of the given international price p_i to the labor input coefficient l_{ij} for $i = 1, 2, \ldots, m$. For this good (say, k) is the dual constraint in (4.44) fulfilled as an equality, whereas

$$w_j^0 > \frac{p_i}{l_{ij}} \quad \text{or} \quad w_j^0 l_{ij} > p_i \quad \text{for } i = 1, 2, \ldots, m \quad \text{and} \quad i \neq k.$$

According to the complementary slackness theorem, if the unit cost of good i ($i = 1, 2, \ldots, m$; $i \neq k$) in country j is greater than the price of this good, then good i will not be produced in country j. In other words, country j will specialize in product k.

This reasoning can be applied to all countries ($j = 1, 2, \ldots, n$) in order to estimate the best product for each country.

The answer to the question whether the best product k is different from the best product of all other countries or many countries will specialize on the same product depends on the international relative prices for the goods.

Due to Theorem 4.4, for the optimal quantities x_{ij}^0 and the optimal wage rates w_j^0, the values of the objective functions are equal:

$$\sum_{i=1}^{m} p_i \sum_{j=1}^{n} x_{ij}^0 = \sum_{j=1}^{n} w_j^0 L_j.$$

In economic terms, the value of world output coincides with total factor income of the world.

The reader may verify that the application of this general model to the simple Ricardian example from Section 1.2.4 leads to the same results as the graphical solution in Figures 4.3 and 4.4.

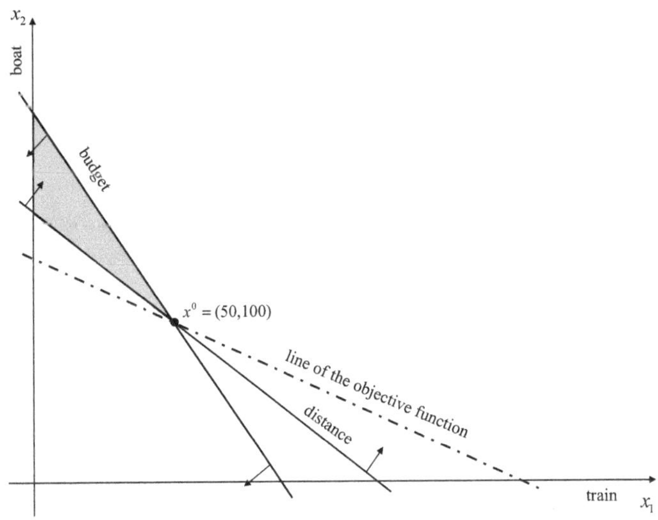

Fig. 4.5. Graphical solution of problem (4.46).

4.6.2 The Giffen Paradox

According to the letter from A. Marshall to F. Y. Edgeworth quoted in Section 1.2.5, the following linear programming problem of a traveler in Holland has been formulated:

$$\begin{aligned} \text{minimize} \quad & c_1x_1 + c_2x_2 \\ \text{subject to} \quad & x_1 + x_2 \geqq 150 \quad \text{(distance)}, \\ & 2x_1 + x_2 \leqq 200 \quad \text{(budget)}, \\ & x_1 \geqq 0, \quad x_2 \geqq 0, \end{aligned} \tag{4.46}$$

where x_1, x_2 denote the distance traveled by train and boat, respectively, and c_1, c_2 are the times required per kilo covered by the two vehicles ($c_1 < c_2$, the speed of the train is higher than that of the boat).

In Figure 4.5, a graphical solution of problem (4.46) supports the travelers' decision, described in the aforementioned letter, to go 100 km by boat and 50 km by train.

The optimal solution $x_1^0 = 50$, $x_2^0 = 100$ is the intersection point of the two constraints:

$$\begin{aligned} x_1 + x_2 &= 150, \\ 2x_1 + x_2 &= 200. \end{aligned}$$

After the increased boat charge, we get the modified problem:

$$\begin{array}{ll} \text{minimize} & c_1x_1 + c_2x_2 \\ \text{subject to} & x_1 + x_2 \geqq 150 \quad \text{(distance)}, \\ & 2x_1 + \frac{5}{4}x_2 \leqq 200 \quad \text{(budget)}, \\ & x_1 \geqq 0, \quad x_2 \geqq 0. \end{array}$$

The higher boat fare implies the shift of the budget constraint in Figure 4.5 upward and consequently a new intersection point as the solution of the equation system

$$\begin{aligned} x_1 + x_2 &= 150, \\ 2x_1 + \frac{5}{4}x_2 &= 200. \end{aligned}$$

The new solution is $x_1^0 = 16\frac{2}{3}$ and $x_2^0 = 133\frac{1}{3}$. In spite of the higher boat fare, the distance traveled by boat will increase because the traveler—under the same budget constraint—can afford less the more expensive mean of transportation. What is the reason for this Giffen paradox (higher price will actually lead to a large purchase of this good)?

The problem that confronts us now is about the directions in which variables tend to change in response to changes in the data. (This type of analysis is in economics known as comparative statics, or sensitivity analysis, in linear programming.)

Let us consider the linear programming problem

$$\begin{array}{ll} \text{maximize} & f_0(\mathbf{x}) = \mathbf{c}'\mathbf{x} \\ \text{subject to} & A\mathbf{x} \leqq \mathbf{b}, \\ & \mathbf{x} \geqq \mathbf{0}. \end{array}$$

Now suppose some arbitrary changes

$$\begin{aligned} A &\to A + \Delta A, \\ \mathbf{b} &\to \mathbf{b} + \Delta\mathbf{b}, \\ \mathbf{c} &\to \mathbf{c} + \Delta\mathbf{c} \end{aligned}$$

in the data, such that the new problem still has a solution, giving the changes

$$\begin{aligned} \mathbf{x} &\to \mathbf{x} + \Delta\mathbf{x}, \\ \mathbf{u} &\to \mathbf{u} + \Delta\mathbf{u} \end{aligned}$$

in the primal and dual solutions, respectively. The resulting changes in the primal solution $\mathbf{x}$ and in the dual solution $\mathbf{u}$ must fulfill the following condition (see [5, p. 233] or [27, pp. 329–330]):

$$(\Delta\mathbf{c}' - \mathbf{u}'\Delta A)\Delta\mathbf{x} - \Delta\mathbf{u}'(\Delta\mathbf{b} - \Delta A\mathbf{x}) \geqq \mathbf{0}. \tag{4.47}$$

Assuming that $\Delta A = 0$ and $\Delta\mathbf{c} = \mathbf{0}$, inequality (4.47) yields

$$\Delta \mathbf{u}' \Delta \mathbf{b} \leqq 0.$$

If the quantity of one of the available resources changes, then its shadow price changes in the opposite direction. In other words, the marginal value of a resource increases if its amount is reduced, and vice versa. This property, known as the principle of Le Chatelier, was first applied in economics by Samuelson [30]. It was used in thermodynamics for a long time and can be formulated in the following way: "If the external conditions of a thermodynamic system are altered, the equilibrium of the system will tend to move in such a direction as to oppose the change in external conditions" [13, p. 111]. The Le Chatelier principle in linear programming was introduced by Samuelson [31] and generalized for convex programming by Leblanc and Moeseke [21].

Assuming $\Delta A = 0$ and $\Delta \mathbf{b} = 0$, inequality (4.47) yields

$$\Delta \mathbf{c}' \Delta \mathbf{x}' \geqq 0.$$

In particular, in the case of a change in a single coefficient, $\Delta c_k \neq 0$ and $\Delta c_j = 0$ for $j \neq k$,

$$\Delta c_k \Delta x_k \geqq 0.$$

The level of activity increases if the corresponding coefficient in the objective function (the price of the commodity or the profit per unit of the activity) increases. In the paper by Leblanc and Moeseke [21], this property is called the Le Chatelier principle II.

In our example (4.46) described above, we consider the case in which only one element (a_{22}) in the matrix A is changed and all coefficients of the vectors $\mathbf{b}$ and $\mathbf{c}$ remain constant ($\Delta \mathbf{c} = \mathbf{0}$ and $\Delta \mathbf{b} = \mathbf{0}$). From (4.47), we obtain

$$-u_2 \Delta a_{22} \Delta x_2 + \Delta u_2 \Delta a_{22} x_2 \geqq 0,$$

or

$$\Delta a_{22}(\Delta u_2 x_2 - u_2 \Delta x_2) \geqq 0. \tag{4.48}$$

First, we will show that the higher boat fare implies higher shadow price u_2 for the budget constraint, which can be interpreted as the marginal utility of money. For this purpose, we write the problem (4.46) as a maximization problem:

$$\begin{array}{ll} \text{maximize} & -c_1 x_1 - c_2 x_2 \\ \text{subject to} & -x_1 - x_2 \leqq -150, \\ & 2x_1 + x_2 \leqq 200, \\ & x_1 \geqq 0, \quad x_2 \geqq 0. \end{array}$$

The corresponding dual problem has the following form:

$$\begin{array}{ll} \text{minimize} & -150u_1 + 200u_2 \\ \text{subject to} & -u_1 + 2u_2 \geqq -c_1, \\ & -u_1 + u_2 \geqq -c_2, \\ & u_1 \geqq 0, \quad u_2 \geqq 0. \end{array}$$

Using the complementary slackness theorem from Section 4.3, we obtain the shadow price for the budget constraint,

$$u_2^0 = c_2 - c_1 > 0 \quad (\text{because of } c_1 < c_2).$$

Increasing boat fare from 1 cent to $1\frac{1}{4}$ cents per kilo leads to the following dual problem:

$$\begin{array}{ll} \text{minimize} & -150u_1 + 200u_2 \\ \text{subject to} & \qquad -u_1 + 2u_2 \geqq -c_1, \\ & \qquad -u_1 + \frac{5}{4}u_2 \geqq -c_2, \\ & \qquad\qquad u_1 \geqq 0, \quad u_2 \geqq 0. \end{array}$$

The reader may verify that the solution of this model yields

$$u_2^* = \frac{4}{3}(c_2 - c_1) > u_2^0.$$

The increasing boat fare ($\Delta a_{22} > 0$) implies increasing marginal utility of money ($\Delta u_2 > 0$). The reader may verify that this result is valid for any train and boat fares p_1 and p_2, respectively, assuming that $p_1 > p_2$ and $c_2 > c_1$ (the boat is slower and cheaper).

From (4.48), we conclude for $\Delta a_{22} > 0$ that

$$\Delta u_2 x_2 - u_2 \Delta x_2 \geqq 0,$$

or

$$\frac{\Delta u_2}{u_2} \geqq \frac{\Delta x_2}{x_2}. \tag{4.49}$$

As Beckmann [5] has pointed out, inequality (4.49) provides a sufficient condition for the Giffen paradox. The distance traveled by boat can increase as a result of the higher boat fare only if the marginal utility of money has been increased at least at the same proportion.

Rewriting inequality (4.49) as

$$\Delta x_2 - x_2 \frac{\Delta u_2}{u_2} \leqq 0, \tag{4.50}$$

we can observe the relation to the Slutsky equation in the neoclassical theory of households. The Slutsky equation shows the response of a utility-maximizing consumer to a change in price. It decomposes the total effect of a change in price on demand into two components: first, a response to price change holding the consumer on the original indifference curve (a *substitution effect*), and second, an *income effect*, where income is changed, holding prices constant, to reach a tangency on the new indifference curve.

Denoting the purchases of the good j by x_j, the price of the good j by p_j, the given income available to the household (or to the traveler) by M, and the utility function (in terms of travel time) by U, we write the Slutsky equation as

$$\frac{\partial x_i^M}{\partial p_i} = \frac{\partial x_i^U}{\partial p_i} - x_i^M \frac{\partial x_i^M}{\partial M}, \tag{4.51}$$

where $x_i = x_i^M(p_1, p_2, M)$ is the "money income held constant" demand curve and $x_i = x_i^U(p_1, p_2, U)$ is the "utility held constant" or "income-compensated" demand curve. (For the modern derivation of the Slutsky equation, see [32, pp. 282–286].) The first term on the right side of (4.51) describes the substitution effect, and the second term the income effect.

Silberberg and Suen [32] show that along a given indifference curve, as some prices change, the change in the marginal utility of income is related in a very simple manner to the income effect of the good whose price has changed:

$$\frac{\partial \lambda^H}{\partial p_i} = -\lambda^M \frac{\partial x_i^M}{\partial M},$$

where the Lagrange multiplier λ^M represents the marginal utility of income and λ^H is a "compensated" marginal utility of income, showing the response in this value as one moved along a single indifference curve,

$$\lambda^H(p_1, p_2, U^0) = \lambda^M(p_1, p_2, M^*(p_1, p_2, U^0)).$$

According to the interpretation of the multiplier u_2 as the marginal utility of money, the term $\frac{\Delta u_2}{u_2}$ in (4.50) represents its change and describes the income effect of the increasing boat fare. Condition (4.49) points directly at the source of the Giffen paradox, namely, sufficiently high income effect.

4.6.3 Leontief Pollution Model

In Section 1.2.8, the basic Leontief's input–output model has been introduced. With increasing pollution as a by-product of regular economic activities, there arises the need to incorporate environmental effects in an input–output framework. In the well-known paper [22], Leontief extended the input–output model in two ways. First, he added rows to show the output of pollutants by industries. Second, he introduced a pollution abatement "industry" with a specific technology for the elimination of each pollutant. With respect to the exogenously given level of tolerated pollution, two formulations of the model can be found in his paper.

In the first version, an exogenously given vector of tolerated level of pollutants or environmental standards is treated as a negative variable on the right-hand side of the model (see also [23]). It consists of the following equations:

$$(E - A_{11})\mathbf{x}_1 - A_{12}\mathbf{x}_2 = \mathbf{y}_1, \tag{4.52a}$$

$$-A_{21}\mathbf{x}_1 + (E - A_{22})\mathbf{x}_2 = -\mathbf{y}_2, \tag{4.52b}$$

where

$\mathbf{x}_1$ is the n-dimensional vector of gross industrial outputs;
$\mathbf{x}_2$ is the k-dimensional vector of abatement (or antipollution) activity levels;
A_{11} is the $n \times n$ matrix of conventional input coefficients, showing the input of good i per unit of the output of good j (produced by sector j);
A_{12} is the $n \times k$ matrix with a_{ig} representing the input of good i per unit of the eliminated pollutant g (eliminated by abatement activity g);
A_{21} is the $k \times n$ matrix that shows the output of pollutant g per unit of good i (produced by sector i);
A_{22} is the $k \times k$ matrix that shows the output of pollutant g per unit of eliminated pollutant h (eliminated by abatement activity h);
E is the identity matrix;
$\mathbf{y}_1$ is the n-dimensional vector of final demands for economic commodities;
$\mathbf{y}_2$ is the k-dimensional vector of the net generation of pollutants which remain untreated. The gth element of this vector represents the environmental standard of pollutant g and indicates the tolerated level of net pollution.

From (4.52a), we can see that one part of the industrial output is used as an input in the other sectors of the economy ($A_{11}\mathbf{x}_1$), another part is used as an input for the abatement activities ($A_{12}\mathbf{x}_2$), and another part is devoted for the final demand ($\mathbf{y}_1$). The balance equations for the pollutants or for the undesirable outputs are given by (4.52b). The total amount of pollution consists of pollution generated by production of desirable goods ($A_{21}\mathbf{x}_1$) and by the abatement activities themselves ($A_{22}\mathbf{x}_2$). One part of the gross pollution will be eliminated ($\mathbf{x}_2$), and the amount $\mathbf{y}_2$ remains untreated because it is tolerated.

The solution of (4.52) for given levels of final demand $\mathbf{y}_1$ and given pollution standards $\mathbf{y}_2$ can be obtained by inverting the augmented Leontief matrix such that

$$\begin{pmatrix} \mathbf{x}_1 \\ \mathbf{x}_2 \end{pmatrix} = \begin{pmatrix} E - A_{11} & -A_{12} \\ -A_{21} & E - A_{22} \end{pmatrix}^{-1} \begin{pmatrix} \mathbf{y}_1 \\ -\mathbf{y}_2 \end{pmatrix}.$$

The sufficient conditions for the existence of a nonnegative solution of the systems in (4.52) are given in [25].

The price model corresponding to the model (4.52) has the form

$$\mathbf{p}'(E - A_{11}) - \mathbf{r}'A_{21} = \mathbf{v}_1', \tag{4.53a}$$

$$-\mathbf{p}'A_{12} + \mathbf{r}'(E - A_{22}) = \mathbf{v}_2', \tag{4.53b}$$

with $\mathbf{p}'$ the $(1 \times n)$ vector of commodity prices, and $\mathbf{r}'$ the $(1 \times k)$ vector of prices (= cost per unit) for eliminating pollutants. $\mathbf{v}_1'$ and $\mathbf{v}_2'$ are the exogenously given $(1 \times n)$ and $(1 \times k)$ vectors of primary input values per unit of production and per unit level of abatement activities, respectively.

Equation (4.53a) shows that the commodity prices $\mathbf{p}'$ must be such that they cover the costs of inputs from other sectors of the economy ($\mathbf{p}'A_{11}$), the costs of primary factors $\mathbf{v}_1'$, and the pollution costs ($\mathbf{r}'A_{21}$). Equation (4.53b) determines the prices of pollutant $\mathbf{r}'$ from abatement cost ($\mathbf{p}'A_{12}$), costs of primary inputs per unit level of abatement activities $\mathbf{v}_2'$, and the pollution costs of the abatement activities themselves ($\mathbf{r}'A_{22}$).

The solution of the price or of the dual model is then

$$(\mathbf{p}', \mathbf{r}') = (\mathbf{v}_1', \mathbf{v}_2') \begin{pmatrix} E - A_{11} & -A_{12} \\ -A_{21} & E - A_{22} \end{pmatrix}^{-1}.$$

In the second version of the Leontief pollution model, the environmental standard has been defined—in the one-pollutant case—as the ratio of eliminated pollution x_{2g} to the gross pollution, which is the sum of net pollution and abatement activity $(x_{2g} + y_{2g})$. Denoting the $(k \times k)$ diagonal matrix of proportions of abated gross pollutants by $\hat{S}$, we have

$$\hat{S}(\mathbf{x}_2 + \mathbf{y}_2) = \mathbf{x}_2.$$

Then the quantity model can be formulated as (see also [33, 23, 1])

$$(E - A_{11})\mathbf{x}_1 - A_{12}\mathbf{x}_2 = \mathbf{y}_1, \tag{4.54a}$$

$$-\hat{S}A_{21}\mathbf{x}_1 + (E - \hat{S}A_{22})\mathbf{x}_2 = \mathbf{0}. \tag{4.54b}$$

Equation (4.54b) determines the level of abatement activity $\mathbf{x}_2$ as the sum of abated pollution generated by the production $(\hat{S}A_{21}\mathbf{x}_1)$ and by the antipollution activities themselves $(\hat{S}A_{22}\mathbf{x}_2)$. Obviously, if $\mathbf{y}_2 = \mathbf{0}$, then $\hat{S} = E$. This is the case of complete abatement (no pollution is tolerated), where the models (4.52) and (4.54) coincide.

The corresponding price model is

$$\mathbf{p}_s'(E - A_{11}) - \mathbf{r}_s'\hat{S}A_{21} = \mathbf{v}_1', \tag{4.55a}$$

$$-\mathbf{p}_s'A_{12} + \mathbf{r}_s'(E - \hat{S}A_{22}) = \mathbf{v}_2'. \tag{4.55b}$$

Note that prices in this model are subscripted by s.

According to (4.55a) the commodity prices $\mathbf{p}_s'$ include the costs of intermediate inputs $(\mathbf{p}_s'A_{11})$, the costs of the primary inputs $(\mathbf{v}_1')$, and the pollution abatement costs $(\mathbf{r}_s'\hat{S}A_{21})$. The interpretation of (4.55b) for the pollutant prices $\mathbf{r}_s'$ is similar.

The solution of the price model (4.55) is

$$(\mathbf{p}_s', \mathbf{r}_s') = (\mathbf{v}_1', \mathbf{v}_2') \begin{pmatrix} E - A_{11} & -A_{12} \\ -\hat{S}A_{21} & E - \hat{S}A_{22} \end{pmatrix}^{-1}.$$

The given environmental standards or the tolerated level of the net pollution $\mathbf{y}_2$, the corresponding elements of the diagonal matrix $\hat{S}$ can be chosen such that the models (4.52) and (4.54) share the same solution (for the levels of production and abatement). Even in this case the commodity prices $\mathbf{p}_s$ and the prices for eliminating pollutants $\mathbf{r}_s$ are smaller than or equal to the prices determined by the model (4.53) for any nonnegative vector $(\mathbf{v}_1', \mathbf{v}_2')$ when some net pollution is left untreated [26, Theorem 1, p. 267].

Already in the paper by Lowe [23] the price solutions of both models were compared. He showed that only the prices $\mathbf{p}_s$ and $\mathbf{r}_s$ were the appropriate industrial prices and effluent charges because they were consistent with financial viability and national income–expenditure balance. That means that all chosen activities could be met from revenue. At the other end, the prices determined by model (4.53) only possessed the property of opportunity costs of environmental restriction in terms of extra value-added or lost final demand. Differences in the two sets of prices appear, because untreated net pollution is discharged free to final consumers. The question that arises by imposing emission charges (effluent taxes) for untreated pollution is that of how to estimate the level of emission charges in both models, such that the prices for both models—providing the same level of production and of net pollution—are the same?

For this purpose, the augmented Leontief model (4.52) is formulated as an optimization model with the net generation of pollutants $\mathbf{y}_2$ as endogenous variables that are limited to specified amounts $\overline{\mathbf{y}}_2$. However, untreated pollutants are not discharged free in a receiving medium, but the polluters have to pay effluent charge on every untreated unit. Denoting by $\mathbf{t}$ the vector of effluent taxes levied per unit of residual pollutants, the environmental costs $\mathbf{t}'\mathbf{y}_2$ will be added to the costs of primary factors required by industrial production $\mathbf{x}_1$ and abatement activities $\mathbf{x}_2$. The gross national product (GNP) at factor costs, including the environmental costs, should be minimized for a given level of final demand $\overline{\mathbf{y}}_1$. The resulting optimization model (with a possibility of including alternative techniques of industrial production and pollution abatement), denoted as Model I [26, p. 269], is then

$$\text{minimize} \quad V(\mathbf{x}_1, \mathbf{x}_2, \mathbf{y}_2) = \mathbf{v}_1'\mathbf{x}_1 + \mathbf{v}_2'\mathbf{x}_2 + \mathbf{t}'\mathbf{y}_2 \tag{4.56}$$

$$\text{subject to} \quad (E - A_{11})\mathbf{x}_1 - A_{12}\mathbf{x}_2 \geqq \overline{\mathbf{y}}_1, \tag{4.57}$$

$$-A_{21}\mathbf{x}_1 + (E - A_{22})\mathbf{x}_2 + \mathbf{y}_2 \geqq \mathbf{0}, \tag{4.58}$$

$$-\mathbf{y}_2 \geqq -\overline{\mathbf{y}}_2, \tag{4.59}$$

$$\mathbf{x}_1 \geqq \mathbf{0}, \quad \mathbf{x}_2 \geqq \mathbf{0}, \quad \mathbf{y}_2 \geqq \mathbf{0}. \tag{4.60}$$

The inequalities in (4.57) express the requirement that a given bill of goods $\overline{\mathbf{y}}_1$ for final demand be provided. According to the expressions (4.58) and (4.59), the actual amount of pollutants $\mathbf{y}_2$ that remain untreated after abatement activity does not exceed the environmental standards $\overline{\mathbf{y}}_2$.

The subject of our interest is the dual or price model corresponding to the model (4.56)–(4.60), i.e.,

$$\text{maximize} \quad W(\mathbf{p}, \mathbf{r}, \mathbf{s}) = \mathbf{p}'\overline{\mathbf{y}}_1 - \mathbf{s}'\overline{\mathbf{y}}_2 \tag{4.61}$$

$$\text{subject to} \quad \mathbf{p}'(E - A_{11}) - \mathbf{r}'A_{21} \leqq \mathbf{v}_1', \tag{4.62}$$

$$-\mathbf{p}'A_{12} + \mathbf{r}'(E - A_{22}) \leqq \mathbf{v}_2', \tag{4.63}$$

$$\mathbf{r}' - \mathbf{s}' \leqq \mathbf{t}', \tag{4.64}$$

$$\mathbf{p}' \geqq \mathbf{0}, \quad \mathbf{r}' \geqq \mathbf{0}, \quad \mathbf{s}' \geqq \mathbf{0}, \tag{4.65}$$

where $\mathbf{s}'$ is a $(1 \times k)$ vector of dual variables related to the environmental constraints (4.59).

For positive levels of gross industrial outputs $\mathbf{x}_1$ and of abatement activities $\mathbf{x}_2$, the constraints (4.62) and (4.63) are fulfilled as equalities (due to the complementary slackness theorem):

$$\mathbf{p}' = \mathbf{p}'A_{11} + \mathbf{v}_1' + \mathbf{r}'A_{21}, \tag{4.66}$$
$$\mathbf{r}' = \mathbf{p}'A_{12} + \mathbf{v}_2' + \mathbf{r}'A_{22}. \tag{4.67}$$

These equations correspond to (4.53a)–(4.53b) and provide the economic foundation to the "polluter pays principle."

A positive level of net pollution $\mathbf{y}_2 > \mathbf{0}$ implies an equality in expression (4.64), $\mathbf{r}' - \mathbf{s}' = \mathbf{t}'$ or $\mathbf{r}' = \mathbf{t}' + \mathbf{s}'$, where s_g indicates the increase of GNP at factor costs, by tightening the environmental standard $\bar{y}_{2g}$ $(g = 1, 2, \ldots, k)$ by a small unit. If the amount of untreated pollutant g is below the tolerated level $\bar{y}_{2g}$, then the corresponding dual variable s_g is equal to zero and the price of pollutants g is determined by the effluent tax t_g levied on residual pollutant. When the environmental constraint (4.59) is binding, the shadow price $\mathbf{s}'$ can be positive, and the prices of pollutants $\mathbf{r}'$ and commodity prices $\mathbf{p}'$ will rise to include the additional environmental cost $\mathbf{s}'$ caused by the obligation to meet the standards. The higher environmental quality is paid for by increasing commodity prices and prices for eliminating pollutants.

The modification of model (4.54) by imposing effluent taxes $\mathbf{t}_s'$ per unit of untreated pollution leads to the following optimization model, denoted in [26, p. 271] as Model II:

$$\begin{aligned}
\text{minimize} \quad & V_s(\mathbf{x}_1, \mathbf{x}_2, \mathbf{y}_2) = \mathbf{v}_1'\mathbf{x}_1 + \mathbf{v}_2'\mathbf{x}_2 + \mathbf{t}_s'\mathbf{y}_2 && (4.68)\\
\text{subject to} \quad & (E - A_{11})\mathbf{x}_1 - A_{12}\mathbf{x}_2 \geqq \bar{\mathbf{y}}_1, && (4.69)\\
& -\hat{S}A_{21}\mathbf{x}_1 + (E - \hat{S}A_{22})\mathbf{x}_2 \geqq \mathbf{0}, && (4.70)\\
& -(E - \hat{S})A_{21}\mathbf{x}_1 - (E - \hat{S})A_{22}\mathbf{x}_2 + \mathbf{y}_2 \geqq \mathbf{0}, && (4.71)\\
& \mathbf{x}_1 \geqq \mathbf{0}, \quad \mathbf{x}_2 \geqq \mathbf{0}, \quad \mathbf{y}_2 \geqq \mathbf{0}. && (4.72)
\end{aligned}$$

Note that we use subscript s to distinguish the variables or parameters of both models. The objective function (4.68), apart from the possible differences in the level of effluent taxes $\mathbf{t}_s'$ and $\mathbf{t}'$, respectively, is the same as the objective function (4.56). Furthermore, there is no difference in the constraints (4.57) and (4.69). The expression (4.70) requires that the levels of abatement activities $\mathbf{x}_2$ must at least meet the given proportions of gross pollution. Because of the objective function (4.68) in the optimal solution of Model II, condition (4.71) will be fulfilled under equality. Thereupon constraint (4.71) describes the levels of untreated pollution $\mathbf{y}_2$ for which the specific effluent taxes are levied.

Again, the subject of our analysis is the price model of Model II, i.e.,

$$\begin{aligned}
\text{maximize} \quad & W_s(\mathbf{p}, \mathbf{r}, \mathbf{s}) = \mathbf{p}_s'\bar{y}_1 && (4.73)\\
\text{subject to} \quad & \mathbf{p}_s'(E - A_{11}) - \mathbf{r}_s'\hat{S}A_{21} - \mathbf{s}_s'(E - \hat{S})A_{21} \leqq \mathbf{v}_1', && (4.74)\\
& -\mathbf{p}_s'A_{12} + \mathbf{r}_s'(E - \hat{S}A_{22}) - \mathbf{s}_s'(E - \hat{S})A_{22} \leqq \mathbf{v}_2', && (4.75)\\
& \mathbf{s}_s' \leqq \mathbf{t}_s', && (4.76)\\
& \mathbf{p}_s' \geqq \mathbf{0}, \quad \mathbf{r}_s' \geqq \mathbf{0}, \quad \mathbf{s}_s' \geqq \mathbf{0}. && (4.77)
\end{aligned}$$

Assuming again positive levels of industrial production $\mathbf{x}_1$ and of abatement activities $\mathbf{x}_2$, the constraints (4.74) and (4.75) can be written as equalities:

$$\mathbf{p}_s' = \mathbf{p}_s' A_{11} + \mathbf{v}_1' + \mathbf{r}'\hat{S}A_{21} + \mathbf{s}_s'(E - \hat{S})A_{21}, \tag{4.78}$$

$$\mathbf{r}_s' = \mathbf{p}_s' A_{12} + \mathbf{v}_2' + \mathbf{r}'\hat{S}A_{22} + \mathbf{s}_s'(E - \hat{S})A_{22}. \tag{4.79}$$

For positive levels of untreated pollution $\mathbf{y}_2$, the dual variables $\mathbf{s}_s'$ are equal to the effluent taxes $\mathbf{t}_s'$. Then the price equations (4.78) and (4.79) get a clear economic meaning. Compared with the price equation (4.66), the price equation (4.78) takes into account not only the costs of intermediate inputs $\mathbf{p}_s' A_{11}$ and the primary inputs $\mathbf{v}_1'$ but also the pollution abatement cost $\mathbf{r}'\hat{S}A_{21}$ and the charges for untreated pollution, given by $\mathbf{t}_s'(E - \hat{S})A_{22}$, as well. The interpretation of (4.79) for the pollutant prices $\mathbf{r}_s'$ is similar.

The answer to the question how to avoid the differences in the prices for the models (4.53) and (4.55) is given in the following.

Proposition 4.1 (see [26, p. 272, including the proof]). *If* $\mathbf{t}_s' = \mathbf{t}' + \mathbf{s}'$ *for given levels of the effluent taxes* $\mathbf{t}'$*, and if the model* (4.56)–(4.60) *and the model* (4.68)–(4.72) *share the same optimal solution, then the commodity prices and the prices for eliminating pollutants for both programs, i.e.,* (4.61)–(4.65) *and* (4.73)–(4.77), *respectively, are equal.*

In this way, the prices are consistent with financial viability and can be interpreted as opportunity cost variables. The shadow prices $\mathbf{s}$ provide the appropriate rates for the effluent taxes to be charged on untreated pollutions. If these charges are lower than the shadow prices of environmental standards, then production and abatement activities will exactly meet the standards. For effluent taxes higher than the shadow prices, the pollutants will be cleaned up completely. It is cheaper to abate than to pay taxes. The switch between completely protected economy ($\mathbf{y}_2 = \mathbf{0}$) and polluting up to the standards ($\mathbf{y}_2 = \bar{\mathbf{y}}_2$) follows from the linearity of the model.

References and Further Reading

[1] J. Arrous, The Leontief pollution model: A systematic formulation, *Econ. Syst. Res.*, **6** (1994), 105–107.

[2] K. J. Arrow, George Dantzig in the development of economic analysis, *Discrete Optim.*, **5** (2008), 159–167.

[3] M. L. Balinski and W. J. Baumol, The dual in nonlinear programming and its economic interpretation, *Rev. Econ. Stud.*, **35**-3 (1968), 237–256.

[4] W. J. Baumol and R. C. Bushnell, Error produced by linearization in mathematical programming, *Econometrica*, **35**-3–4 (1967), 447–471.

[5] M. J. Beckmann, Comparative statics in linear programming and the Giffen paradox, *Rev. Econ. Stud.*, **23** (1956), 232–235.

[6] A. Charnes, S. Duffuaa, and M. Ryan, The more for less paradox in linear programming, *Eur. J. Oper. Res.*, **31** (1987), 194–197.

[7] A. Charnes and D. Klingman, The more-for-less paradox in the distribution model, *Cahiers Centre Études Rech. Opér.*, **13**-1 (1971), 11–22.

[8] H. B. Chenery and W. J. Raduchel, Substitution in planning models, in H. B. Chenery, ed., *Studies in Development Planning*, Harvard University Press, Cambridge, MA, 1971.

[9] M. Chobot and F. Turnovec, O jednom ekonomickom paradoxe v modeloch lineárneho programovania (One economic paradox in linear programming models), *Ekon.-Mat. Obzor*, **4** (1974), 376–386.
[10] V. Chvátal, *Linear Programming*, W. H. Freeman, New York, 1983.
[11] G. B. Dantzig and W. Orchard-Hays, The product form for the inverse in the simplex method, *Math. Tables Other Aids Comput.*, **8** (1954), 64–67.
[12] R. Dorfman, P. A. Samuelson, and R. M. Solow, *Linear Programming and Economic Analysis*, McGraw–Hill, New York, 1958.
[13] E. Fermi, *Thermodynamics*, Dover, New York, 1937.
[14] G. Gandolfo, *International Trade Theory and Policy*, Springer-Verlag, Berlin, 1998.
[15] D. Gale, *The Theory of Linear Economic Models*, McGraw–Hill, New York, 1960.
[16] S. I. Gass, *Linear Programming: Methods and Applications*, 3rd ed., McGraw–Hill, New York, 1969.
[17] S. I. Gass, *Linear Programming: Methods and Applications*, 4th ed., McGraw–Hill, New York, 1975.
[18] N. Karmarkar, A new polynomial-time algorithm for linear programming, *Combinatorica*, **4** (1984), 373–395.
[19] L. G. Khachian, A polynomial algorithm in linear programming, *Dokl. Akad. Nauk SSSR*, **224** (1979), 1093–1096 (in Russian); *Sov. Math. Dokl.*, **20** (1979), 191–194 (English translation).
[20] V. Klee and G.J. Minty, How good is the simplex algorithm?, in O. Shisha, ed., *Inequalities*, Vol. III, Academic Press, New York, 1972, 159–175.
[21] G. Leblanc and P. V. Moeseke, The Le Chatelier principle in convex programming, *Rev. Econ. Stud.*, **43**-133 (1976), 143–147.
[22] W. W. Leontief, Environmental repercussions and the economic structure: An input–output approach, *Rev. Econ. Statist.*, **52** (1970), 262–271.
[23] P. D. Lowe, Pricing problems in an input–output approach to environmental protection, *Rev. Econ. Statist.*, **61** (1979), 110–117.
[24] M. Luptáčik, A note on the "more-for-less" paradox, in P. Kischka, U. Leopold-Wildburger, R. H. Möhring, and F. J. Rademacher, eds., *Models, Methods and Decision Support for Management: Essays in Honor of Paul Stähly*, Physica-Verlag, Würtzburg, Germany, 2001.
[25] M. Luptáčik and B. Böhm, Reconsideration of non-negative solutions for the augmented Leontief model, *Econ. Syst. Res.*, **6** (1994), 167–170.
[26] M. Luptáčik and B. Böhm, A consistent formulation of the Leontief pollution model, *Econ. Syst. Res.*, **11** (1999), 263–275.
[27] Y. Murata, *Mathematics for Stability and Optimization of Economic Systems*, Academic Press, New York, 1977.
[28] D. Ricardo, *Principles of Political Economy and Taxation*, John Murray, London, 1817; reprint of the 3rd ed., Bell and Sons, London, 1924.
[29] C. Ross, T. Terlaky, and J. Ph. Vial, *Theory and Algorithms for Linear Optimization: An Interior Point Approach*, Wiley, New York, 1997.
[30] P. A. Samuelson, *Foundations of Economic Analysis*, Harvard University Press, Cambridge, MA, 1947.
[31] P. A. Samuelson, *The Le Chatelier Principle in Linear Programming*, Research memorandum, Rand Corporation, Santa Monica, CA, 1949.
[32] E. Silberberg and W. Suen, *The Structure of Economics: A Mathematical Analysis*, 3rd ed., McGraw–Hill, Singapore, 2001.
[33] A. E. Steenge, Environmental repercussions and the economic structure: Further comments, *Rev. Econ. Statist.*, **60** (1978), 482–486.

[34] A. Takayama, *International Trade: An Approach to the Theory*, Holt, Reinhart, and Winston, New York, 1972.
[35] Th. ten Raa, *Linear Analysis of Competitive Economics*, LSE Handbooks in Economics, Harvester Wheatsheaf, London, 1995.
[36] Th. ten Raa, *The Economics of Input–Output Analysis*, Cambridge University Press, New York, 2005.
[37] D. C. Vandermeulen, *Linear Economic Theory*, Prentice–Hall, Englewood Cliffs, NJ, 1971.

5

Data Envelopment Analysis

In Section 1.2.9, the original model of data envelopment analysis (DEA), developed by Charnes, Cooper, and Rhodes [8], was introduced. With their study, DEA began as a new approach for efficiency and productivity analysis. They described DEA as a "mathematical programming model applied to observational data [that] provides a new way of obtaining empirical estimates of extremal relationships such as the production functions and/or efficient production possibility surfaces that are a cornerstone of modern economics" [34, p. 8].

There are two fundamental approaches for the estimation of frontiers in economics—the parametric and nonparametric approaches.

The parametric approach, described in [29] and [4], requires the imposition of a specific functional form (e.g., a regression equation, a production function, etc.) relating the independent variables to the dependent variables. The functional form selected also requires specific assumptions about the distribution of the error terms (e.g., independently and identically normally distributed). As a result, one can derive some conclusions about the underlying production processes by evaluating marginal products, partial elasticities, marginal costs, or elasticities of substitution.

DEA represents the nonparametric approach for frontier estimation in the sense that it does not require any assumption about the functional form. That is, it does not assume that the underlying technology "belongs to a certain class of functions of a specific functional form which depend on a finite number of parameters, such as the well-known Cobb–Douglas functional form" [15, p. 131]. DEA is also "nonstatistical" because it makes no explicit assumption on the probability distribution of "errors" (i.e., the efficiency residuals) in the production function. In DEA, any deviation from the frontier is treated as inefficiency, and there is no provision for random shocks. DEA provides a single measure of efficiency even when dealing with multiple inputs and outputs, and it obviates the need to assign prespecified weights to either. It measures the efficiency of a decision-making unit (DMU) relative to all other DMUs with the simple restriction that all DMUs lie on or below the efficient frontier. Each DMU not on the frontier (an inefficient DMU) is scaled against a linear or a convex combination of the DMUs on the frontier faced closest to it. For each

M. Luptáčik, *Mathematical Optimization and Economic Analysis*,
Springer Optimization and Its Applications 36, DOI 10.1007/978-0-387-89552-9_5,

inefficient unit, DEA identifies the sources and level of inefficiency for each of the inputs and outputs.

At present, DEA—with the great number and variety of applications in the fields of public services, banking, hospital management, agriculture, education, industrial production, environmental economics, and so on, and with important new developments in concepts and methodology—has become one of the most widely used tools for efficiency analysis. The recent book by Cooper, Seiford, and Tone [12, p. xxxi] cites *A Bibliography of Data Envelopment Analysis* (1978–2001) by G. Tavares[1] as referencing more than 3,600 papers, books, etc., by more than 1,600 authors in 42 countries. According to Ray [31, p.1], an Internet search for DEA produces no fewer than 12,700 entries. One can easily agree with Bouyssou [5] that "DEA can safely be considered as one of the recent 'success stories' in operations research." For further reading on efficiency and productivity analysis containing the new developments in the methodology as well as the empirical applications, the reader is referred to the very recent book by Fried, Lovell, and Schmidt [23].

In this chapter, we discuss DEA as one of the most important recent applications of mathematical programming in economics. In Section 5.1, the concepts of efficiency and productivity will be presented. They are closely related but different measures of a firm's resource-utilization performance. Section 5.2 provides a description of the basic DEA models and their classification with respect to the type of envelopment surface efficiency measurement, the orientation, and the effect of scale changes. Section 5.3 relates the DEA to the economic theory of production and to the Pareto–Koopmans notion of efficiency. A novel application of DEA for measuring of ecoefficiency—using data for industry in 16 OECD countries—is presented in Section 5.4.

5.1 Productivity and Technical and Allocative Efficiency

Production is a process of transforming inputs (labor, capital, materials, etc.) into outputs (goods or services). An input–output combination is a feasible production plan if the output quantity can be produced from the associated input quantity. The technology available to a firm at a given point in time defines which input–output combinations are feasible. For simplicity, let us consider a simple production process in which a single input (x) is used to produce a single output (y). The curve OP in Figure 5.1 represents the maximum output attainable from each input level. Using x_A units of the input (x), the maximum y_A units of the output (y) can be produced. The line OP represents a production frontier because it sets a bound on the range of all feasible input–output combinations. The set consisting of all points between the production frontier OP and the X-axis (inclusive of these bounds) describes a feasible production set. Thus production may take place at or below the frontier, but at no points above it. Hence the production frontier reflects the current state of technology in the industry. Firms in this industry operating on that frontier (like the points A

[1] gtavares@rutcor.rutgers.edu.

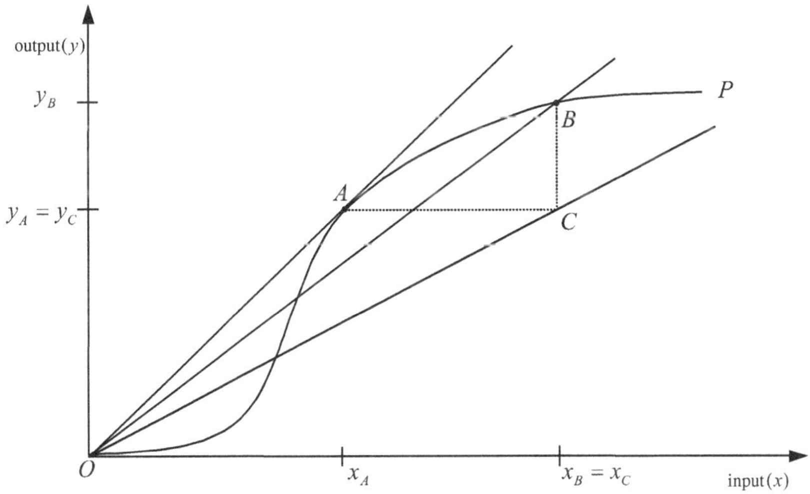

Fig. 5.1. Production frontier, technical efficiency, and productivity.

and B in Figure 5.1) are *technically efficient*: The higher output of the firm B (y_B)—compared with the output of the firm A (y_A)—has been achieved by increasing input ($x_B > x_A$). A firm operating at point C is inefficient because it produces the same amount of output with more input than the firm operating at point A (or it produces a smaller amount of output than the firm operating at point B although both firms use the same amount of input). The amounts by which a firm lies below its production frontier can be regarded as measures of relative technical efficiency.

To illustrate the distinction between technical efficiency and productivity, we use Figure 5.1 again. The *productivity* of a firm is defined as a ratio of the output that it produces to the input that it uses:[2]

$$\text{productivity} = \frac{\text{output}}{\text{input}}.$$

In our figure, we use a ray through the origin to measure the productivity at a particular data point. The slope of this ray is $\frac{y}{x}$ and hence provides a measure of productivity. The firms operating at points A and B are technically efficient. If the firm operating at point B were to move to point A, the slope of the ray would be greater, implying higher productivity at point A. This movement is an example of exploiting *scale economies*: To the right of point A, the output increases more weakly than the input; to the left of point A, the output increases more strongly than the input. Point A is the point of (technically) optimal scale. Operation at any other point on the production frontier results in lower productivity. If the technically inefficient firm operating at point C were to move to the technically efficient point A (or B), the slope of the ray would be greater, implying higher productivity at point A (or B).

[2] When multiple inputs and/or outputs are involved, a method for aggregating these inputs and/or outputs must be used.

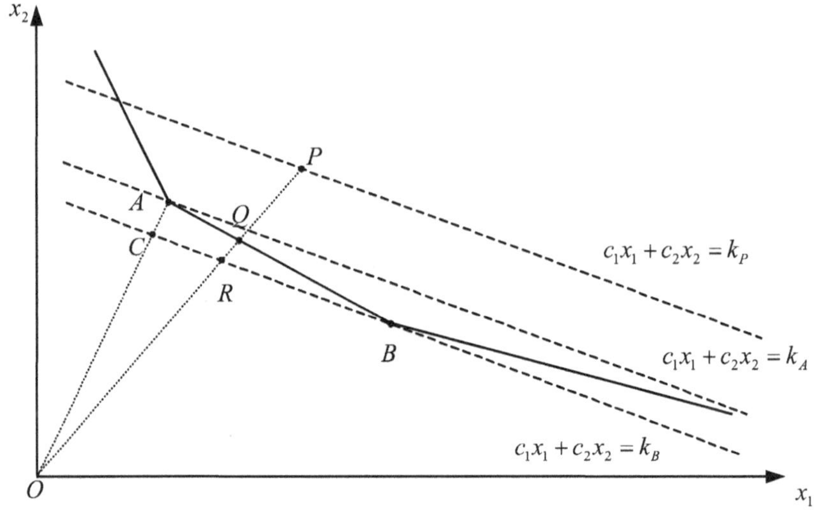

Fig. 5.2. Technical, allocative, and overall efficiency.

Summarizing, a firm may be technically efficient but may still be able to improve its productivity by exploiting scale economies. Productivity and technical efficiency are equivalent only when the technology exhibits constant returns to scale.

To explain the notion of allocative efficiency, we move to Figure 5.2. The solid lines in this figure are segments of an isoquant that represents all possible combinations of the input amounts (x_1, x_2) that are needed to produce a prescribed amount (e.g., one unit) of a single output. All points at this piecewise linear curve are *technically efficient*: The amount of the first input can be reduced only if the amount of the second input increases. P is a point in the interior of the production possibility set representing the activity of a DMU that produces this same amount (one unit) of output but with greater amounts of both inputs. Hence the corresponding DMU is technically inefficient. To evaluate the performance of P, we will use the measure of radial efficiency, represented as

$$0 < \frac{OQ}{OP} \leqq 1.$$

This measure can be interpreted as the distance from O to Q relative to the distance from O to P. The technical efficiency (TE) is the ratio of the potential to actual input consumption. The components of this ratio lie on the dotted line from the origin through Q to P.

Technical efficiency as represented by points A and B per se is not sufficient to minimize cost. Denoting the input prices by c_1 and c_2, the broken line $c_1x_1 + c_2x_2 = k_P$ passing through P contains all input combinations with cost k_P. However, this cost can be reduced by moving this line parallel in a downward direction. The coordinates of A then give $c_1x_1^A + c_2x_2^A = k_A$, where $k_A < k_P$, showing that total cost is reduced. Further parallel movement in a downward direction leads to the point B with the minimal cost at the given output level. The point B is technically and *allocatively* efficient (AE), while the point A is technically but not allocatively efficient.

The ratio

$$0 < \frac{OC}{OA} \leqq 1$$

is commonly referred to as *allocative efficiency*. It provides a measure of the extent to which the technically efficient DMU, represented by point A, falls short of achieving minimal cost because of a failure to make the substitutions involved in moving from A to B along the efficiency frontier.

The measure for *overall efficiency* can be represented in ratio form as

$$0 < \frac{OR}{OP} \leqq 1.$$

This is a measure of the extent to which the originally observed values of P have fallen short of achieving minimal cost. According to Farrel [18], *overall efficiency* (OE) can be decomposed into two multiplicative components:

$$\text{OE} \equiv \frac{OR}{OP} = \underbrace{\frac{OQ}{OP}}_{\text{TE}} \cdot \underbrace{\frac{OR}{OQ}}_{\text{AE}}.$$

Overall efficiency is equal to the product of *technical* times *allocative* efficiency. Overall efficiency (i.e., OE $=$ 1) requires simultaneous *technical* and *allocative efficiency* (TE $=$ AE $=$ 1), which is achieved at B in Figure 5.2.

5.2 Basic DEA Models

Once we step outside the simplified world of single-input, single-output production, we must use a method for aggregating the inputs and/or outputs into a single index of inputs and/or outputs to estimate a ratio measure of productivity and efficiency. Charnes, Cooper, and Rhodes [8] introduced the method of DEA to measure the efficiency of DMUs with multiple inputs and multiple outputs in the absence of market prices.

As introduced in Section 1.2.9, we consider n DMUs ($j = 1, 2, \ldots, n$), each using a varying amount of m different inputs ($i = 1, 2, \ldots, m$) to produce s different outputs ($r = 1, 2, \ldots, s$). Specifically, DMU$_j$ consumes amount x_{ij} of input i and produces amount y_{rj} of output r. We assume that $x_{ij} \geqq 0$ and $y_{rj} \geqq 0$ and further that each DMU has at least one positive input and one positive output value. DEA treats the observed inputs $\mathbf{x}_j$ and outputs $\mathbf{y}_j$ ($j = 1, 2, \ldots, n$) as given constants and chooses values of the input and output weights for a particular DMU$_0$ such that the efficiency—defined as the ratio of weighted sum of outputs to the weighted sum of inputs—will be maximized subject to the less-than-unity constraints. These constraints ensure that the optimal weights for DMU$_0$ in the objective function do not imply an efficiency higher than unity either for itself or for any of the other DMUs. The indicated maximization then accords the evaluated DMU_0 the most favorable weighting that the constraints allow.

There are two basic types of DEA models with respect to envelopment surfaces, referred to as *constant returns-to-scale* (CRS) and *variable returns-to-scale* (VRS) surfaces. The assumption of constant returns to scale implies that if an activity $(\mathbf{x}, \mathbf{y})$ is feasible, then for every positive scalar t, the activity $(t\mathbf{x}, t\mathbf{y})$ is also feasible. This postulate is in the theory of production denoted as "ray unboundedness." The appropriateness of a particular surface is frequently determined by economic and other assumptions regarding the data set to be analyzed. The choice of envelopment surface implies the selection of the particular DEA mathematical programming model.

5.2.1 The Input-Oriented Model under a Constant Returns-to-Scale Assumption

The first measure of the efficiency of any DMU proposed by by Charnes, Cooper, and Rhodes [8] leads to the following fractional programming problem (see (1.27)) for DMU_0:

$$\begin{aligned}
&\underset{\mathbf{u},\mathbf{v}}{\text{maximize}} && h_0(\mathbf{u},\mathbf{v}) = \frac{\sum_{r=1}^{s} y_{r0}u_r}{\sum_{i=1}^{m} x_{i0}v_i} \\
&\text{subject to} && \frac{\sum_{r=1}^{s} y_{rj}u_r}{\sum_{i=1}^{m} x_{ij}v_i} \leqq 1 \quad (j = 1, 2, \ldots, n), \\
& && u_r \geqq 0 \quad (r = 1, 2, \ldots, s), \\
& && v_i \geqq 0 \quad (i = 1, 2, \ldots, m),
\end{aligned}$$

where u_r is the weight for output r $(r = 1, 2, \ldots, s)$ and v_i is the weight given to input i $(i = 1, 2, \ldots, m)$. An efficiency of unity implies that DMU_0 lies on the efficient frontier; the observed and the potential performance coincide. In this case, DMU_0 is said to be "best practice." If the efficiency h_0 is less than one, DMU_0 is relatively inefficient. Its performance is poorer than that of some of its peer DMUs.

The fractional program can be thought of as the conceptual DEA model, but for its nonlinearity and nonconvexity it is not used for actual computation of the efficiency scores. Fortunately, using the transformation of variables [7]:

$$\begin{aligned}
\mu_r &= t u_r \quad (r = 1, 2, \ldots, s), \\
\nu_i &= t v_i \quad (i = 1, 2, \ldots, m), \\
t &= \frac{1}{\sum_{i=1}^{m} x_{i0}v_i}
\end{aligned}$$

the fractional programming problem (1.27) can be converted into an ordinary linear program:

$$
\begin{aligned}
\underset{\boldsymbol{\mu},\boldsymbol{\nu}}{\text{maximize}} \quad & \omega_0(\boldsymbol{\mu}) = \sum_{r=1}^{s} y_{r0}\mu_r \\
\text{subject to} \quad & \sum_{r=1}^{s} y_{rj}\mu_r - \sum_{i=1}^{m} x_{ij}\nu_i \leqq 0 \quad (j = 1, 2, \ldots, n), \\
& \sum_{i=1}^{m} x_{i0}\nu_i = 1, \\
& \mu_r \geqq 0 \quad (r = 1, 2, \ldots, s), \\
& \nu_i \geqq 0 \quad (i = 1, 2, \ldots, m).
\end{aligned}
\tag{5.1}
$$

The measures of efficiency described by problems (1.27) and (5.1) are "units invariant," i.e., they are independent of the units of measurement used. The following theorem can be proved [11, pp. 24 and 39].

Theorem 5.1. *The optimal values of* $\max h_0 = h^0$ *in* (1.27) *and* $\max w_0 = w^0$ *are independent of the units in which the inputs and the outputs are measured provided these units are the same for every DMU.*

In the first paper by Charnes, Cooper, and Rhodes [8], only the nonnegativity condition for the input and output weights (ν_i and μ_r, respectively) has been imposed. However, they showed in a correction to this paper [8] that under the nonnegativity constraints, obviously inefficient units can appear as efficient. Therefore, they restricted the input and output weights such that

$$
\nu_i \geqq \epsilon, \qquad \mu_r \geqq \epsilon,
$$

where ϵ is an infinitesimal or non-Archimedean constant.[3]

Therefore, the linear programming equivalent of the Charnes–Cooper–Rhodes (CCR) ratio model that we will use in what follows is

[3] Solving the non-Archimedean models as linear programs, with an explicit value for ϵ, can lead to inaccurate results. If the value of ϵ used is "small enough," then accurate results would be obtained. However, what "small enough" is depends on the particular data set. As shown by Ali and Seiford [2], if

$$
\epsilon \geqq \min_{j=1,\ldots,n} \frac{1}{\sum_{i=1}^{m} x_{ij}},
$$

then the linear program (5.1) has unbounded objective function values. For a discussion of this issue and other computational aspects of DEA, we refer to Ali [1].

$$\begin{aligned}
&\underset{\boldsymbol{\mu},\boldsymbol{\nu}}{\text{maximize}} && w_0(\boldsymbol{\mu}) = \sum_{r=1}^{s} y_{r0}\mu_r \\
&\text{subject to} && \sum_{r=1}^{s} y_{rj}\mu_r - \sum_{i=1}^{m} x_{ij}\nu_i \leqq 0 \qquad (j = 1, 2, \ldots, n), \\
& && \sum_{i=1}^{m} x_{i0}\nu_i = 1, \\
& && -\mu_r \leqq -\epsilon \qquad (r = 1, 2, \ldots, s), \\
& && -\nu_i \leqq -\epsilon \qquad (i = 1, 2, \ldots, m),
\end{aligned} \tag{5.2}$$

whose dual problem is

$$\begin{aligned}
&\underset{\theta,\boldsymbol{\lambda},\mathbf{s}^-,\mathbf{s}^+}{\text{minimize}} && g_0(\theta, \mathbf{s}^-, \mathbf{s}^+) = \theta - \epsilon\left(\sum_{i=1}^{m} s_i^- + \sum_{r=1}^{s} s_r^+\right) \\
&\text{subject to} && \theta x_{i0} - \sum_{j=1}^{n} x_{ij}\lambda_j - s_i^- = 0 \qquad (i = 1, 2, \ldots, m), \\
& && \sum_{j=1}^{n} y_{rj}\lambda_j - s_r^+ = y_{r0} \qquad (r = 1, 2, \ldots, s), \\
&\text{(weights on DMUs)} && \lambda_j \geqq 0 \qquad (j = 1, 2, \ldots, n), \\
&\text{(input slacks)} && s_i^- \geqq 0 \qquad (i = 1, 2, \ldots, m), \\
&\text{(output slacks)} && s_r^+ \geqq 0 \qquad (r = 1, 2, \ldots, s).
\end{aligned} \tag{5.3}$$

Leaving the data for DMU_0 (the DMU being evaluated) in the constraints guarantees that solutions exist for both problems (5.2) and (5.3), respectively. By the duality theory of linear programming, it follows that they will have finite and equal optimal values.

The condition for "DEA efficiency" now becomes

$$\sum_{r=1}^{s} y_{r0}\mu_r = 1$$

for problem (5.2). Because of the duality theory of linear programming, it must hold for the optimal solution (denoted by o) of the dual problem (5.3):

$$\theta^o - \epsilon\left(\sum_{i=1}^{m} s_i^{-0} + \sum_{r=1}^{s} s_r^{+0}\right) = 1.$$

Evidently, the following two statements are equivalent:

1. A DMU is efficient if and only if the following two conditions are satisfied:

$$\begin{aligned} &\text{(a)} \quad \theta^0 = 1, \\ &\text{(b)} \quad s_i^{-0} = s_r^{+0} = 0 \quad \text{for all } i \text{ and } r. \end{aligned} \tag{5.4}$$

2. A DMU is efficient if and only if $w_0^0 = g_0^0 = 1$.

The (scalar) variable θ_0^0 gives the proportion of all inputs of DMU_0 that must be sufficient—compared with the units on the efficient frontier—to achieve the given output levels. In other words, $1 - \theta_0^0$ gives the necessary proportional reduction of all inputs of DMU_0 being evaluated in order to achieve the efficient frontier. This reduction is applied simultaneously to all inputs and results in a radial movement toward the envelopment surface.

However, a DMU can be a boundary point ($\theta^0 = 1$) and be inefficient, as already shown by Charnes, Cooper, and Rhodes [9].

Because of the presence of the non-Archimedean ϵ in the dual objective function, θ is to be preemptively minimized, after which the sum of the slacks in (5.3) is to be maximized. Thus the computation proceeds in two stages: with maximal reduction of inputs being achieved first, via the optimal θ^0; then in the second stage, substituting θ^0 in the dual constraints, movement onto the efficient frontier is achieved via the slack variables ($\mathbf{s}^-$ and $\mathbf{s}^+$). In this way, we do not specify the non-Archimedean constant $\epsilon > 0$ explicitly.

The nonzero slacks and/or the value of $\theta^0 < 1$ identify the sources and amounts of inefficiency in each input and output of the DMU being evaluated.

As pointed out by Cooper, Thompson, and Thrall [13], the objective function in (5.3) implies that two types of inefficiencies can be distinguished in (5.4). A value of $\theta^0 < 1$ shows *radial inefficiencies* in the form of excessive use of *all* inputs without altering their proportions. The positive slacks indicate that further reduction can be made, which will necessarily alter the proportions used, and hence they show *mix inefficiencies*.

Furthermore, (5.3) seeks values of λ_j to construct a composite unit, with outputs $\sum_j y_{rj}\lambda_j$ $(r = 1, 2, \ldots, s)$ and inputs $\sum_j x_{ij}\lambda_j$ $(i = 1, 2, \ldots, m)$. The dual constraints with respect to inputs imply that even after the proportional reduction of all inputs, the inputs of the evaluated DMU_0 cannot be lower than the inputs of the composite unit. Similarly, the outputs of DMU_0 cannot be higher than the outputs of the composite unit. DMU_0 will be efficient when it has proved impossible to construct a composite unit that outperforms DMU_0. Conversely, if DMU_0 is inefficient, the optimal values of λ_j form a composite unit outperforming DMU_0 and providing targets for DMU_0. In order words, the positive values of λ_j provide the linear combination of the DMUs on the efficiency frontier faced closest to DMU_0 (the peer group for DMU_0). In this way, the dual problem (5.3) constructs the piecewise linear envelopment surface and is called the *envelopment problem*, whereas the primal problem (5.2) seeking the values for the weights is the *multiplier problem*.

Because of the focus on maximal movement toward the frontier through proportional reduction of inputs, models (5.2)–(5.3) are denoted *input-oriented CCR models*.

Table 5.1. Example with one input and one output.

DMU	1	2	3	4	5	6	7
Input	2	3	5	7	3	6	8
Output	2	4	6	7	1	5	7

To illustrate, we consider an example consisting of seven DMUs each consuming a single input to produce a single output (see Table 5.1).

We can evaluate the efficiency of DMU_1 from the data in Table 5.1 by solving the linear programming problem (5.2) below:

$$\begin{aligned} \underset{\mu,\nu}{\text{maximize}} \quad & w_0(\mu) = 2\mu & \\ \text{subject to} \quad & 2\mu - 2\nu \leqq 0, & (\text{DMU}_1) \\ & 4\mu - 3\nu \leqq 0, & (\text{DMU}_2) \\ & 6\mu - 5\nu \leqq 0, & (\text{DMU}_3) \\ & 7\mu - 7\nu \leqq 0, & (\text{DMU}_4) \\ & \mu - 3\nu \leqq 0, & (\text{DMU}_5) \\ & 5\mu - 6\nu \leqq 0, & (\text{DMU}_6) \\ & 7\mu - 8\nu \leqq 0, & (\text{DMU}_7) \\ & 2\nu = 1, & \end{aligned}$$

where all variables are constrained to be nonnegative. The optimal solution, easily obtained by simple ratio calculations (or graphically), is given by $\mu^0 = 0.375$, $\nu^0 = 0.5$, and $w_{01}^0 = 0.75$.

The corresponding dual problem (5.3) for DMU_1 is stated as

$$\begin{aligned} \underset{\lambda,\theta}{\text{minimize}} \quad & \theta \\ \text{subject to} \quad & 2\lambda_1 + 4\lambda_2 + 6\lambda_3 + 7\lambda_4 + \lambda_5 + 5\lambda_6 + 7\lambda_7 \geqq 2, \\ & -2\lambda_1 - 3\lambda_2 - 5\lambda_3 - 7\lambda_4 - 3\lambda_5 - 6\lambda_6 - 8\lambda_7 + 2\theta \geqq 0, \\ & \lambda_j \geqq 0 \quad (j = 1, 2, \ldots, 7), \\ & \theta \text{ free.} \end{aligned}$$

The optimal solution is $\lambda_2^0 = 0,5$; $\lambda_1^0 = \lambda_3^0 = \lambda_4^0 = \lambda_5^0 = \lambda_6^0 = \lambda_7^0 = 0$; and $\theta_1^0 = 0,75$.

In a similar way, the linear programming problems (5.2) and (5.3) for every DMU are calculated with the optimal solution values given in Table 5.2.

Under constant returns to scale, the only efficient unit is DMU_2 ($\theta_2^0 = 1$, $s^+ = s^- = 0$), and it is the reference set of all the other DMUs. The efficient frontier must pass through the origin, and it is given by the ray $\{\lambda(x_2, y_2)/\lambda \geqq 0\}$ as illustrated in Figure 5.3.

The interpretation of the envelopment problem (5.3) for DMU_1 with $P_1 = (2; 2)$ is the selection of a point at the efficient frontier that allows the maximal input reduction

Table 5.2. Optimal solution values for (5.2) and (5.3) with one input and one output.

DMU	θ^0	s^+	s^-	λ	w_0^0	μ	ν
1	0.75	0	0	$\lambda_2 = \frac{1}{2}$	0.75	$\frac{3}{8}$	$\frac{1}{2}$
2	1	0	0	$\lambda_2 = 1$	1	$\frac{1}{4}$	$\frac{1}{3}$
3	0.90	0	0	$\lambda_2 = \frac{3}{2}$	0.90	$\frac{3}{20}$	$\frac{1}{5}$
4	0.75	0	0	$\lambda_2 = \frac{7}{4}$	0.75	$\frac{3}{28}$	$\frac{1}{7}$
5	0.25	0	0	$\lambda_2 = \frac{1}{4}$	0.25	$\frac{1}{4}$	$\frac{1}{3}$
6	0.625	0	0	$\lambda_2 = \frac{5}{4}$	0.625	$\frac{1}{8}$	$\frac{1}{6}$
7	0.656	0	0	$\lambda_2 = \frac{7}{4}$	0.656	$\frac{3}{32}$	$\frac{1}{8}$

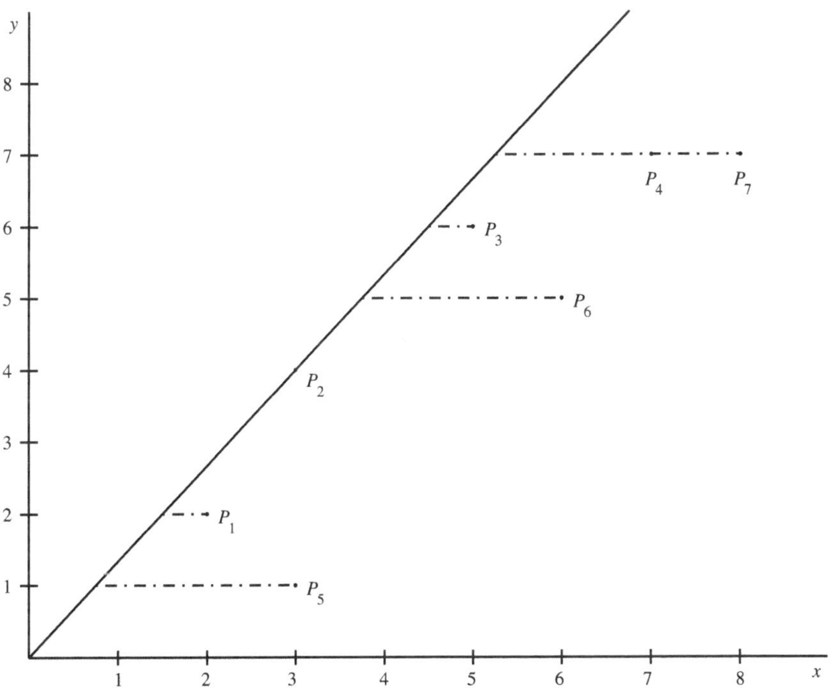

Fig. 5.3. Projections for the input-oriented CCR model from Table 5.1.

or minimal input necessary to produce two units of output. This is $\theta_1^0 = 0.75$; the input of DMU_1 should be reduced by one quarter. Hence P_1 is projected into the boundary point $(1.5; 2) = \lambda_2 P_2 = 0.5 \times (3; 4)$.

The input of DMU_5 should be reduced to 0.75, and therefore P_5 is projected into the boundary point $(0.75; 1) = \lambda_2 P_2 = 0.25 \times (3; 4)$.

It is apparent that each DMU would be projected onto this ray through P_2, whose slope is defined by the ratio of the multipliers μ and ν. From Figure 5.3, it is easy to see that the slope of the efficiency frontier is $\frac{4}{3}$, which corresponds exactly to the ratio of the multipliers $\frac{\nu_j}{\mu_j} = \frac{4}{3}$ for $j = 1, 2, \ldots, 7$.

In a similar way, the results for the remaining DMUs can be interpreted.

To move to multiple inputs and outputs and their treatment, we turn to Table 5.3, which lists the performance of seven DMUs each with two inputs and one output. The input values are normalized to values for getting one unit of output.

Table 5.3. Example with two inputs and two output.

DMU	1	2	3	4	5	6	7
Input 1	2	3	5	7	8	5	4
Input 2	5	3	2	2	3	4	6
Output	1	1	1	1	1	1	1

The linear program (5.2) for DMU_1 is

$$\begin{aligned}
\underset{\mu,\nu}{\text{maximize}} \quad & w_0(\mu) = \mu \\
\text{subject to} \quad & \mu - 2\nu_1 - 5\nu_2 \leqq 0, && (\text{DMU}_1) \\
& \mu - 3\nu_1 - 3\nu_2 \leqq 0, && (\text{DMU}_2) \\
& \mu - 5\nu_1 - 2\nu_2 \leqq 0, && (\text{DMU}_3) \\
& \mu - 7\nu_1 - 2\nu_2 \leqq 0, && (\text{DMU}_4) \\
& \mu - 8\nu_1 - 3\nu_2 \leqq 0, && (\text{DMU}_5) \\
& \mu - 5\nu_1 - 4\nu_2 \leqq 0, && (\text{DMU}_6) \\
& \mu - 4\nu_1 - 6\nu_2 \leqq 0, && (\text{DMU}_7) \\
& 2\nu_1 + 5\nu_2 = 1, \\
& \mu \geqq 0, \quad \nu_1 \geqq 0, \quad \nu_2 \geqq 0.
\end{aligned}$$

This problem can be solved by the simplex method or by simply deleting ν_2 from the inequalities by inserting $\nu_2 = \frac{1-2\nu_1}{5}$ and then solving graphically. An optimal solution is $\nu_1^0 = 0.222$, $\nu_2^0 = 0.111$, $\mu^0 = 1$, and $w_0^0 = 1$.

The envelopment problem (5.3) for DMU_1 becomes the following:

$$\begin{aligned}
\underset{\lambda,\theta}{\text{minimize}} \quad & \theta \\
\text{subject to} \quad & \lambda_1 + \lambda_2 + \lambda_3 + \lambda_4 + \lambda_5 + \lambda_6 + \lambda_7 \geqq 1, \\
& -2\lambda_1 - 3\lambda_2 - 5\lambda_3 - 7\lambda_4 - 8\lambda_5 - 5\lambda_6 - 4\lambda_7 + 2\theta \geqq 0, \\
& -5\lambda_1 - 3\lambda_2 - 2\lambda_3 - 2\lambda_4 - 3\lambda_5 - 4\lambda_6 - 6\lambda_7 + 5\theta \geqq 0, \\
& \lambda_j \geqq 0 \quad (j = 1, 2, \ldots, 7), \\
& \theta \text{ free.}
\end{aligned}$$

Table 5.4. Results of the example from Table 5.3.

DMU	v_1^0	v_2^0	μ^0	θ^0	s^{+0}	s_1^{-0}	s_2^{-0}	Reference set	λ_j^0
1	0.222	0.111	1	1	0	0	0	DMU_1	$\lambda_1^0 = 1$
2	0.222	0.111	1	1	0	0	0	DMU_2	$\lambda_2^0 = 1$
3	0.111	0.222	1	1	0	0	0	DMU_3	$\lambda_3^0 = 1$
4	0	0.5	1	1	0	2	0	DMU_3	$\lambda_3^0 = 1$
5	0	0.333	0.666	0.666	0	0.333	0	DMU_3	$\lambda_3^0 = 1$
6	0.0769	0.153846	0.6923	0.6923	0	0	0	DMU_2, DMU_3	$\lambda_2^0 = 0.769$, $\lambda_3^0 = 0.231$
7	0.1428	0.0714	0.64285	0.6428	0	0	0	DMU_1, DMU_2	$\lambda_1^0 = 0.4286$, $\lambda_2 = 0.5714$

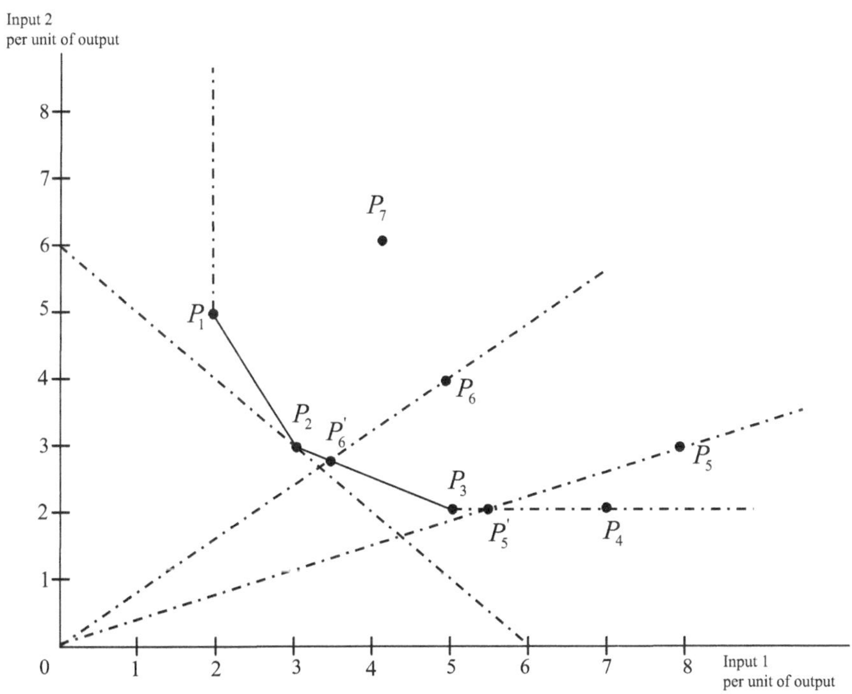

Fig. 5.4. Efficiency frontier for the example from Table 5.3.

The (unique) optimal solution is $\lambda_1^0 = 1$; $\lambda_2^0 = \lambda_3^0 = \lambda_4^0 = \lambda_5^0 = \lambda_6^0 = \lambda_7^0 = 0$; $\theta^0 = 1$; and all slack variables s^+, s_1^-, s_2^- equal zero. DMU_1 is CCR efficient.

We can proceed in a similar way with the other DMUs. The results are summarized in Table 5.4, and Figure 5.4 portrays the situation geometrically. The efficient units are DMU_1, DMU_2, and DMU_3. The efficient frontier represented by the solid line passes through the points P_1, P_2, and P_3 corresponding to these units.

The optimal solution for DMU_4 yields $\theta^0 = 1$, DMU_4 looks efficient. By comparing DMU_4 with DMU_3, one can see that both units use the same amount of input 2, but DMU_4 has two units of excess in input 1. This inefficiency is concealed because the optimal solution forces the weight of input 1 to zero ($\nu_1^0 = 0$). This is exactly the situation described by Charnes, Cooper, and Rhodes [9]. By assigning a small positive value ϵ to ν_1 (or using two-phase optimization—see [1] and [11, Section 3.6]), it is possible to identify the inefficiency of DMU such as DMU_4 in our example. Looking at the solution of the envelopment model, such an excess in an input is explicitly given by the nonzero value of the slack variable ($s_1^{-0} = 2$). Hence condition (b) in statement (5.4) is not satisfied, so DMU_4 did not achieve efficiency in its performance.

A DMU such as DMU_4, with $\theta^0 = 1$ and with an excess in inputs, is called *radially* efficient but *mix* inefficient (or weakly efficient). Using the duality theory of linear programming, the reader can prove that the following definition of CCR efficiency gives the same efficiency characterization as obtained from (5.4).

Definition 5.1 (CCR efficiency). DMU_0 is efficient if $\theta^0 = 1$ and there exists at least one optimal solution $(\boldsymbol{\mu}^0, \boldsymbol{\nu}^0)$ with $\boldsymbol{\mu}^0 > \mathbf{0}$ and $\boldsymbol{\nu}^0 > 0$. Otherwise, DMU_0 is inefficient.

Thus CCR inefficiency means that either (i) $\theta^0 < 1$ or (ii) $\theta^0 = 1$ and at least one element of $(\boldsymbol{\mu}^0, \boldsymbol{\nu}^0)$ is zero for every optimal solution of (5.2). The multiplier ν_1^0 for DMU_4 is zero.

Now let us analyze the solution for the inefficient DMU, e.g., DMU_6 as given in Table 5.4. $\theta^0 = 0.69$ indicates that DMU_6 can simultaneously reduce both inputs (without altering the proportions in which they are utilized) by 31% in order to achieve the efficiency frontier. In other words, 69% of both inputs must be sufficient to produce one unit of output. Hence DMU_6 is technically inefficient. No mix efficiencies are present because all slacks are zero. $\frac{0P_6'}{0P_6} = 0.69$ corresponds to radial efficiency of DMU_6. Also, we have

- input 1 of $P_6' = 0,6923 \times 5$(input 1 of DMU_6) $= 3,4615$;
- input 2 of $P_6' = 0,6923 \times 4$(input 2 of DMU_6) $= 2,7692$.

Point P_6' gives us the projection for P_6 to the efficiency frontier, and it is a point on the line segment (P_2, P_3). In other words, P_6' can be described as a linear combination of the points P_2 (it represents DMU_2) and P_3 (it represents DMU_3). Therefore, the reference set for DMU_6 is created by DMU_2 and DMU_3. The solution of the envelopment problem (5.3) for DMU_6 yields $\lambda_2^0 = 0,76923$, $\lambda_3^0 = 0,23076$, and $\lambda_1^0 = \lambda_4^0 = \lambda_5^0 = \lambda_6^0 = \lambda_7^0 = 0$. Consequently, the values of point P_6' are calculated as $P_6' = \lambda_2^0 P_2 + \lambda_3^0 P_3$:

- input 1 of $P_6' = 0,76923 \times 3$(input 1 of DMU_2) $+ 0,23076 \times 5$(input 1 of DMU_3) $= 3,461$;
- input 2 of $P_6' = 0,76923 \times 3$(input 2 of DMU_2) $+ 0,23076 \times 2$(input 2 of DMU_3) $= 2,7692$.

Comparing these results we see that the coordinates of P_6', the (virtual) DMU used to evaluate DMU_6, can be derived in either of these two ways.

From the magnitude of coefficients λ_2^0 and λ_3^0, DMU_6 has more similarity to DMU_2 than DMU_3 (as can be seen from Table 5.3 and Figure 5.4).

The multipliers or weights μ, ν_1, and ν_2 are the dual variables to the envelopment problem (5.3) and therefore have a role as a measure of the sensitivity of efficiency score θ with respect to variations in input items. The higher optimal weight ν_2^0 implies that a reduction in input 2 has a bigger effect on efficiency than does a reduction in input 1. In other words, the contribution of input 2 to the efficiency of DMU_6 is higher than the contribution of input 1. The ratio $\frac{\nu_2^0}{\nu_1^0} = \frac{0.153846}{0.0769} = 2$ indicates that it is advantageous for DMU_6 to weight input 2 two times more than input 1 in order to maximize the efficiency. Half a unit of input 2 is as good as one unit of input 1. The marginal rate of substitution of input 2 for input 1 along the line segment (P_2, P_3)—where point P_6 is projected—is two. As can be seen from Figure 5.4, the slope of the line segment (P_2, P_3) is a half, which is exactly the ratio $\frac{\nu_1^0}{\nu_2^0}$. In this way, the ratio $\frac{\nu_1^0}{\nu_2^0}$ expresses the slope of the line segment created by the reference set (or the peer group) to DMU_6.

In a similar way, the results for the inefficient unit DMU_7 can be interpreted.

The interpretation for the inefficient unit DMU_5 is slightly different. Since $\lambda_3 > 0$, the reference set for DMU_5 consists of DMU_3 only. One plan for the improvement of P_5 is to reduce both input values by multiplying them by 0.666 (projection to the point P_5' on the line segment (P_3, P_4)) and further subtracting 0.333 from input 1. When this is done, the thusly altered values coincide with the coordinates of P_3. Geometrically, then, the projection for P_5 is

$$
\begin{aligned}
x_1' &= \theta^0 x_1 - s_1^{-0} = 0.666 \times 8 - 0,333 = 5 && \text{(37.5\% reduction)},\\
x_2' &= \theta^0 x_2 - s_2^{-0} = 0.666 \times 3 - 0 = 2 && \text{(33\% reduction)},\\
y' &= y = 1 && \text{(no change)},
\end{aligned}
$$

which are the same values as for DMU_3 (see Table 5.3).

However, we should be careful with the interpretation of the optimal weights for an efficient unit. From Figure 5.4, we can see that P_2 (representing the efficient unit DMU_2) lies on the line segment (P_1, P_2) as well as on the line segment (P_2, P_3). Thus the optimal weights for an efficient DMU need not be unique, in the case of DMU_2, they are restricted to an interval corresponding to the different slope of (both) the line segments containing the efficient unit DMU_2.

Summarizing the results in Table 5.4, only DMU_1, DMU_2, and DMU_3 are strong (or fully) efficient in the sense of statement (5.4). DMU_6 and DMU_7 fail because $\theta^0 < 1$. DMU_4 has a value of $\theta^0 = 1$ because it is on the frontier. However, this portion of frontier is not efficient because of the nonzero slack s_1^{-0}. Finally, DMU_5 fails to be efficient both because $\theta^0 < 1$ and the positive slack s_1^{-0} is involved in the projection of P_5 to efficiency frontier.

5.2.2 The Output-Oriented Model under a Constant Returns-to-Scale Assumption

Up to this point, we have been dealing with a model whose objective is to minimize inputs while producing at least the given output levels. Alternately, one could focus on maximal movement via proportional augmentation of outputs (for given inputs). In other words, we minimize the inefficiency of DMU_0 given by the ratio of virtual input to virtual output under the constraints that the so-defined inefficiency cannot be lower than one for itself or for any of the other DMUs. The required optimization problem is

$$\begin{aligned}
\underset{\mathbf{u},\mathbf{v}}{\text{minimize}} \quad & z_0(\mathbf{u},\mathbf{v}) = \frac{\sum_{i=1}^m x_{i0}v_i}{\sum_{r=1}^s y_{r0}u_r} \\
\text{subject to} \quad & \frac{\sum_{i=1}^m x_{ij}v_i}{\sum_{r=1}^s y_{rj}u_r} \geqq 1 \quad (j = 1, 2, \ldots, n), \\
& v_i \geqq 0 \quad (i = 1, 2, \ldots, m), \\
& u_r \geqq 0 \quad (r = 1, 2, \ldots, s).
\end{aligned}$$

Again the Charnes–Cooper [7] transformation for linear fractional programming yields the linear programming model (with lower bounds for the multipliers)

$$\begin{aligned}
\underset{\mu,\nu}{\text{minimize}} \quad & f_0(\boldsymbol{\nu}) = \sum_{i=1}^m x_{i0}\nu_i \\
\text{subject to} \quad & -\sum_{r=1}^s y_{rj}\mu_r + \sum_{i=1}^m x_{ij}\nu_i \geqq 0 \quad (j = 1, 2, \ldots, n), \\
& \sum_{r=1}^s y_{r0}\mu_r = 1 \\
& \mu_r \geqq \epsilon \quad (r = 1, 2, \ldots, s), \\
& \nu_i \geqq \epsilon \quad (i = 1, 2, \ldots, m),
\end{aligned} \tag{5.5}$$

whose dual problem is

$$\begin{aligned}
\underset{\varphi,\boldsymbol{\lambda},\mathbf{s}^+,\mathbf{s}^-}{\text{maximize}} \quad & q_0(\varphi,\mathbf{s}^+,\mathbf{s}^-) = \varphi + \epsilon\left(\sum_{r=1}^s s_r^+ + \sum_{i=1}^m s_i^-\right) \\
\text{subject to} \quad & \varphi y_{r0} - \sum_{j=1}^n y_{rj}\lambda_j + s_r^+ = 0 \quad (r = 1, 2, \ldots, s), \\
& \sum_{j=1}^n x_{ij}\lambda_j + s_i^- = x_{i0} \quad (i = 1, 2, \ldots, m), \\
& \lambda_j \geqq 0 \quad (j = 1, 2, \ldots, n), \\
& s_r^+ \geqq 0 \quad (r = 1, 2, \ldots, s), \\
& s_i^- \geqq 0 \quad (i = 1, 2, \ldots, m).
\end{aligned} \tag{5.6}$$

Models (5.5) and (5.6) are called *output-oriented CCR* models. The objective now is to maximize output production while not exceeding the given input levels. Model (5.6) maximizes on φ to achieve proportional output augmentation. The variable φ_0^0 yields the proportion by which all the DMU_0's outputs should be produced (under the given input levels) if DMU_0 is efficient. It should be clear from the construction of the output-oriented model that the optimal value φ_0^0 cannot be smaller than 1. Thereupon $\varphi_0^0 - 1$ indicates the proportional increase of all DMU_0's outputs in order to achieve the frontier. A proportional increase is possible until at least one of the output slack variables is reduced to zero. As for the input-oriented model (5.3), a DMU is efficient if and only if $\varphi^0 = 1$ and all slack variables s_r^+ and s_i^- are equal to zero (or the multipliers μ_r for $r = 1, 2, \ldots, s$ and ν_i for $i = 1, 2, \ldots, m$ are positive).

The constraints in the envelopment problem (5.6) with respect to outputs indicate that even after the proportional increase of all outputs, the outputs of the evaluated DMU_0 cannot be higher than the outputs of the composite unit. Similarly, the inputs of DMU_0 cannot be lower than the inputs of the composite unit. Again, the objective function in (5.6) allows us to distinguish between mix and radial inefficiencies.

To illustrate the output-oriented model and for its comparison to the input-oriented model, we consider the same example with one input and one output presented in Table 5.1.

The linear program (5.5) for DMU_1 is

$$
\begin{array}{lll}
\underset{\mu,\nu}{\text{minimize}} & f_0(\nu) = 2\nu & \\
\text{subject to} & -2\mu + 2\nu \geqq 0, & (\text{DMU}_1)\\
& -4\mu + 3\nu \geqq 0, & (\text{DMU}_2)\\
& -6\mu + 5\nu \geqq 0, & (\text{DMU}_3)\\
& -7\mu + 7\nu \geqq 0, & (\text{DMU}_4)\\
& -\mu + 3\nu \geqq 0, & (\text{DMU}_5)\\
& -5\mu + 6\nu \geqq 0, & (\text{DMU}_6)\\
& -7\mu + 8\nu \geqq 0, & (\text{DMU}_7)\\
& 2\mu = 1, & \\
& \mu \geqq 0, \quad \nu \geqq 0. &
\end{array}
$$

The optimal solution is $\mu^0 = 0,5$, $\nu^0 = 0,666$, and $f_0^0 = 1,333$.

The corresponding envelopment problem (5.6) is stated as

$$
\begin{array}{ll}
\underset{\lambda,\varphi}{\text{maximize}} & \varphi \\
\text{subject to} & -2\lambda_1 - 4\lambda_2 - 6\lambda_3 - 7\lambda_4 - \lambda_5 - 5\lambda_6 - 7\lambda_7 + 2\varphi \leqq 0,\\
& 2\lambda_1 + 3\lambda_2 + 5\lambda_3 + 7\lambda_4 + 3\lambda_5 + 6\lambda_6 + 8\lambda_7 \leqq 2,\\
& \lambda_j \geqq 0 \quad (j = 1, 2, \ldots, 7),\\
& \varphi \text{ free.}
\end{array}
$$

Table 5.5. Optimal solutions for (5.5) and (5.6) with the data set from Table 5.1.

DMU	φ^*	s^{+*}	s^{-*}	λ^*	f_0^*	μ^*	ν^*
1	1.3333	0	0	$\lambda_2 = 0.6666$	1.3333	0.5	0.6666
2	1	0	0	$\lambda_2 = 1$	1	0.25	0.3333
3	1.1111	0	0	$\lambda_2 = 1.6666$	1.1111	0.1666	0.2222
4	1.3333	0	0	$\lambda_2 = 2.3333$	1.3333	0.1428	0.1905
5	4	0	0	$\lambda_2 = 1$	4	1	1.3333
6	1.6	0	0	$\lambda_2 = 2$	1.6	0.2	0.2666
7	1.5238	0	0	$\lambda_2 = 2.6666$	1.5238	0.1428	0.1905

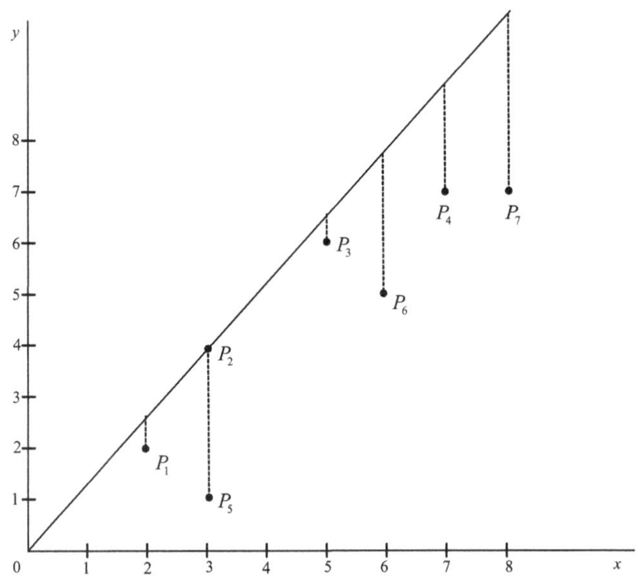

Fig. 5.5. Projections for the output-oriented CCR model from Table 5.1.

The optimal solution is given by $\varphi^0 = 1,3333$, $\lambda_2^0 = 0,666$, $\lambda_1^0 = \lambda_3^0 = \lambda_4^0 = \lambda_5^0 = \lambda_6^0 = \lambda_7^0 = 0$. DMU_1 can increase its output by 33.33% while using no more than the observed amount of input. P_1 is projected into the boundary point $(2; 2.6666) = \lambda_2 P_2 = 0.6666 \times (3; 4) = (2; 2.6666)$.

In a similar way, the linear programming problems (5.5)—under the nonnegativity constraints for the multipliers—and (5.6) for every DMU can be calculated. The results are summarized in Table 5.5. Figure 5.5 depicts the efficiency frontier and the projections of the inefficient DMUs to the efficiency frontier.

As seen in Figures 5.3 and 5.5 and looking at the results for both models (Tables 5.2 and 5.5), the two orientations yield identical envelopment surfaces and identical sets of efficient and inefficient DMUs. The following theorems [34, pp. 23–24] provide the correspondence between solutions for the input-oriented CCR and output-oriented CCR models.

Theorem 5.2. *Let* $(\theta^0, \boldsymbol{\lambda}^0)$ *be an optimal solution to model* (5.3). *Then*

$$(\varphi^*, \boldsymbol{\lambda}^*) = \left(\frac{1}{\theta^0}, \frac{1}{\theta^0}\boldsymbol{\lambda}^0\right)$$

is optimal for (5.6), *and the mapping*

$$(\theta, \boldsymbol{\lambda}) \rightarrow \left(\frac{1}{\theta}, \frac{1}{\theta}\boldsymbol{\lambda}\right)$$

is a $1-1$ *correspondence between optimal solutions of* (5.3) *and* (5.6).

Theorem 5.3. *The mapping*

$$(\boldsymbol{\mu}^*, \boldsymbol{\nu}^*) \rightarrow \frac{1}{w_0^0}(\boldsymbol{\mu}^0, \boldsymbol{\nu}^0)$$

is a $1-1$ *correspondence between optimal solutions of* (5.2) *and* (5.5), *and the respective optimal values* w_0^0, f_0^0 *satisfy the condition* $w_0^0 \cdot f_0^0 = 1$.

The results in Tables 5.2 and 5.5 confirm the simple rule by which the solutions of the input- and output-oriented CCR models are linked. In the same way, the slack variables of the output-oriented model $(\mathbf{s}^{-*}, \mathbf{s}^{+*})$ are related to the slacks of the input-oriented model $(\mathbf{s}^{-0}, \mathbf{s}^{+0})$:

$$\mathbf{s}^{-*} = \frac{1}{\theta^0}\mathbf{s}^{-0}, \qquad \mathbf{s}^{+*} = \frac{1}{\theta^0}\mathbf{s}^{+0}.$$

The reader should note that an inefficient DMU will be projected to different points on the frontier under the input and output orientations, as illustrated in Figures 5.3 and 5.5.

5.2.3 The Additive Model under a Constant Returns-to-Scale Assumption

To further clarify what is being accommodated in DEA, we now introduce another type, known as "additive models." In the previous models, one set of variables (inputs or outputs) preempts the other in proportional (radial) movement toward the frontier. However, the projected point can be equivalently represented in terms of the vector of output slacks $\mathbf{s}^+$ and the vector of excess inputs $\mathbf{s}^-$. In other words, the movement toward the efficiency frontier is not radial but via augmenting particular outputs *and* reducing particular inputs.

Generally speaking, an envelopment surface consists of portions of supporting hyperplanes in R^{m+s} that form particular facets of the convex hull of the points $(\mathbf{y}_j, \mathbf{x}_j)$ for $j = 1, 2, \ldots, n$. The general equation for a hyperplane in R^{m+s} is given by

$$\sum_{r=1}^{s} y_r \mu_r - \sum_{i=1}^{m} x_i \nu_i + \mu_0 = 0. \tag{5.7}$$

Such a hyperplane is a supporting hyperplane (and forms a facet of the envelopment surface) if and only if all the points $(\mathbf{y}_j, \mathbf{x}_j)$ lie on or beneath the hyperplane and, additionally, the hyperplane passes through at least one of the points.

Under the assumption of constant returns to scale, all supporting hyperplanes pass through the origin (see Figure 5.1 for one input and one output). Thus $\mu_0 = 0$, and (5.7) reduces to

$$\sum_{r=1}^{s} y_r \mu_r - \sum_{i=1}^{m} x_i \nu_i = 0.$$

Such a hyperplane forms a facet of the CRS envelopment surface if and only if

$$\begin{aligned} &\sum_{r=1}^{s} y_{rj} \mu_r - \sum_{i=1}^{m} x_{ij} \nu_i \leqq 0 \quad (j = 1, 2, \ldots, n), \\ &\sum_{r=1}^{s} y_{rk} \mu_r - \sum_{i=1}^{m} x_{ik} \nu_i = 0 \quad \text{for some } k. \end{aligned} \tag{5.8}$$

The set of constraints (5.8) ensures that all points lie on or below the envelopment surface. The maximization of the efficiency of DMU_0 selects a hyperplane that minimizes the distance from DMU_0 to this envelopment surface.

The formulation of the additive CRS multiplier program (see [2]) that follows is a direct consequence of the above conditions for the CRS envelopment surface:

$$\begin{aligned} \underset{\boldsymbol{\mu},\boldsymbol{\nu}}{\text{maximize}} \quad & w_0(\boldsymbol{\mu}, \boldsymbol{\nu}) = \sum_{r=1}^{s} y_{r0} \mu_r - \sum_{i=1}^{m} x_{i0} \nu_i \\ \text{subject to} \quad & \sum_{r=1}^{s} y_{rj} \mu_r - \sum_{i=1}^{m} x_{ij} \nu_i \leqq 0 \quad (j = 1, 2, \ldots, n), \\ & -\mu_r \leqq -1 \quad (r = 1, 2, \ldots, s), \\ & -\nu_i \leqq -1 \quad (i = 1, 2, \ldots, m) \end{aligned} \tag{5.9}$$

with $\epsilon = 1$.

Note that the objective function value is nonpositive, and we are maximizing a nonpositive quantity. Hence an optimal value of zero indicates that DMU_0 lies on the frontier. Inefficient units lie below the closest supporting hyperplane and thus correspond to nonzero objective values. As in the oriented models, each of the n optimal solutions given by the sets of values $(\boldsymbol{\mu}_j^0, \boldsymbol{\nu}_j^0)$ for $j = 1, 2, \ldots, n$ is the set of coefficients for a supporting hyperplane which defines a facet of the envelopment surface.

The CRS envelopment model, which is the dual problem to the above multiplier program, can be expressed as follows:

$$
\begin{aligned}
&\underset{\boldsymbol{\lambda},\mathbf{s}^-,\mathbf{s}^+}{\text{minimize}} && g_0(\boldsymbol{\lambda},\mathbf{s}^-,\mathbf{s}^+) = -\sum_{r=1}^{s} s_r^+ - \sum_{i=1}^{m} s_i^- \\
&\text{subject to} && \sum_{j=1}^{n} y_{rj}\lambda_j \quad s_r^+ = y_{r0} \qquad (r - 1,2,\ldots,s), \\
& && -\sum_{j=1}^{n} x_{ij}\lambda_j - s_i^- = -x_{i0} \qquad (i = 1,2,\ldots,m), \\
& && \lambda_j \geqq 0 \qquad (j = 1,2,\ldots,n), \\
& && s_r^+ \geqq 0 \qquad (r = 1,2,\ldots,s), \\
& && s_i^- \geqq 0 \qquad (i = 1,2,\ldots,m).
\end{aligned}
\tag{5.10}
$$

From the complementary slackness conditions, we know that for each $\lambda_j^0 > 0$, the corresponding constraint in the multiplier problem is binding, that is, $\sum_{r=1}^{s} y_{rj}\mu_r^0 - \sum_{i=1}^{m} x_{ij}\nu_i^0 = 0$. Thus the DMU$_j$ with $\lambda_j^0 > 0$ is efficient and lies on the hyperplane which defines a facet of the frontier.

The duality theory of linear programming implies that $w_0^0 = g_0^0$. Thus DMU$_0$ is efficient if and only if $w_0^0 = g_0^0 = 0$. In other words, in the additive model DMU$_0$ is efficient if and only if $\mathbf{s}_0^{+^0} = \mathbf{0}$ and $\mathbf{s}_0^{-^0} = \mathbf{0}$. The DMU$_0$ is inefficient if it does not lie on the frontier, i.e., if any component of the slack variables $\mathbf{s}^+$ or $\mathbf{s}^-$ is positive; the values of these nonzero components identify the sources and amounts of inefficiency in the corresponding outputs and inputs.

As in the oriented models for an inefficient DMU$_0$, the vector $\boldsymbol{\lambda}^0$ defines the projected point on the efficient frontier as

$$
(\mathbf{y}_0', \mathbf{x}_0') = \left(\sum_{j=1}^{n} \lambda_j^0 \mathbf{y}_j, \sum_{j=1}^{n} \lambda_j^0 \mathbf{x}_j \right).
$$

An important distinction between the additive model and the oriented models lies in the improvement to an efficient activity. In an input-oriented model, $(1 - \theta^0)$ represents reductions that can be achieved without altering the proportions in which inputs are used. $(\varphi^0 - 1)$ plays a similar role for output expansions that do not alter the proportions in which outputs are produced.

The projection in terms of the slack variables $\mathbf{s}^{+^0}$ and $\mathbf{s}^{-^0}$ is by the additive model represented as

$$
\mathbf{y}_0' = \mathbf{y}_0 + \mathbf{s}^{+^0} \quad \text{and} \quad \mathbf{x}_0' = \mathbf{x}_0 - \mathbf{s}^{-^0}.
$$

For the example data set in Table 5.3, the multiplier form of the CRS model for DMU$_1$ is given by

$$
\begin{aligned}
&\underset{\mu,\boldsymbol{\nu}}{\text{maximize}} && w_1(\mu,\boldsymbol{\nu}) = \mu - 2\nu_1 - 5\nu_2 \\
&\text{subject to} && \mu - 2\nu_1 - 5\nu_2 \leqq 0,
\end{aligned}
$$

$$
\begin{aligned}
\mu - 3\nu_1 - 3\nu_2 &\leqq 0,\\
\mu - 5\nu_1 - 2\nu_2 &\leqq 0,\\
\mu - 7\nu_1 - 2\nu_2 &\leqq 0,\\
\mu - 8\nu_1 - 3\nu_2 &\leqq 0,\\
\mu - 5\nu_1 - 4\nu_2 &\leqq 0,\\
\mu - 4\nu_1 - 6\nu_2 &\leqq 0,\\
-\mu &\leqq -1,\\
-\nu_1 &\leqq -1,\\
-\nu_2 &\leqq -1,
\end{aligned}
$$

while the corresponding envelopment problem is formulated as

$$
\begin{aligned}
\underset{\boldsymbol{\lambda}, s^+, s^-}{\text{minimize}} \quad & g_1(\boldsymbol{\lambda}, s^+, \mathbf{s}^-) = -s^+ - s_1^- - s_2^-\\
\text{subject to} \quad & \lambda_1 + \lambda_2 + \lambda_3 + \lambda_4 + \lambda_5 + \lambda_6 + \lambda_7 - s^+ = 1,\\
& -2\lambda_1 - 3\lambda_2 - 5\lambda_3 - 7\lambda_4 - 8\lambda_5 - 5\lambda_6 - 4\lambda_7 - s_1^- = -2,\\
& -5\lambda_1 - 3\lambda_2 - 2\lambda_3 - 2\lambda_4 - 3\lambda_5 - 4\lambda_6 - 6\lambda_7 - s_2^- = -5,\\
& \lambda_j \geqq 0 \quad (j = 1, 2, \ldots, 7),\\
& s^+ \geqq 0,\\
& s_i^- \geqq 0 \quad (i = 1, 2).
\end{aligned}
$$

The optimal solution to the multiplier problem is $\mu^0 = 9$, $\nu_1^0 = 2$, $\nu_2^0 = 1$, and $w_1^0 = 0$. The optimal envelopment solution $(\boldsymbol{\lambda}^0, s^{+^0}, \mathbf{s}^{-^0})$ is given by $\lambda_1^0 = 1$, $\lambda_j^0 = 0$ $(j = 2, 3, \ldots, 7)$, $s^{+^0} = 0$, $s_1^{-^0} = 0$, $s_2^{-^0} = 0$. Thus DMU_1 is efficient.

Table 5.6 reports optimal solutions for the additive CRS model for each DMU, and Figure 5.6 depicts the efficient frontier and the projections for the inefficient

Table 5.6. Results for the additive model.

DMU	ν_1^0	ν_2^0	μ	w^0	s^{+^0}	$s_1^{-^0}$	$s_2^{-^0}$	Reference set	λ_j^0
1	2	1	9	0	0	0	0	DMU_1	$\lambda_1^0 = 1$
2	1	1	6	0	0	0	0	DMU_2	$\lambda_2^0 = 1$
3	1	2	9	0	0	0	0	DMU_3	$\lambda_3^0 = 1$
4	1	2	9	2	0	2	0	DMU_3	$\lambda_3^0 = 1$
5	1	1	6	5	0	5	0	DMU_2	$\lambda_2^0 = 1$
6	1	1	6	3	0	2	1	DMU_2	$\lambda_2^0 = 1$
7	1	1	6	4	0	1	3	DMU_2	$\lambda_2^0 = 1$

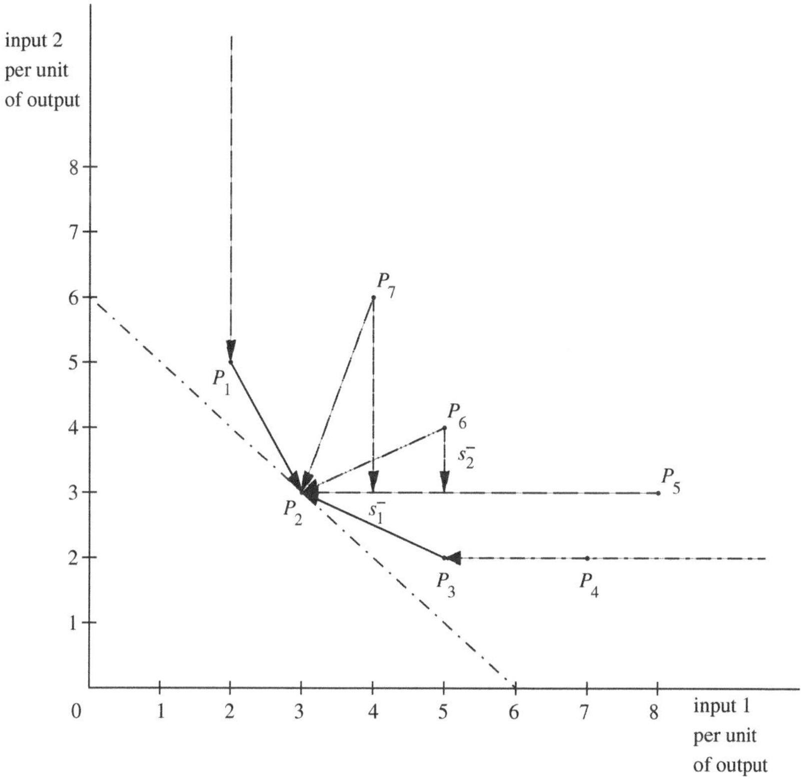

Fig. 5.6. Efficiency frontier and the projections for the additive model with the data from Table 5.3.

DMUs. (For this simple example with one output, whose value is unitized to 1, we can describe the situation using a two-dimensional graph.)

The efficient units are DMU_1, DMU_2, and DMU_3 because all components of the slack variables, s^+ and $\mathbf{s}^-$, are equal to zero. The piecewise linear envelopment surface passes through the points P_1, P_2, and P_3 representing DMU_1, DMU_2, and DMU_3, respectively. It is easily verified that both DMU_1 and DMU_2 lie on the facet of CRS envelopment surface defined by the hyperplane

$$9y - 2x_1 - x_2 = 0.$$

Similarly, both DMU_2 and DMU_3 lie on the facet of efficient frontier expressed by the equation

$$9y - x_1 - 2x_2 = 0. \tag{5.11}$$

The solution of the linear programming problem (5.9) for DMU_2 is not unique. DMU_2 lies on two above-defined facets of the envelopment surface. From the row for DMU_2 in Table 5.6, a supporting hyperplane with equation

$$6y - x_1 - x_2 = 0 \tag{5.12}$$

is derived. In Figure 5.6, it is shown by the dotted line (with slope -1) passing through the point P_2.

DMU_4, DMU_5, DMU_6, and DMU_7 are obviously inefficient because of excess inputs $\mathbf{s}^-$. For these DMUs, the objective function value, w_0, measures the distance to the closest supporting hyperplane, i.e., $\boldsymbol{\mu}\mathbf{y}_0 - \boldsymbol{\nu}\mathbf{x}_0 = w_0$ with maximal w_0, where $w_0 \leqq 0$. For example, the supporting hyperplane (5.11) for DMU_3 and the parallel hyperplane

$$9y - x_1 - 2x_2 = -2$$

that passes through P_4 are an L1 distance of 2 units apart. The supporting hyperplane (5.12) and the parallel hyperplane

$$6y - x_1 - x_2 = -5$$

passing through P_5 are an L1 distance of 5 units apart. The multipliers ν_1, ν_2, and μ obtained from the rows in Table 5.6 are the coefficients of supporting hyperplanes for efficient DMUs, which serve as the closest supporting hyperplane for an inefficient DMU. The projections on the efficient frontier for these units illustrate the difference to the (input-)oriented model.

While in the input-oriented model the improvement to an efficient activity for DMU_5 is obtained first by the proportional reduction of both inputs by 33% (see point P_5' in Figure 5.4), and additionally input 1 was reduced by 0,333, in the additive model efficiency is achieved by reducing input 1 by 5 units. In this way, point P_5 in Figure 5.6 is projected to point P_2 and the reference DMU for DMU_5 is DMU_2.

The arrows s_1^- and s_2^- in Figure 5.6 denote the efficiency improvement for the inefficient DMU_6 represented by point P_6. In arriving at point P_2 on the efficient frontier, which is most distant from P_6, DMU_6 will reduce input 1 by 2 units and input 2 by 1 unit. Consequently, the proportion in which inputs are used will change.

5.2.4 DEA Models under a Variable Returns-to-Scale Assumption

In the preceding sections, we discussed DEA models built on the assumption of constant returns to scale of activities. This means that a proportional increase in inputs leads to the same proportional increase in outputs. Geometrically speaking, all supporting hyperplanes for a CRS-efficient frontier pass through the origin (see Figure 5.3 for the single-input and single-output case). However, this assumption can be modified or relaxed in order to "restrict our attention to production inefficiencies at the given level of operations for each DMU, and thus develop an efficiency measurement procedure that assigns an efficiency rating of one to a DMU if and only if the DMU lies on the efficient production surface, even when it may not be operating at the most efficient scale size. This identification of the efficient production surface will also allow us to determine whether increasing, constant, or decreasing returns to scale prevail in different segments of the production surface" [6, p. 1084]. The extension of the CCR model proposed by Banker, Charnes, and Cooper [6] allows us to analyze situations in which increasing inputs imply more (or less) than proportional increasing outputs at points on the efficient production surface and separate them from output increases resulting from the elimination of technical inefficiencies.

Let us start with the simple example from the previous section (Table 5.1) with one input and one output. The efficient frontier of the CCR model in Figure 5.7 is the dotted line that passes through P_2 from the origin. The output increases proportionally to input. Under the assumption of variable returns to scale, the efficient frontier consists of the bold lines connecting P_1, P_2, P_3, and P_4. DMU_1, DMU_2, DMU_3, and DMU_4 are on the frontier and efficient under variable returns-to-scale assumption. However, only DMU_2 is CCR efficient.

Reading values from this graph, the Banker–Charnes–Cooper (BCC) efficiency (or the VRS efficiency) of DMU_6, represented by P_6, is evaluated by

$$\frac{QS}{QP_6} = \frac{4}{6} = 0,6666,$$

while its CCR efficiency (or CRS efficiency) is smaller, with value

$$\frac{QR}{QP_6} = \frac{3.75}{6} = 0.625.$$

The BCC efficiency of DMU_1, represented in Figure 5.7 by P_1, is given by

$$\frac{LP_1}{LP_1} = 1,$$

while its CCR efficiency is smaller than one because

$$\frac{LM}{LP_1} = \frac{1.5}{2} = 0.75.$$

DMU_1 is BCC efficient but CCR inefficient.

Generally, the CRS efficiency does not exceed the VRS efficiency.

In the output orientation, the VRS efficiency of DMU_6 is determined by

$$\frac{VT}{P_6T} = \frac{6.5}{5} = 1.3.$$

In order to achieve the VRS efficiency frontier in Figure 5.7, DMU_6 should increase its output (under the given input level) from its observed value to $1.3 \times 5 = 6.5$ units, or by 30%.

CRS efficiency is evaluated by

$$\frac{WT}{P_6T} = \frac{8}{5} = 1.6,$$

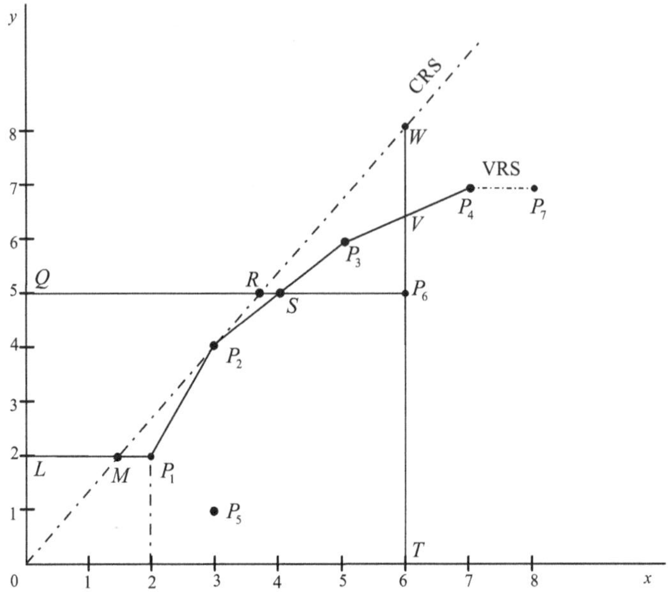

Fig. 5.7. Production frontier under CRS and VRS.

which means that the achievement of efficiency under a constant returns-to-scale assumption would require augmenting DMU_6's output to $1.6 \times 5 = 8$ units, or by 60%. As Figure 5.7 makes clear, under a CRS assumption a still-greater augmentation is needed to achieve efficiency. According to Theorem 5.2, the CRS efficiency in the output-oriented model (1.6) is the reciprocal of its input efficiency (0.625). The example of DMU_6 illustrates that this simple "reciprocal relation" between input- and output-oriented models is not valid under the variable returns-to-scale assumption.

As already mentioned and as illustrated in Figure 5.7, all supporting hyperplanes for a CRS envelopment surface pass through the origin.

Under the variable returns-to-scale assumption, the equation for a hyperplane is given by (5.7). Consequently, the conditions for a facet of the VRS envelopment surface may be formulated as

$$\sum_{r=1}^{s} y_{rj}\mu_r - \sum_{i=1}^{m} x_{ij}v_i + \mu_0 \leqq 0 \quad (j = 1, 2, \ldots, n),$$
$$\sum_{r=1}^{s} y_{rk}\mu_r - \sum_{i=1}^{m} x_{ik}v_i + \mu_0 = 0 \quad \text{for some } k.$$

The difference between CRS and VRS models is present in the free variable μ_0 (it may be positive or negative or zero), which describes the shift of the supporting hyperplanes from the origin.

The input-oriented VRS or BCC model in the multiplier form is expressed as

$$
\begin{aligned}
&\underset{\boldsymbol{\mu},\boldsymbol{v},\mu_0}{\text{maximize}} && w_0(\boldsymbol{\mu}, z) = \sum_{r=1}^{s} y_{r0}\mu_r + \mu_0 \\
&\text{subject to} && \sum_{r=1}^{s} y_{rj}\mu_r - \sum_{i=1}^{m} x_{ij}\nu_i + \mu_0 \leqq 0 \quad (j = 1, 2, \ldots, n), \\
&&& \sum_{i=1}^{m} x_{i0}\nu_i = 1, \\
&&& \mu_r \geqq 0 \quad (r = 1, 2, \ldots, s), \\
&&& \nu_i \geqq 0 \quad (i = 1, 2, \ldots, m), \\
&&& \mu_0 \text{ free in sign.}
\end{aligned}
\tag{5.13}
$$

The equivalent BCC fractional programming problem from which the linear program (5.13) can be obtained (by a transformation of the variables as in the CCR model) is stated as

$$
\begin{aligned}
&\underset{\mathbf{u},\mathbf{v},u_0}{\text{maximize}} && h_0(\mathbf{u}, \mathbf{v}, u_o) = \frac{\sum_{r=1}^{s} y_{r0}u_r + u_0}{\sum_{i=1}^{m} x_{i0}v_i} \\
&\text{subject to} && \frac{\sum_{r=1}^{s} y_{rj}u_r + u_0}{\sum_{i=1}^{m} x_{ij}v_i} \leqq 1 \quad (j = 1, 2, \ldots, n), \\
&&& u_r \geqq 0 \quad (r = 1, 2, \ldots, s), \\
&&& v_i \geqq 0 \quad (i = 1, 2, \ldots, m), \\
&&& w_0 \text{ free.}
\end{aligned}
$$

The envelopment problem as the dual form of the linear program (5.13) is given by

$$
\begin{aligned}
&\underset{\theta\rightarrow}{\text{minimize}} && g_0(\theta) = \theta \\
&\text{subject to} && \sum_{j=1}^{n} y_{rj}\lambda_j \geqq y_{r0} \quad (r = 1, 2, \ldots, s), \\
&&& -\sum_{j=1}^{n} x_{ij}\lambda_j + \theta x_{i0} \geqq 0 \quad (i = 1, 2, \ldots, m), \\
&&& \sum_{j=1}^{n} \lambda_j = 1, \\
&&& \lambda_j \geqq 0 \quad (j = 1, 2, \ldots, n), \\
&&& \theta \text{ free.}
\end{aligned}
\tag{5.14}
$$

According to the duality theory of linear programming, the free variable μ_0 in the multiplier problem (5.13) is the dual variable associated with the constraint $\sum_{j=1}^{n} \lambda_j = 1$, which does not appear in the CCR model.

From the economic interpretation point of view, the BCC model assumes the convex combination of the observed DMUs as the production possibility set, and the BCC score is called *local pure technical efficiency* (PTE). The constant returns-to-scale assumption (without the convexity condition $\sum_{j=1}^{n} \lambda_j = 1$) implies that the radial expansion and reduction of all observed DMUs, and their nonnegative combinations are possible and the CCR score is called *global technical efficiency* (TE). Therefore, comparisons of the CCR and BCC scores provide deeper insight into the sources of inefficiency that a DMU might have.

Returning to our example in Figure 5.7, DMU_1 is operating locally efficiently but not globally efficiently due to its scale size. Thereupon it is an interesting subject to decompose the inefficiency of a DMU into its component parts. Based on the CCR and BCC scores, *scale efficiency* is defined by the ratio of the following two scores.

Definition 5.2 (scale efficiency). Let θ^0_{CCR} and θ^0_{BCC} denote the CCR and BCC scores of a DMU. The scale efficiency is defined by

$$\text{SE} = \frac{\theta^0_{\text{CCR}}}{\theta^0_{\text{BCC}}} = \frac{\text{TE}}{\text{PTE}}. \tag{5.15}$$

Using the relationship (5.15), the (global) technical efficiency (TE) of a DMU is decomposed as

$$\text{TE} = \text{PTE} \times \text{SE}\,.$$

The global or overall inefficiency of a DMU is explained by inefficient operation (PTE) or by the scale effect (SE) or by both.

For the BCC-efficient DMU_1 in Figure 5.7, its scale efficiency is given by

$$\text{SE}(\text{DMU}_1) = \frac{\theta^0_{\text{CCR}}(\text{DMU}_1)}{\theta^0_{\text{BCC}}(\text{DMU}_1)} = \frac{LM}{LP_1} = 0.75,$$

which implies that the overall inefficiency (TE) is caused by the scale inefficiency. The scale efficiency for DMU_6 is

$$\begin{aligned}\text{SE}(\text{DMU}_6) &= \frac{\theta^0_{\text{CCR}}(\text{DMU}_6)}{\theta^0_{\text{BCC}}(\text{DMU}_6)} = \frac{\frac{QR}{QP_6}}{\frac{QS}{QP_6}} \\ &= \frac{QR}{QP_6} \cdot \frac{QP_6}{QS} = \frac{QR}{QS} = \frac{3.75}{4} = 0.9375.\end{aligned}$$

The global technical inefficiency of DMU_6 can be decomposed as

$$\text{TE}(\text{DMU}_6) = \text{PTE}(\text{DMU}_6) \times \text{SE}(\text{DMU}_6) = 0.6666 \times 0.9375 = 0.62499.$$

Thus the overall inefficiency of DMU_6 is primarily caused by its inefficient operation but also caused—but to a lesser extent—by the disadvantageous conditions under which DMU_6 is operating.

For computational purposes—similar to the CCR case—the lower bound ϵ for the multiplier μ_r $(r = 1, 2, \ldots, s)$ and ν_i $(i = 1, 2, \ldots, m)$ in the linear program (5.13) are introduced, and the corresponding envelopment problem (5.14) is solved using a two-phase procedure. In the first phase, we minimize θ, and in the second phase, we maximize the sum of the slack variables, keeping $\theta = \theta^0$ (the optimal objective function value). The BCC efficiency is then given by the following.

Definition 5.3 (BCC efficiency). A DMU is BCC efficient if and only if the following two conditions are fulfilled:

(a) $\theta^0 = 1$;
(b) $s_i^{-0} = s_r^{+0} = 0$ for all i and r.

An interesting property of BCC efficiency is provided by the next theorem [11, p. 90].

Theorem 5.4. *A DMU that has a minimum input value for any input item, or a maximum output value for any output item, is BCC efficient.*

The multiplier problem of the output-oriented BCC model has the form

$$\begin{aligned}
\text{minimize} \quad & f_0(\nu, \nu_0) = \sum_{i=1}^{m} x_{i0}\nu_i + \nu_0 \\
\text{subject to} \quad & -\sum_{r=1}^{s} y_{rj}\mu_r + \sum_{i=1}^{m} x_{ij}\nu_i + \nu_0 \geqq 0 \quad (j = 1, 2, \ldots, n), \\
& \sum_{r=} y_{r0}\mu_r = 1, \\
& \mu_r \geqq \epsilon \quad (r = 1, 2, \ldots, s), \\
& \nu_i \geqq \epsilon \quad (i = 1, 2, \ldots, m), \\
& \nu_0 \text{ free in sign.}
\end{aligned} \tag{5.16}$$

The dual (envelopment) problem associated with the linear program (5.16) is expressed as

$$\begin{aligned}
\underset{\varphi,\boldsymbol{\lambda},\mathbf{s}^+,\mathbf{s}^-}{\text{maximize}} \quad & q_0(\varphi, \mathbf{s}^+, \mathbf{s}^-) = \varphi + \epsilon \left(\sum_{r=1}^{s} s_r^+ + \sum_{i=1}^{m} s_i^- \right) \\
\text{subject to} \quad & \varphi y_{r0} - \sum_{j=1}^{n} y_{rj}\lambda_j + s_r^+ = 0 \quad (r = 1, 2, \ldots, s), \\
& \sum_{i=1}^{m} x_{ij}\lambda_j + s_i^- = x_{i0} \quad (i = 1, 2, \ldots, m), \\
& \sum_{j=1}^{n} \lambda_j = 1, \\
& \lambda_j \geqq 0 \quad (j = 1, 2, \ldots, n), \\
& s_r^+ \geqq 0 \quad (r = 1, 2, \ldots, s), \\
& s_i^- \geqq 0 \quad (i = 1, 2, \ldots, m).
\end{aligned} \tag{5.17}$$

Table 5.7. Results for the input-oriented BCC model with the data set from Table 5.1.

DMU	θ^0	s^{+0}	s^{-0}	λ^0	μ^0	ν^0	μ_0^0
1	1	0	0	$\lambda_1 = 1$	0.25	0.5	0.5
2	1	0	0	$\lambda_2 = 1$	0.25	0.3333	0
3	1	0	0	$\lambda_3 = 1$	0.2	0.2	−0.2
4	1	0	0	$\lambda_4 = 1$	0.2857	0.1428	−1
5	0.666	1	0	$\lambda_1 = 1$	0	0.3333	0.6666
6	0.666	0	0	$\lambda_2 = \lambda_3 = 0.5$	0.1666	0.1666	−0.1666
7	0.875	0	0	$\lambda_4 = 1$	0.25	0.125	−0.875

Table 5.8. Results for the output-oriented BCC model with the data set from Table 5.1.

DMU	φ^0	s^{+0}	s^{-0}	λ^0	μ^0	ν^0	ν_0^0
1	1	0	0	$\lambda_1 = 1$	0.5	1	1
2	1	0	0	$\lambda_2 = 1$	0.25	0.3333	0
3	1	0	0	$\lambda_3 = 1$	0.1666	0.1666	−0.1666
4	1	0	0	$\lambda_4 = 1$	0.1428	0.0714	−0.5
5	4	0	0	$\lambda_2 = 1$	1	1.333	0
6	1.3	0	0	$\lambda_3 = \lambda_4 = 0.5$	0.2	0.1	−0.7
7	1	0	1	$\lambda_4 = 1$	0.1428	0	−1

To illustrate and to compare with the CCR models, we return to the example in Table 5.1. The results for the input-oriented BCC model are summarized in Table 5.7 and for the output-oriented BCC model in Table 5.8.

As explained above, the comparison of the CCR and BCC scores allows us to decompose the inefficiency of a DMU into the pure technical efficiency and the scale efficiency. In our example, the global technical inefficiency of DMU_1, DMU_3, and DMU_4 is due to their scale inefficiency.

The scale efficiency of DMU_5 is given by

$$\mathrm{SE}(\mathrm{DMU}_5) = \frac{\theta^0_{\mathrm{CCR}}(\mathrm{DMU}_5)}{\theta^0_{\mathrm{BCC}}(\mathrm{DMU}_5)} = \frac{0.25}{0.666} = 0.375.$$

Thus the overall inefficiency of DMU_5 is more attributable to SE(0.375) than to PTE(0.666). As already shown, the global inefficiency of DMU_6 is primarily caused by its PTE (0.666); the scale efficiency is relatively high (0.9375).

The overall inefficiency of DMU_7 (0.656) is explained by its PTE (0.875) and SE (0.75).

We conclude this section by formulating the additive model under the assumption of variable returns to scale. As in the previous input- and output-oriented models, we need only to add the convexity condition $\sum_{j=1}^{n} \lambda_j = 1$ to the constraints of the additive CRS model (5.10). The additive VRS envelopment model can then be written as follows:

$$
\begin{aligned}
\underset{\boldsymbol{\lambda},\mathbf{s}^+,\mathbf{s}^-}{\text{minimize}} \quad & g_0(\boldsymbol{\lambda}, \mathbf{s}^+, \mathbf{s}^-) = -\sum_{r=1}^{s} s_r^+ - \sum_{i=1}^{m} s_i^- \\
\text{subject to} \quad & \sum_{j=1}^{n} y_{rj}\lambda_j - s_r^+ = y_{r0} \qquad (r = 1, 2, \ldots, s), \\
& -\sum_{j=1}^{n} x_{ij}\lambda_j - s_i^- = -x_{i0} \qquad (i = 1, 2, \ldots, m), \\
& \sum_{j=1}^{n} \lambda_j = 1, \\
& \lambda_j \geqq 0 \qquad (j = 1, 2, \ldots, n), \\
& s_r^+ \geqq 0 \qquad (r = 1, 2, \ldots, s), \\
& s_i^- \geqq 0 \qquad (i = 1, 2, \ldots, m).
\end{aligned}
\tag{5.18}
$$

The dual (multiplier) problem associated with the linear program (5.18) is expressed as

$$
\begin{aligned}
\underset{\boldsymbol{\mu},\boldsymbol{\nu}}{\text{maximize}} \quad & w_0(\boldsymbol{\mu}, \boldsymbol{\nu}, \mu_0) = \sum_{r=1}^{s} y_{r0}\mu_r - \sum_{i=1}^{m} x_{i0}\nu_i + \mu_0 \\
\text{subject to} \quad & \sum_{r=1}^{s} y_{rj}\mu_r - \sum_{i=1}^{m} x_{ij}\nu_i + \mu_0 \leqq 0 \qquad (j = 1, 2, \ldots, n), \\
& -\mu_r \leqq -1 \qquad (r = 1, 2, \ldots, s), \\
& -\nu_i \leqq -1 \qquad (i = 1, 2, \ldots, m), \\
& \mu_0 \text{ free.}
\end{aligned}
\tag{5.19}
$$

As in the CRS additive model, DMU_0 is efficient if and only if $\mathbf{s}^{+^0} = \mathbf{0}$ and $\mathbf{s}^{-^0} = \mathbf{0}$. As an illustration, the results of the additive model under the VRS assumption for the data set given in Table 5.1 are summarized in Table 5.9.

Table 5.9. Results for the additive VRS model with the data set from Table 5.1.

DMU	g_0^0	s^+	s^-	$\boldsymbol{\lambda}^0$	w_0^0	μ^0	ν^0	μ_0^0
1	0	0	0	$\lambda_1 = 1$	0	1	2	2
2	0	0	0	$\lambda_2 = 1$	0	1	1	−1
3	0	0	0	$\lambda_3 = 1$	0	1	1	−1
4	0	0	0	$\lambda_4 = 1$	0	2	1	−7
5	−3	3	0	$\lambda_2 = 1$	−3	1	1	−1
6	−2	0	2	$\lambda_2 = \lambda_3 = \frac{1}{2}$	−2	1	1	−1
7	−1	0	1	$\lambda_4 = 1$	−1	2	1	−7

A comparison of the results in Table 5.9 with those for oriented BCC models (see Tables 5.7 and 5.8) shows that the set of efficient units is the same. The following theorem [3] is valid.

Theorem 5.5. DMU_0 *is additive-efficient if and only if it is BCC efficient.*

The additive model does not provide a *scalar* efficiency score like θ^0 in the input-oriented model. The objective function in (5.18) reflects inefficiencies present in *both* inputs and outputs, whereas θ^0 reflects purely technical efficiency. Therefore, in the second phase of the optimization procedure, the slacks are maximized in order to identify further inefficiences. Any positive slack variable necessarily implies a change in the input or output proportions, or in other words, in the mix. In this way, the CCR and BCC models distinguish between purely technical and mix inefficiencies.

Another property of the additive models is known as "translation invariance." In many applications, the assumption of semipositivity for some inputs or outputs may be not fulfilled. Then the data set (X, Y) can be translated by introducing arbitrary constants α_i $(i = 1, 2, \ldots, m)$ and β_r $(r = 1, 2, \ldots, s)$ to obtain new data

$$\bar{x}_{ij} = x_{ij} + \alpha_i \quad (i = 1, 2, \ldots, m;\; j = 1, 2, \ldots, n),$$
$$\bar{y}_{rj} = y_{rj} + \beta_r \quad (r = 1, 2, \ldots, s;\; j = 1, 2, \ldots, n).$$

Will the optimal solution for the envelopment problem change because of the data translation? Cooper, Seiford, and Tone [11, p. 94] define translation invariance as follows:

Definition 5.4. Given any problem, a DEA model is said to be translation invariant if translating the original input and/or output values results in a new problem that has the same optimal solution for the envelopment form as the old one.

Turning to Figure 5.7, we can see that for the input-oriented BCC model, the efficiency of DMU_6 described by the ratio $\frac{QS}{QP_6}$ remains the same even if we shift the output value by changing the origin, e.g., by two units. Thus the input-oriented BCC model is translation invariant with respect to outputs (but not inputs). Similarly, the output-oriented BCC model is translation invariant with respect to inputs (but not outputs). For the additive model, it can be shown that the efficiency evaluation is not affected if the origin of the coordinate system is shifted.

Theorem 5.6 (see [2]). *The additive model* (5.18) *is translation invariant.*

It should be noted that the convexity condition $\sum_{j=1}^{n} \lambda_j = 1$ is crucial for the proof of this theorem.

As mentioned above the additive DEA model lacks a one-dimensional efficiency measure like θ^0 or φ^0. Because the slacks are expressed in different units, the value of the objective function g_0 in (5.18) cannot be used as an efficiency score.

In order to overcome such disadvantage of the additive model, Cooper, Seiford, and Tone [11] proposed its modification by introducing a measure in the form of a single scalar (with values between zero and one) called a "slack-based measure" (SBM), which has the following properties:

1. The measure is invariant with respect to the unit of measurement of each input and output item (units invariant).
2. The measure is monotone decreasing in each input and output slack.

SBM accounts for all inefficiencies and can be expressed as a product of input and output inefficiencies. For details, see [11, Section 4.4].

5.3 Production Technologies and Efficiency Measurement

As mentioned in the introduction, DEA provides a new (nonparametric) way for a description of a production possibility set and, in particular, for estimation of an efficient subset or efficiency envelope based on observed input and output data. The various parametric production functions (e.g., the Cobb–Douglas, constant elasticity of substitution (CES), and translog functions) that have been widely used in empirical work are restricted to the single-output situation. The advantage of an alternative description of the technology as a list of all of the technologically feasible combinations of inputs and outputs is that it encompasses the situation of multiple outputs. Shephard [36] introduced the concept of a distance function that can model multioutput multiinput technologies and at the same time represents them with convenient functional forms. In addition, this concept can be related to the important pioneering work of Farrel [18] in measuring efficiency directly from observational data, at least in the single-output case.

Our objective now is to relate DEA models developed in the previous section to the theory of production and, in particular, to the input and output distance functions. There are some papers and books in the literature dealing with these deeper economic aspects of DEA, including Debreu–Farrel efficiency, Pareto–Koopmans efficiency, and more general technical efficiency axiomatic approaches [14, 17, 6, 10, 32, 22, 31].

In what follows, we follow Färe and Primont [22, Chapter 2], with the notation of the variables as introduced in Section 5.1, and relate the oriented DEA models to Shephard's concept of a distance function.

We denoted by $\mathbf{x} = (x_1, x_2, \ldots, x_m)$ the vector of m observed inputs and by $\mathbf{y} = (y_1, y_2, \ldots, y_s)$ the vector of s observed outputs.

The technology set as the set of all feasible input–output vectors is represented by

$$T = \{(\mathbf{x}, \mathbf{y}) | \mathbf{x} \in R^m_+,\ \mathbf{y} \in R^s_+,\ \mathbf{y} \text{ can be produced from } \mathbf{x}\}.$$

For a given T and for a single output, the production function

$$F : R^m_+ \to R_+$$

is defined by

$$F(\mathbf{x}) = \max_y \{y | (\mathbf{x}, y) \in T\}. \tag{5.20}$$

It represents the maximum output that can be produced for any specified input vector.

Starting with a production function F, the technology set is then defined by

$$T^* = \{(\mathbf{x}, y) | F(\mathbf{x}) \geqq y,\ y \in R_+\}. \tag{5.21}$$

If F is defined from T by (5.20) and if T^* is defined from F using (5.21), then $T^* = T$. The output distance function for the single-output case [37, 22] is defined by

$$D_O(\mathbf{x}, y) = \min_{\varphi} \left\{ \varphi | F(\mathbf{x}) \geqq \frac{y}{\varphi} \right\}.$$

It follows that

$$D_O(\mathbf{x}, y) = \frac{y}{F(\mathbf{x})}.$$

The output distance function defined from the technology set is

$$D_O(\mathbf{x}, y) = \min_{\varphi} \left\{ \varphi > 0 \middle| \left(\mathbf{x}, \frac{y}{\varphi} \right) \in T \right\}. \tag{5.22}$$

The advantage of definition (5.22) is that it remains valid even for a multioutput case.

Following Shephard [37] and Färe and Primont [22], we define the output possibility set (the set of feasible outputs) $P(\mathbf{x})$, for each $\mathbf{x}$, as

$$P(\mathbf{x}) = \{\mathbf{y} | (\mathbf{x}, \mathbf{y}) \in T\}. \tag{5.23}$$

Then an alternative and equivalent definition of the output distance function in terms of the output sets is given by

$$D_O(\mathbf{x}, \mathbf{y}) = \min_{\varphi} \left\{ \varphi > 0 \middle| \left(\frac{\mathbf{y}}{\varphi} \right) \in P(\mathbf{x}) \right\} \quad \text{for all } \mathbf{x} \in R^m_+. \tag{5.24}$$

The illustration of definition (5.24) for two outputs is provided in Figure 5.8.

A given input vector $\mathbf{x}^*$ determines the output possibility set, $P(\mathbf{x}^*)$. An output vector $\mathbf{y}^*$ is arbitrarily chosen. The value $D_O(\mathbf{x}^*, \mathbf{y}^*)$ puts $\frac{\mathbf{y}^*}{D_O(\mathbf{x}^*, \mathbf{y}^*)}$ on the boundary of $P(\mathbf{x}^*)$ and on the ray through $\mathbf{y}^*$. In Figure 5.8, $\mathbf{y}^*$ is an interior point of $P(\mathbf{x}^*)$, and thus $D_O(\mathbf{x}^*, \mathbf{y}^*) < 1$.

The following two properties for the output possibility set, $P(\mathbf{x})$, are introduced.

Postulate 5.1. $\mathbf{0}_s \in P(\mathbf{x})$ *for all* $\mathbf{x} \in R^m_+$.

Postulate 5.2. *For all* $(\mathbf{x}, \mathbf{y})$ *in* R^{m+s}_+, *if* $\mathbf{y} \in P(\mathbf{x})$ *and* $0 < \varphi \leqq 1$, *then* $\varphi \mathbf{y} \in P(\mathbf{x})$.

The first assumption implies that given any input vector, it is always possible to produce nothing ($\mathbf{y} = \mathbf{0}_s$). The second postulate is interpreted as the weak disposability of outputs. This means that if $\mathbf{x}$ can produce $\mathbf{y}$, then $\mathbf{x}$ can produce any proportional reduction of $\mathbf{y}$.

It follows from Figure 5.8 that $\mathbf{y} \in P(\mathbf{x})$ if and only if $D_O(\mathbf{x}, \mathbf{y}) \leqq 1$.

The interesting implication of the above result is that the technology specified by the output possibility set can be given an equivalent specification in terms of the output distance function.

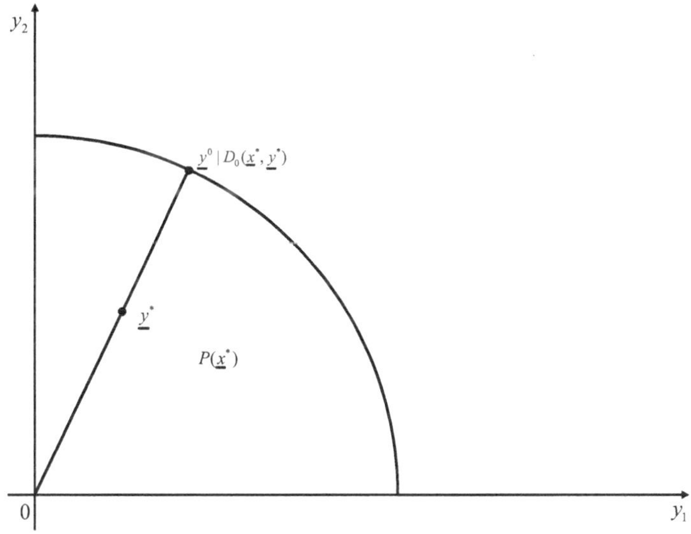

Fig. 5.8. The output set and the output distance function.

For the analysis of input changes, Shephard [37] introduced the input distance function. First, he defined the input possibility set $L(\mathbf{y})$, for each $\mathbf{y}$, as

$$L(\mathbf{y}) = \{\mathbf{x} | (\mathbf{x}, \mathbf{y}) \in T\}, \tag{5.25}$$

where T is the technology set. It can be seen from a comparison of (5.23) and (5.25) that output sets and input sets are "inverse" in the following sense:

$$\mathbf{y} \in P(\mathbf{x}) \quad \text{if and only if } \mathbf{x} \in L(\mathbf{y}).$$

Analogously to the definition of the output distance function (5.24), the input distance function in terms of the input sets is defined as

$$D_i(\mathbf{y}, \mathbf{x}) = \max_{\theta} \left\{ \theta > 0 \middle| \left(\frac{\mathbf{x}}{\theta} \right) \in L(\mathbf{y}) \text{ for all } \mathbf{y} \in R_+^s \right\}.$$

The illustration of the input distance function for two inputs is provided in Figure 5.9.

A given output vector $\mathbf{y}^*$ determines the input set, $L(\mathbf{y}^*)$. An input vector, $\mathbf{x}^*$, is arbitrarily chosen. The value of $D_i(\mathbf{y}^*, \mathbf{x}^*)$ puts $\frac{\mathbf{x}^*}{D_i(\mathbf{y}^*, \mathbf{x}^*)}$ on the boundary of $L(\mathbf{y}^*)$ and on the ray through $\mathbf{x}^*$. In Figure 5.9, $\mathbf{x}^*$ lies in the interior of the input set $L(\mathbf{y}^*)$, and thus $D_i(\mathbf{y}^*, \mathbf{x}^*) > 1$, for if $\mathbf{x}^*$ had been outside $L(\mathbf{y}^*)$, the value of the input distance function would have been less than one.

In order to characterize the input possibility set by the input distance function, we assume that inputs are weakly disposable, i.e.,

$$\text{if } \mathbf{x} \in L(\mathbf{y}) \quad \text{and} \quad \theta \geqq 1, \quad \text{then } \theta\mathbf{x} \in L(\mathbf{y}). \tag{5.26}$$

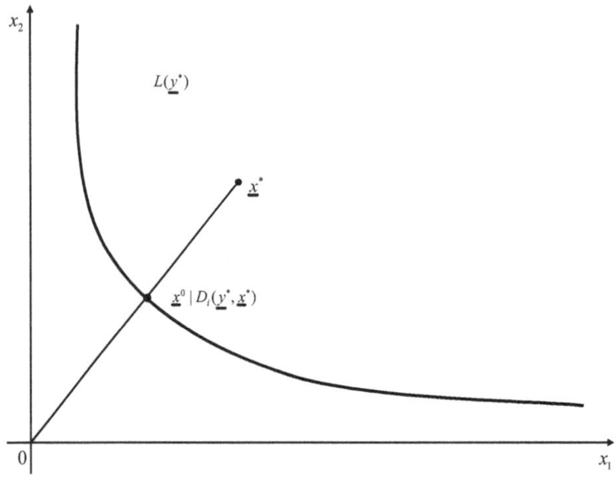

Fig. 5.9. The input set and the input distance function.

The economic interpretation is straightforward: If **y** can be produced from **x**, then any proportional increase in **x** can produce **y**.

Using property (5.26), Färe and Primont [22, p. 22] proved the following.

Proposition 5.1. *Inputs are weakly disposable if and only if*

$$L(\mathbf{y}) = \{\mathbf{x} | D_i(\mathbf{y}, \mathbf{x}) \geqq 1\}.$$

We can summarize as follows: Weak disposability of outputs is necessary and sufficient for the validity of

$$\mathbf{y} \in P(\mathbf{x}) \quad \text{if and only if } D_O(\mathbf{x}, \mathbf{y}) \leqq 1,$$

and weak disposability of inputs is necessary and sufficient for the validity of

$$\mathbf{x} \in L(\mathbf{y}) \quad \text{if and only if } D_i(\mathbf{y}, \mathbf{x}) \geqq 1.$$

Moreover, if both inputs and outputs are weakly disposable, the following inverse relation between the two distance functions holds [22, p. 23]:

$$\begin{aligned} D_O(\mathbf{x}, \mathbf{y}) &= \min_{\varphi} \left\{ \varphi \middle| D_i \left(\frac{\mathbf{y}}{\varphi}, \mathbf{x} \right) \geqq 1 \right\}, \\ D_i(\mathbf{y}, \mathbf{x}) &= \max_{\theta} \left\{ \theta \middle| D_O \left(\frac{\mathbf{x}}{\theta}, \mathbf{y} \right) \leqq 1 \right\}. \end{aligned} \tag{5.27}$$

An interesting result can be derived if the inverse relation (5.27) is applied to the technology with constant returns to scale (CRS). A formal definition of CRS is the following.

Definition 5.5. A technology described by T exhibits constant returns to scale globally if $T = kT$ for all $k > 0$.

For global CRS, a technology set T is a cone with vertex at $(\mathbf{0}_m, \mathbf{0}_s)$. In terms of output and input sets, the condition that $P(\theta\mathbf{x}) = \theta P(\mathbf{x})$ for all $\theta > 0$ is equivalent to the condition that $L(\varphi\mathbf{y}) = \varphi L(\mathbf{y})$ for all $\varphi > 0$, which is equivalent to global CRS. The interesting result is contained in the next proposition, proved by Färe and Primont [22, pp. 24–25].

Proposition 5.2. *The technology exhibits global constant returns to scale if and only if* $D_O(\mathbf{x}, \mathbf{y}) = \frac{1}{D_i(\mathbf{y}, \mathbf{x})}$ *for all* $(\mathbf{x}, \mathbf{y}) \in R_+^{m+s}$.

In order to relate the input- and output-oriented DEA measures of technical efficiency to the input and output distance functions, we introduce the following definitions of the input and output efficiency, respectively [22, pp. 28–29].

Definition 5.6. A feasible production plan $(\mathbf{x}, \mathbf{y})$ is input efficient if any reduction in one or more of the inputs will render $\mathbf{y}$ an infeasible output vector. Thus the input-efficient subset of $L(\mathbf{y})$ is defined by

$$\text{Eff } L(\mathbf{y}) = \{\mathbf{x} | \mathbf{x} \in L(\mathbf{y}),\ \mathbf{x}' \leqq \mathbf{x} \text{ and } \mathbf{x}' \neq \mathbf{x} \Rightarrow \mathbf{x}' \notin L(\mathbf{y})\}, \quad \mathbf{y} \geq \mathbf{0}_s.$$

Definition 5.7. A feasible production plan $(\mathbf{x}, \mathbf{y})$ is input-isoquant efficient if any proportional reduction of the inputs will render $\mathbf{y}$ an infeasible output vector. Thus the input-isoquant-efficient subset of $L(\mathbf{y})$ is defined by

$$\text{Isoq } L(\mathbf{y}) = \{\mathbf{x} | \mathbf{x} \in L(\mathbf{y}),\ \theta < 1 \Rightarrow \theta\mathbf{x} \notin L(\mathbf{y})\}, \quad \mathbf{y} \geq \mathbf{0}_s.$$

The notion of input-isoquant efficiency is weaker in the sense that an input-efficient production plan is input-isoquant efficient, but not conversely. The reader can verify that these notions correspond to the notion of efficiency and strong efficiency, respectively, in DEA models.

Similarly, on the output side we introduce the following.

Definition 5.8. A feasible production plan $(\mathbf{x}, \mathbf{y})$ is output efficient if $\mathbf{y}$ belongs to

$$\text{Eff } P(\mathbf{x}) = \{\mathbf{y} | \mathbf{y} \in P(\mathbf{x}),\ \mathbf{y}' \geqq \mathbf{y} \text{ and } \mathbf{y}' \neq \mathbf{y} \Rightarrow \mathbf{y}' \notin P(\mathbf{x})\}, \quad \mathbf{x} \geqq \mathbf{0}_m.$$

For a given input vector $\mathbf{x}$, there does not exist an output vector $\mathbf{y}' \in P(\mathbf{x})$ such that the output at least of one good is higher than the output described by $\mathbf{y}$.

Definition 5.9. A feasible production plan $(\mathbf{x}, \mathbf{y})$ is output-isoquant efficient if $\mathbf{y}$ belongs to

$$\text{Isoq } P(\mathbf{x}) = \{\mathbf{y} | \mathbf{y} \in P(\mathbf{x}),\ \varphi > 1 \Rightarrow \varphi\mathbf{y} \notin P(\mathbf{x})\}, \quad \mathbf{x} \geqq \mathbf{0}_m.$$

The set Isoq $P(\mathbf{x})$ is related to the set of strong efficient points in a DEA model.

Now the Debreu [14] and Farrel [18] input-oriented measures of technical efficiency can be calculated by

$$\left(\frac{1}{D_i(\mathbf{y},\mathbf{x})}\right) = \min_{\theta}\{\theta \,|\, \theta\mathbf{x} \in L(\mathbf{y})\}. \tag{5.28}$$

We get an efficiency measure that lies between zero and one (due to $D_i(\mathbf{y},\mathbf{x}) \geqq 1$) and whose higher value implies higher input efficiency of a production plan, $(\mathbf{x},\mathbf{y})$. Using a set of observed inputs and outputs $(\mathbf{x}_j,\mathbf{y}_j)$ for different DMUs $(j = 1, 2, \ldots, n)$, the Debreu–Farrel efficiency score can be obtained by solving a linear programming problem.

We assume—in addition to our assumption in Section 5.1 (each DMU has at least one positive input and one positive output value)—the following:

(i)
$$\sum_{j=1}^{n} x_{ij} > 0 \quad (i = 1, 2, \ldots, m).$$
A positive amount of each input is used by at least one DMU.

(ii)
$$\sum_{j=1}^{n} y_{rj} > 0 \quad (r = 1, 2, \ldots, s).$$
A positive amount of each output is produced by at least one DMU.

(iii)
$$(\mathbf{x}_j,\mathbf{y}_j) \in T \quad \text{and} \quad (\lambda_j\mathbf{x}_j,\lambda_j\mathbf{y}_j) \in T,$$
for all $\lambda_j \geqq 0$, where λ_j denotes the intensity level of activity of DMU j. We specify a technology T that exhibits global constant returns to scale.

(iv) If $(\lambda_j\mathbf{x}_j,\lambda_j\mathbf{y}_j) \in T$ $(j = 1, 2, \ldots, n)$, then
$$\left(\sum_{j=1}^{n}\lambda_j\mathbf{x}_j, \sum_{j=1}^{n}\lambda_j\mathbf{y}_j\right) \in T.$$
We postulate the additivity property of the activities.

(v) If $\mathbf{y} \in P(\mathbf{x})$ and $\mathbf{y}' \leqq \mathbf{y}$, then $\mathbf{y}' \in P(\mathbf{x})$; i.e., if $(\mathbf{x},\mathbf{y}) \in T$ and $\mathbf{y}' \leqq \mathbf{y}$, then $(\mathbf{x},\mathbf{y}') \in T$. We assume strong disposability of outputs.

(vi) Next, we postulate strong disposability of inputs: If $\mathbf{x} \in L(\mathbf{y})$ and $\mathbf{x}' \geqq \mathbf{x}$, then $\mathbf{x}' \in L(\mathbf{y})$; i.e., if $(\mathbf{x},\mathbf{y}) \in T$ and $\mathbf{x}' \geqq \mathbf{x}$, then $(\mathbf{x}',\mathbf{y}) \in T$.

These last two assumptions are often denoted as an inefficiency postulate. They imply that if $(\sum_{j=1}^{n}\lambda_j\mathbf{x}_j, \sum_{j=1}^{n}\lambda_j\mathbf{y}_j) \in T$, $\mathbf{x} \geqq \sum_{j=1}^{n}\lambda_j\mathbf{x}_j$, and $\mathbf{y} \leqq \sum_{j=1}^{n}\lambda_j\mathbf{y}_j$, then $(\mathbf{x},\mathbf{y}) \in T$.

Then the production possibility set, as the smallest convex cone that contains all of the data points, is

$$T = \left\{(\mathbf{x},\mathbf{y}) \,\middle|\, \sum_{j=1}^{n}\lambda_j\mathbf{y}_j \geqq \mathbf{y} \geqq \mathbf{0}_s,\ \sum_{j=1}^{n}\lambda_j\mathbf{x}_j \leqq \mathbf{x},\ \lambda_j \geqq 0,\ j = 1, 2, \ldots, n\right\}.$$

The input possibility set that corresponds to T is described by

$$L(\mathbf{y}) = \left\{ \mathbf{x} \,\middle|\, \sum_{j=1}^{n} \lambda_j y_{rj} \geqq y_r \quad (r = 1, 2, \ldots, s), \right.$$
$$\sum_{j=1}^{n} \lambda_j x_{ij} \leqq x_i \quad (i = 1, 2, \ldots, m), \tag{5.29}$$
$$\left. \lambda_j \geqq 0 \quad (j = 1, 2, \ldots, n) \right\}.$$

The specification of the input possibility set (5.29) can be applied to any observation (or any DMU). In order to estimate the input-oriented measure of technical efficiency for one specific DMU, denoted by subscript 0, we solve the following linear programming problem:

$$\left[\frac{1}{D_i(\mathbf{y}_0, \mathbf{x}_0)}\right] = \min_{\theta, \boldsymbol{\lambda}} \{\theta | \theta \mathbf{x}_0 \in L(\mathbf{y}_0)\}$$
$$= \min_{\theta, \boldsymbol{\lambda}} \left\{ \theta \,\middle|\, \sum_{j=1}^{n} \lambda_j y_{rj} \geqq y_{r0} \quad (r = 1, 2, \ldots, s), \right.$$
$$\sum_{j=1}^{n} \lambda_j x_{ij} \leqq \theta x_{i0} \quad (i = 1, 2, \ldots, m), \tag{5.30}$$
$$\left. \lambda_j \geqq 0 \quad (j = 1, 2, \ldots, n) \right\}.$$

We obtained in the CCR input-oriented model (5.30)—without the input and output slacks—from the previous section. The efficiency score θ^* indicates that the observed input vector, $\mathbf{x}_0$, of DMU$_0$ could (at most) be proportionally reduced to $\theta^* \mathbf{x}_0$ while still producing $\mathbf{y}_0$; i.e., $\theta^* \mathbf{x}_0 \in L(\mathbf{y}_0)$ and $\theta \mathbf{x}_0 \notin L(\mathbf{y}_0)$ for $\theta < \theta^*$.

If the same output can be produced with proportionally less input, then, alternatively, proportionally more output can be produced with the same input:

$$\left[\frac{1}{D_0(\mathbf{x}_0, \mathbf{y}_0)}\right] = \max_{\varphi, \boldsymbol{\lambda}} \{\varphi | \varphi \mathbf{y}_0 \in P(\mathbf{x}_0)\}$$
$$= \max_{\varphi, \boldsymbol{\lambda}} \left\{ \varphi \,\middle|\, \sum_{j=1}^{n} \lambda_j y_{rj} \geqq \varphi y_{r0} \quad (r = 1, 2, \ldots, s), \right.$$
$$\sum_{i=1}^{m} \lambda_j x_{ij} \leqq x_{i0} \quad (i = 1, 2, \ldots, m), \tag{5.31}$$
$$\left. \lambda_j \geqq 0 \quad (j = 1, 2, \ldots, n) \right\}.$$

This alternative view of efficiency leads to the output-oriented CCR model (5.6)—excluding the input and output slacks—from Section 5.2.2.

However, if we relax the assumption of constant returns to scale, then the input- and output-oriented models will provide different information. Supposing that the technology exhibits variable returns to scale (VRS), we postulate the convexity property for the production possibility set, T.

Postulate 5.3. *If* $(\mathbf{x}_j, \mathbf{y}_j) \in T$, $j = 1, 2, \ldots, n$, *and* $\boldsymbol{\lambda}_j \geqq 0$ *are nonnegative scalars such that* $\sum_{j=1}^{n} \lambda_j = 1$, *then* $(\sum_{j=1}^{n} \lambda_j \mathbf{x}_j, \sum_{j=1}^{n} \lambda_j \mathbf{y}_j) \in T$.

A piecewise linear technology that satisfies VRS is given by

$$L(\mathbf{y}) = \left\{ \mathbf{x} \,\middle|\, \begin{array}{ll} \displaystyle\sum_{j=1}^{n} \lambda_j y_{rj} \geqq y_r & (r = 1, 2, \ldots, s), \\ \displaystyle\sum_{j=1}^{n} \lambda_j x_{ij} \leqq x_i & (i = 1, 2, \ldots, m), \\ \lambda_j \geqq 0 & (j = 1, 2, \ldots, n), \\ \displaystyle\sum_{j=1}^{n} \lambda_j = 1 & \end{array} \right\}. \tag{5.32}$$

The VRS technology in (5.32) differs from the CRS technology in (5.29) because of the additional constraint $\sum_{j=1}^{n} \lambda_j = 1$. The input-oriented model for DMU$_0$ under VRS is then

$$\left[\frac{1}{D_i(\mathbf{y}_0, \mathbf{x}_0)}\right] = \min_{\theta, \boldsymbol{\lambda}} \left\{ \theta \,\middle|\, \begin{array}{ll} \displaystyle\sum_{j=1}^{n} \lambda_j y_{rj} \geqq y_{r0} & (r = 1, 2, \ldots, s), \\ \displaystyle\sum_{j=1}^{n} \lambda_j x_{ij} \leqq \theta x_{i0} & (i = 1, 2, \ldots, m), \\ \lambda_j \geqq 0 & (j = 1, 2, \ldots, n), \\ \displaystyle\sum_{j=1}^{n} \lambda_j = 1 & \end{array} \right\}.$$

In the same way, the output-oriented model satisfying VRS can be formulated.

For technology exhibiting nonincreasing returns to scale (NIRS) globally, the restriction $\sum_{j=1}^{n} \lambda_j = 1$ is replaced by $\sum_{j=1}^{n} \lambda_j \leqq 1$, and for nondecreasing returns to scale (NDRS) by $\sum_{j=1}^{n} \lambda_j \geqq 1$.

In economics, the concept of efficiency is intimately related to the idea of Pareto optimality: An input–output bundle is not Pareto optimal if there remains the possibility of any net increase in outputs or net reduction in inputs. The drawback of the radial models presented above is the presence of input and/or output slacks at the optimal solution of these models. A DMU cannot be found to be Pareto efficient as long as there is any slack in any input or output. Charnes et al. [10] showed how DEA can be related to the Pareto notion of efficiency, and in this way the more general additive DEA model can be derived.

Using the notation $\mathbf{x}_j$, $\mathbf{y}_j$ for the observed vectors of the inputs and outputs of the jth DMU, Charnes et al. [10] defined the "empirical production set" T_E as the convex hull of these points, i.e.,

$$T_E = \left\{ (\mathbf{x}, \mathbf{y}) | \mathbf{x} = \sum_{j=1}^{n} \mathbf{x}_j \mu_j,\ \mathbf{y} = \sum_{j=1}^{n} \mathbf{y}_j \mu_j,\ \mu_j \geqq 0,\ \sum_{j=1}^{n} \mu_j = 1 \right\}.$$

They extended it to the empirical production possibility set $\bar{T}_E$ by adding to T_E all points with inputs in T_E and outputs not greater than some output in T_E:

$$\bar{T}_E = \{(\mathbf{x}, \mathbf{y}) | \mathbf{x} = \bar{\mathbf{x}},\ \mathbf{y} \leqq \bar{\mathbf{y}} \text{ for some } (\bar{\mathbf{x}}, \bar{\mathbf{y}}) \in T_E\}.$$

Comparison with the previous studies dealing with axiom systems to characterize the production possibility set of DEA shows that the approach used by Charnes et al. [10] uses fewer axioms than every production possibility set heretofore employed.

Let T_E^r, $\bar{T}_E^r$ denote the sets corresponding to T_E and $\bar{T}_E$ when only the output y_r is considered. A frontier function $f_r(\mathbf{x})$ is then determined by

$$f_r(\mathbf{x}) = \max y_r \quad \text{for } (\mathbf{x}, \mathbf{y}) \in \bar{T}_E.$$

Proposition 5.3. *$f_r(\mathbf{x})$ is a concave, piecewise linear function on T_E.*

It is worth noting that the proof of this proposition given in Charnes et al. [10] does not require the nonnegativity assumption for input and output values.

Charnes et al. [10]—and later other authors [12]—use the designations "Pareto efficiency" and "Pareto–Koopmans efficiency" or "strong efficiency" synonymously (or together) in recognition of Koopmans's work in the adaptation of the Pareto concept of "welfare efficiency" for use in "production economics."

Definition 5.10. A Pareto–Koopmans-efficient (minimum) point for a finite set of functions $g_1(\mathbf{x}), g_2(\mathbf{x}), \ldots, g_k(\mathbf{x})$ is a point $\mathbf{x}_0$ such that there is no other point of these functions such that

$$g_k(\mathbf{x}) \leqq g_k(\mathbf{x}_0) \quad (k = 1, 2, \ldots, K) \tag{5.33}$$

with at least one strict inequality.

The question that now arises is that of how to test a point $\mathbf{x}_0$ for Pareto–Koopmans efficiency. Charnes and Cooper [7] showed that $\mathbf{x}_0$ is Pareto–Koopmans efficient if and only if $\mathbf{x}_0$ is an optimal solution to the mathematical programming problem

$$\begin{aligned} \underset{\mathbf{x}}{\text{minimize}} \quad & \sum_{k=1}^{K} g_k(\mathbf{x}) \\ \text{subject to} \quad & g_k(\mathbf{x}) \leqq g_k(\mathbf{x}_0), \quad k = 1, 2, \ldots, K. \end{aligned} \tag{5.34}$$

Using (5.34), we can determine the Pareto–Koopmans-efficient points from n empirical points. The Pareto–Koopmans empirical frontier function is then defined on the convex hull of their inputs by convex combination of the output values. For efficient production, we wish to maximize on outputs while minimizing on inputs. Therefore, the functions $g_k(\mathbf{x})$ $(k = 1, 2, \ldots, K)$ include both outputs and inputs:

$$\begin{aligned} -g_k(\mathbf{x}) &= y_k \quad (k = 1, 2, \ldots, s), \\ g_k(\mathbf{x}) &= x_k \quad (k = s + i, \ i = 1, 2, \ldots, m) \quad \text{for } (\mathbf{x}, \mathbf{y}) \in \bar{T}_E. \end{aligned}$$

The optimization problem (5.34) need only consider $(\mathbf{x}, \mathbf{y}) \in T_E$ rather than $\bar{T}_E$. Thus the constraint inequalities in (5.33) for a test point $(\mathbf{x}_0, \mathbf{y}_0)$ are

$$\mathbf{y} \geqq \mathbf{y}_0, \qquad \mathbf{x} \leqq \mathbf{x}_0, \tag{5.35}$$

which correspond to the envelopment constraints of DEA for an observed input vector $\mathbf{x}_0$ and related output vector $\mathbf{y}_0$. The following theorem can be proved [10, pp. 96–97].

Theorem 5.7. *The envelopment constraints of DEA in production analysis are the Charnes–Cooper constraints of* (5.34) *for testing Pareto–Koopmans efficiency of an empirical production unit.*

Proof. Given the input–output vector $(\mathbf{x}_0, \mathbf{y}_0)$, the Charnes–Cooper test (5.34) becomes

$$\underset{\boldsymbol{\lambda}}{\text{minimize}} \quad -\mathbf{e}'Y\boldsymbol{\lambda} + \mathbf{e}'X\boldsymbol{\lambda} \tag{5.36}$$

$$\begin{aligned} \text{subject to} \quad Y\boldsymbol{\lambda} &\geqq \mathbf{y}_0, \\ X\boldsymbol{\lambda} &\leqq \mathbf{x}_0, \\ \mathbf{e}'\boldsymbol{\lambda} &= 1, \\ \boldsymbol{\lambda} &\geqq \mathbf{0}. \end{aligned} \tag{5.37}$$

Including the slack variables $\mathbf{s}^- = \mathbf{x}_0 - X\boldsymbol{\lambda}$ and $\mathbf{s}^+ = Y\boldsymbol{\lambda} - \mathbf{y}_0$, the terms $-\mathbf{e}'(Y\boldsymbol{\lambda} - \mathbf{y}_0) + \mathbf{e}'(X\boldsymbol{\lambda} + \mathbf{x}_0)$ differ from (5.36) by only a constant, and we can rewrite the problem (5.36)–(5.37) as

$$\underset{\boldsymbol{\lambda}, \mathbf{s}^+, \mathbf{s}^-}{\text{minimize}} \quad -\mathbf{e}'\mathbf{s}^+ - \mathbf{e}'\mathbf{s}^- \tag{5.38}$$

$$\begin{aligned} \text{subject to} \quad Y\boldsymbol{\lambda} - \mathbf{s}^+ &= \mathbf{y}_0, \\ -X\boldsymbol{\lambda} - \mathbf{s}^- &= -\mathbf{x}_0, \\ \mathbf{e}'\boldsymbol{\lambda} &= 1, \\ \boldsymbol{\lambda} \geqq \mathbf{0}, \quad \mathbf{s}^+ &\geqq \mathbf{0}, \quad \mathbf{s}^- \geqq \mathbf{0}, \end{aligned} \tag{5.39}$$

which is the additive DEA model (5.18) introduced in the previous section. Here the optimal value of the objective function (5.38) is equal to zero if and only if $(\mathbf{y}_0, \mathbf{x}_0)$ is Pareto–Koopmans efficient. As shown by Charnes et al. [10], other variations of $\bar{T}_E$ can be considered easily by simple modification of or additions to the constraints on λ. □

Then the performance of DMU_0 is to be considered Pareto–Koopmans efficient if and only if the performance of other DMUs does not provide evidence that some of the inputs or outputs of DMU_0 could have been improved without worsening some of its other inputs or outputs.

5.4 Technical versus Environmental Efficiency, or How to Measure Ecoefficiency

The new concept in the current state of the public discussion on environmental policy is the concept of ecoefficiency. There is an urgent need for ecoefficient solutions such that goods and services can be produced with less energy and resources and with less waste and emission. But how do we measure "ecoefficiency" in an operational way in order to provide decision support for firms and for economic policy? It is necessary to have new performance indicators for the firm and for the national economy that take into account environmental aspects. From the 1980s onward, there arose in the economic literature a growing interest in the construction of indicators of a firm's environmental behavior (see [38] for a survey).

The main problem in developing of the ecoefficiency indicators is the lack of evaluations like market prices for the waste and emissions (or the undesirable outputs as the by-product of many production processes). According to Theorem 5.1, efficiency indicators obtained by DEA are independent of the units in which the inputs and the outputs are measured. Therefore, DEA can provide an appropriate methodological approach for developing of the ecoefficiency indicators.

The first paper to present a nonparametric approach for multilateral productivity comparison when some outputs are undesirable is by Färe et al. [20]. To treat desirable and undesirable outputs asymmetrically, they use the enhanced hyperbolic output efficiency measure. The resulting nonlinear programming problem is solved by taking a linear approximation of the nonlinear constraints. This methodology was applied to a sample of mills producing paper and pollutants.

Golany, Roll, and Rybak [25] applied DEA to measure the efficiency of power plants, taking into account pollution generated by electricity production. An activity analysis of the environmental performance of firms with an application to fossil fuel–fired electric utilities can be found in papers by Färe, Grosskopf, and Tyteca [21] and Tyteca [39]. The paper by Scheel [33] deals with incorporating undesirable outputs in DEA as outputs with a negative sign. An overview about DEA in ecological context is provided by Dyckhoff and Allen [16]. Hernandez Sancho, Picazo Tadeo, and Reig Martinez [26] extended the Färe et al. [20] methodology in order to adapt their enhanced indices of efficiency to situations in which only a subset of undesirable outputs is constrained by environmental regulations. They applied this method to the analysis of productive efficiency for a sample of Spanish production firms in the wooden goods and furnishings industry.

Korhonen and Luptáčik [28] proposed different variants of DEA models that can be used for estimation of ecoefficiency. The first model (called "Model A") uses negative weights for undesirable outputs, the second (called "Model B") considers the undesirable outputs as inputs, and in the third variant (called "Model C") the ratio of the weighted sum of the desirable outputs minus that of the inputs to that of the undesirable outputs is taken into account. It can be shown that the set of (strongly) ecoefficient DMUs is the same, no matter which model is used.

In the second edition of the book by Cooper, Seiford, and Tone [12], a new chapter dealing with undesirable outputs is added. The slacks-based measure of efficiency mentioned in Section 5.2.4 is modified in order to take undesirable outputs into account.

A new approach to modeling undesirable outputs, based on the zero-sum gains DEA models, is proposed by Gomes and Lins [24]. These models consider the production dependence among the DMUs, including, as an additional restriction, the zero-sum game property. The approach can be used to model CO_2 emission trade, following the Kyoto protocol.

In what follows, we follow [28] and extend the data set of the basic DEA model by p undesirable outputs ($k = 1, 2, \dots, p$) or pollutants (NO_x, SO_2, $CO_2, \dots$) generated by the production of s desirable outputs ($r = 1, 2, \dots, s$). We denote by $X \in R^{m\times n}$, $Y^g \in R^{s\times n}$, and $Y^b \in R^{p\times n}$ the matrices, consisting of nonnegative elements, describing the observed input, desirable outputs ("goods"), and pollutants ("bads") measures for the DMUs.

The problem we are facing now is how to take into account pollutants in order to get an indicator for the ecoefficiency. We can consider two ways to approach the problem.

5.4.1 Composition of Technical and Environmental Efficiency

We decompose the problem into two parts and measure the ecoefficiency in two steps. Beside the standard DEA model (1.27) or (5.2) for measuring *technical efficiency*, we formulate another DEA model for measuring the so-called *environmental efficiency*, defined as a ratio of a weighted sum of (desirable) outputs to the weighted sum of pollutants:

$$
\begin{aligned}
\underset{u,d}{\text{maximize}} \quad & h_0^E = \frac{\sum_r y_{r0}^g u_r}{\sum_k y_{k0}^b d_k} \\
\text{subject to} \quad & \frac{\sum_r y_{rj}^g u_r}{\sum_k y_{kj}^b d_k} \le 1 \quad (j = 1, 2, \dots, n), \\
& u_r \ge \varepsilon \quad (r = 1, 2, \dots, s), \\
& d_k \ge \varepsilon \quad (k = 1, 2, \dots, p), \\
& \varepsilon > 0 \quad (\text{"non-Archimedean" constant}).
\end{aligned}
$$

By substitution of the variables $\mu_r = tu_r$, $\delta_k = td_k$, $t = \frac{1}{\sum_k y_{k0}^b d_k}$, we obtain the following linear programming problem:

$$\begin{aligned}
\underset{\mu,\delta}{\text{maximize}} \quad & h_0^E = \sum_r y_{r0}^g u_r \\
\text{subject to} \quad & \sum_r y_{rj}^g \mu_r - \sum_k y_{kj}^b \delta_k \le 0 \ (j = 1, 2, \ldots, n), \\
& \sum_k y_{k0}^b \delta_k = 1, \\
& \mu_r \ge \varepsilon \ (r = 1, 2, \ldots, s), \\
& \delta_k \ge \varepsilon \ (k = 1, 2, \ldots, p).
\end{aligned} \tag{5.40}$$

The envelopment model for the environmental efficiency is the following:

$$\begin{aligned}
\underset{\theta^E,\lambda,s^g,s^b}{\text{maximize}} \quad & \theta^E - \varepsilon \left(\sum_r s_r^g + \sum_i s_s^b \right) \\
\text{subject to} \quad & \sum_j y_{rj}^g \lambda_j - s_r^g = y_{r0}^g \quad (r = 1, 2, \ldots, s), \\
& \theta^E y_{k0}^b - \sum_j y_{kj}^b \lambda_j - s_k^b = 0 \quad (k = 1, 2, \ldots, p), \\
& \lambda_j \ge 0 \quad (j = 1, 2, \ldots, n), \\
& s_r^g \ge 0 \quad (r = 1, 2, \ldots, s), \\
& s_k^b \ge 0 \quad (k = 1, 2, \ldots, p).
\end{aligned} \tag{5.41}$$

In this way, we obtain for every DMU an indicator for technical efficiency and one for environmental efficiency. We used the input-oriented CCR model, but any other DEA model can be applied as well.

The concept of ecoefficiency described above includes technical as well as environmental efficiency. How do we compose or aggregate the indicators of technical and environmental efficiency in order to get an ecoefficiency performance indicator? Because the DEA model chooses the most favorable weights for a DMU whose performance is being evaluated, the technical and environmental efficiency scores are now the output variables for the new DEA model (with the inputs equal to 1), which yields the indicator for *ecoefficiency*.

To illustrate, we apply this approach to analyzing the relative ecoefficiency of the industry in 16 OECD countries [30]. The data for the year 1993 are described in Table 5.10. The desirable output is the industry production and the undesirable output emissions of CO_2. As the inputs, we consider the labor and the capital stock. (In the case of a lack of data on capital stock for some countries as reported by the OECD in 1997, we used an approximation. These data are denoted by an asterisk.)

Solving the input-oriented CCR model (5.3) and model (5.41), we obtained the measures of the technical and environmental efficiency, respectively, given in Table 5.11. The environmental efficiency model (5.41) in our case is a very simple

Table 5.10. Data for 16 OECD countries. (Source: OECD (1997) and the author's estimation.)

	Labor (in 1,000 employers)	Capital (in billions ATS)	CO_2 (in 1,000 tons)	Industry production (in billions ATS)
Canada	1,647.00	1,867.05	93.60	2,693.85
USA	16,875.00	*30,319.065	694.30	35,481.53
Japan	10,885.00	31,100.01	296.90	30,413.81
Australia	1,009.00	1,039.56	46.60	1,218.09
New Zealand	233.80	*220.00	5.80	247.28
Austria	564.10	*927.407	11.90	1,020.15
Denmark	500.30	770.21	5.00	616.29
Finland	342.80	926.08	14.50	647.15
Germany	7,203.90	11,802.18	152.70	13,066.45
Greece	317.40	*431.747	9.20	237.20
Italy	2,801.00	6,438.52	84.60	4,226.74
Netherlands	719.60	*1,493.169	34.70	1,642.49
Norway	245.30	645.67	6.00	464.61
Spain	1,945.50	*3,100.74	45.30	2,363.80
Sweden	587.50	1,100.39	12.00	964.73
UK	4,379.00	7,178.00	81.80	6,570.72

DEA model with only one input (CO_2) and one output (industry production). The only efficient unit is the industry of Denmark. Austrian industry is environmentally inefficient; for a given level of industry production, the emission of CO_2 should be reduced by approximately 30% in order to be environmentally efficient.

Industries in Canada, the USA, and Japan are technically efficient; Austrian industry is inefficient and should reduce, for a given level of industry production, both inputs—employment and capital stock—by approximately 10% in order to achieve technical efficiency. None of the countries is, under the assumption of constant returns to scale, technically *and* environmentally efficient.

The results of technical and environmental efficiency provide the output variables for the new DEA model (with input equal to 1). The ecoefficiency indicators are given in column 3 of Table 5.11, and the ecoefficiency frontier is drawn in Figure 5.10. Industries in Denmark and Japan are ecoefficient. As can be seen from Figure 5.10, industries in the USA and Canada are weakly ecoefficient because of their environmental inefficiency. Furthermore, for all units lying outside the ecoefficiency cone (e.g., Austria), the indicator of the ecoefficiency is simply the better result from the CCR model and model (5.41), respectively. For Norway, the UK, and Sweden, lying inside the cone, the ecoefficiency indicator is higher than the indicators of technical and environmental efficiency, respectively.

Looking at the solution of the corresponding multiplier model, we can see the importance of technical and environmental efficiency in determining ecoefficiency. For most of the countries, the reason for their ecoefficiency lies in their environmental inefficiency—only Denmark, Norway, Sweden, and the UK have their strengths or

Table 5.11. Ecoefficiency of 16 OECD countries.

	Technical efficiency	Environ-mental efficiency	Ecoefficiency (as a composite of technical and environmental efficiency)	Ecoefficiency under CRS	Ecoefficiency under VRS
Canada	100.00	23.35	100.00	100.00	100.00
USA	100.00	41.46	100.00	100.00	100.00
Japan	100.00	83.11	100.00	100.00	100.00
Australia	81.21	21.21	81.21	84.00	84.92
New Zealand	77.90	34.59	77.90	91.51	100.00
Austria	90.64	69.55	90.64	99.72	100.00
Denmark	64.11	100.00	100.00	100.00	100.00
Finland	70.11	36.21	70.11	70.11	90.48
Germany	91.10	69.42	91.10	100.00	100.00
Greece	41.63	20.92	41.63	46.45	73.66
Italy	62.10	40.53	62.10	63.06	63.28
Netherlands	99.93	38.40	99.93	99.93	100.00
Norway	71.54	62.68	74.12	74.71	100.00
Spain	62.00	43.33	62.00	67.93	68.07
Sweden	76.29	65.22	77.68	85.14	89.55
UK	75.34	65.17	77.30	87.32	87.34

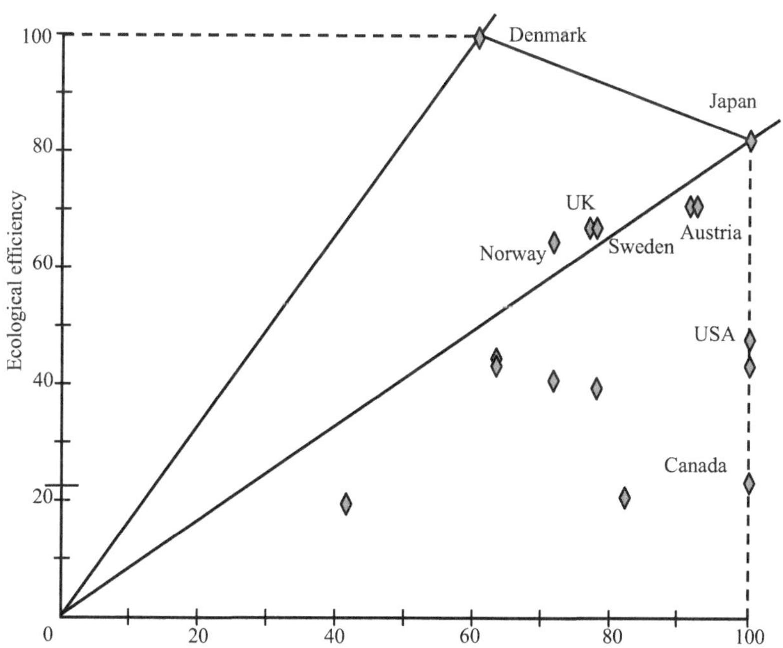

Fig. 5.10. Ecoefficiency frontier for 16 OECD countries.

advantages in environmental efficiency. The contribution of environmental efficiency in determining the ecoefficiency of industry in Denmark was 77% (the contribution of technical efficiency was 23%), in Norway 65% (35%), in Sweden 65% (36%), and in the UK 65% (35%).

5.4.2 Comprehensive Measurement of Ecoefficiency

To provide deeper insight in the causes of the ecoinefficiency and to indicate potential improvements for the particular inputs, desirable outputs, and pollutants, we formulate a model that incorporates all three categories. From the above-mentioned three variants of DEA models proposed by Korhonen and Luptáčik [28], Model B will be used in what follows. In this formulation, the pollutants are treated as the inputs in the sense that we want to produce desirable outputs as much as possible and reduce undesirable outputs and inputs. The corresponding DEA model can be expressed as follows:

$$\begin{aligned} \underset{\mathbf{u},\mathbf{v},\mathbf{d}}{\text{maximize}} \quad & h_0 = \frac{\sum_r y^g_{r0} u_r}{\sum_i x_{i0} v_i + \sum_k y^b_{k0} d_k} \\ \text{subject to} \quad & \frac{\sum_r y^g_{rj} u_r}{\sum_i x_{ij} v_i + \sum_k y^b_{kj} d_k} \leqq 1 \quad (j = 1, 2, \ldots, n), \\ & u_r \geqq \varepsilon \quad (r = 1, 2, \ldots, s), \\ & v_i \geqq \varepsilon \quad (i = 1, 2, \ldots, m), \\ & d_k \geqq \varepsilon \quad (k = 1, 2, \ldots, p). \end{aligned} \tag{5.42}$$

The transformation of the variables leads to the linear multiplier program

$$\begin{aligned} \underset{\boldsymbol{\mu},\mathbf{v},\boldsymbol{\delta}}{\text{maximize}} \quad & h_0 = \sum_r y^g_{r0} \mu_r \\ \text{subject to} \quad & \sum_r y^g_{rj} \mu_r - \sum_k y^b_{kj} \delta_k - \sum_i x_{ij} v_i \leq 0 \qquad (j = 1, 2, \ldots, n), \\ & \sum_i x_{i0} v_i + \sum_k y^b_{k0} \delta_k = 1, \\ & \mu_r \geq \varepsilon \qquad (r = 1, 2, \ldots, s), \\ & v_i \geq \varepsilon \qquad (i = 1, 2, \ldots, m), \\ & \delta_k \geq \varepsilon \qquad (k = 1, 2, \ldots, p) \end{aligned} \tag{5.43}$$

and the envelopment model

$$\begin{aligned} \underset{\theta,\mathbf{s}^g,\mathbf{s}^b,\mathbf{s}^-}{\text{maximize}} \quad & \theta - \varepsilon \left(\sum_r s^g_r + \sum_k s^b_k + \sum_i s^-_i \right) \\ \text{subject to} \quad & \sum_j y^g_{rj} \lambda_j - s^g_r = y^g_{r0} \quad (r = 1, 2, \ldots, s), \end{aligned}$$

$$\begin{aligned}
\theta y_{k0}^b - \sum_j y_{sj}^b \lambda_j - s_k^b &= 0 \qquad (k = 1, 2, \ldots, p), \\
\theta x_{i0} - \sum_j x_{ij} \lambda_j - s_i^- &= 0 \qquad (i = 1, 2, \ldots, m),
\end{aligned} \tag{5.44}$$

$$\begin{aligned}
s_r^g &\geq 0 \qquad (r = 1, 2, \ldots, s), \\
s_k^b &\geq 0 \qquad (k = 1, 2, \ldots, p), \\
s_i^- &\geq 0 \qquad (i = 1, 2, \ldots, m).
\end{aligned}$$

In this model, the DMU simultaneously reduces the inputs and pollutants in order to increase ecoefficiency. Using the data set of 16 OECD countries (Table 5.10), model (5.44) yields the ecoefficiency indicators for the particular countries summarized in column 4 of Table 5.11. Adding the constraint $\sum_{j=1}^{n} \lambda_j = 1$ to the constraints of program (5.44), we obtain ecoefficiency under the assumption of variable returns to scale (the last column in Table 5.11).

Comparison of columns 3 and 4 in Table 5.11 shows in tendency the same results. The reader may verify that the (comprehensive) ecoefficiency measured by model (5.44) (column 4 in Table 5.11) is not lower than the ecoefficiency obtained as a composition of technical and environmental efficiency (column 3 in Table 5.11). The ecoefficient countries Denmark and Japan are ecoefficient again, and in the previous model the weakly ecoefficient industries of Canada and the USA and the ecoinefficient industry of Germany are ecoefficient with respect to model (5.44). The solution of the multiplier problems (5.43) indicates the reasons for the ecoefficiency in the particular countries with respect to particular inputs and outputs. The input variable capital was given an importance rating of 53% and the variable CO_2 an importance rating of 47% in determining of the ecoefficiency in Denmark. The reason for the ecoefficiency of Japan, the USA, and Canada lies in their technical efficiency. The contribution of the input labor in determining the ecoefficiency was 33% for Japan, 40% for the USA, and 51% for Canada. The importance rating given to the input capital was 67% for Japan, 60% in the USA, and 49% in Canada. In Germany, an importance rating of 12% was given to labor, 79% to capital, and 9% to undesirable output, CO_2, in determining its industry ecoefficiency.

Austrian industry is ecoinefficient because of its scale inefficiency:

$$SE_{\text{Austria}} = \frac{\theta_{\text{CCR}}^{E}}{\theta_{\text{BCC}}^{E}} = 0.9972.$$

The solution of the envelopment model (5.44) characterizes the peer group of any ecoinefficient unit. Austrian industry is in the peer group created by Germany and Japan. The linear combination of Germany (by approximately 98%) and of Japan (by 2%) provides the projection of Austrian industry on the ecoefficient frontier.

With regard to these results, the following remark must be added. Obviously, the ecoefficiency of the industry in a particular country depends on its structure, especially on the proportion of steel production in the total industrial production. According to the OECD data for 1993 (reported by the OECD in 1997), the proportion of steel production in the industrial production of ecoefficient Denmark was indeed very low (0.013). However, the proportion of steel production in ecoefficient Japan was approximately the same (0.062) as in ecoinefficient Austria (0.065). Nevertheless, a more interesting analysis can be done for particular industrial sectors rather than for industry as a whole. The subject of our example was determined by the availability of data.

References and Further Reading

[1] A. I. Ali, Computational aspects of DEA, in A. Charnes, W. W. Cooper, A. Lewin, and L. M. Seiford, eds., *Data Envelopment Analysis: Theory, Methodology, and Applications*, Kluwer Academic Publishers, Boston, 1994, 63–88.

[2] A. I. Ali and L. M. Seiford, Translation invariance in data envelopment analysis, *Oper. Res. Lett.*, **9** (1990), 403–405.

[3] T. Ahn, A. Charnes, and W. W. Cooper, Efficiency characterization in different DEA models, *Socio-Econ. Planning Sci.*, **22** (1988), 253–257.

[4] P. W. Bauer, Recent developments in the econometric estimation of frontiers, *J. Econometrics*, **46** (1990), 39–56.

[5] D. Bouyssou, Using DEA as a tool for MCDM: Some remarks, *J. Oper. Res. Soc.*, **50** (1999), 974–978.

[6] R. D. Banker, A. Charnes, and W. W. Cooper, Some models for estimating technical and scale inefficiences in data envelopment analysis, *Manage. Sci.*, **30** (1984), 1078–1092.

[7] A. Charnes and W. W. Cooper, Programming with linear fractional functionals, *Naval Res. Logist. Q.*, **9** (1962), 181–185.

[8] A. Charnes, W. W. Cooper, and E. Rhodes, Measuring the efficiency of decision making units, *Eur. J. Oper. Res.*, **2** (1978), 429–444.

[9] A. Charnes, W. W. Cooper, and E. Rhodes, Short communication: Measuring the efficiency of decision making units, *Eur. J. Oper. Res.*, **3** (1979), 339.

[10] A. Charnes, W. W. Cooper, B. Golany, L. M. Seiford, and J. Stutz, Foundations of data envelopment analysis for Pareto–Koopmans efficient empirical production functions, *J. Econometrics*, **30** (1985), 91–107.

[11] W. W. Cooper, L. M. Seiford, and K. Tone, *Data Envelopment Analysis: A Comprehensive Text with Models, Applications, References, and DEA-Solver Software*, 2nd printing, Kluwer Academic Publishers, Boston, 2000.

[12] W. W. Cooper, L. M. Seiford, and K. Tone, *Data Envelopment Analysis: A Comprehensive Text with Models, Applications, References, and DEA-Solver Software*, 2nd ed., Springer, New York, 2007.

[13] W. W. Cooper, R. G. Thompson, and R. M. Thrall, Extensions and new development in DEA, *Ann. Oper. Res.*, **66** (1996), 3–45.

[14] G. Debreu, The coefficient of resource utilization, *Econometrica*, **19** (1951), 273–292.

[15] W. E. Diewert and C. Parkan, Linear programming tests of regularity conditions for production frontiers, in W. Eichorn, R. Henn, K. Neumann, and R. W. Shephard, eds., *Quantitative Studies in Production and Prices*, Physica-Verlag, Würzburg, Germany, 1983.

[16] H. Dyckhoff and K. Allen, Measuring ecological efficiency with data envelopment analysis (DEA), *Eur. J. Oper. Res.*, **132** (2001), 312–325.

[17] R. Färe and C. A. K. Lovell, Measuring the technical efficiency of production, *J. Econ. Theory*, **19** (1978), 150–162.

[18] M. J. Farrel, The measurement of productive efficiency, *J. R. Statist. Soc. Ser.* A, **120**-Part 3 (1957), 253–281.

[19] R. Färe, S. Grosskopf, and C. A. K. Lovell, *The Measurement of Efficiency of Production*, Kluwer Academic Publishers, Boston, 1985.

[20] R. Färe, S. Grosskopf, C. A. K. Lovell, and C. Pasurka, Multilateral productivity comparison when some outputs are undesirable: A nonparametric approach, *Rev. Econ. Statist.*, **71** (1989), 90–98.

[21] R. Färe, S. Grosskopf, and D. Tyteca, An activity analysis model of the environmental performance of firms: Application to fossil-fuel-fired electric utilities, *Ecol. Econ.*, **18** (1996), 161–175.

[22] R. Färe and D. Primont, *Multi-Output Production and Duality: Theory and Applications*, Kluwer Academic Publishers, Boston, 1997.

[23] H. O. Fried, C. A. K. Lovell, and S. S. Schmidt, eds., *The Measurement of Productive Efficiency and Productivity Growth*, Oxford University Press, New York, 2008.

[24] E. G. Gomes and M. P. E. Lins, Modeling undesirable outputs with zero sum gains data envelopment analysis models, *J. Oper. Res. Soc.*, **59**-5 (2008), 616–623; advance online publication, 2007.

[25] B. Golany, Y. Roll, and D. Rybak, Measuring efficiency of power plants in Israel by data envelopment analysis, *IEEE Trans. Eng.*, **41**-3 (1994), 291–301.

[26] F. Hernandez Sancho, A. Picazo Tadeo, and E. Reig Martinez, Efficiency and environmental regulation, *Environ. Resource Econ.*, **15** (2000), 365–378.

[27] T. C. Koopmans, Analysis of production as an efficient combination of activities, in T. C. Koopmans, ed., *Activity Analysis of Production and Allocation*, Cowles Commission Monograph 13, Wiley, New York, 1951, 33–97.

[28] P. J. Korhonen and M. Luptáčik, Eco-efficiency analysis of power plants: An extension of data envelopment analysis, *Eur. J. Oper. Res.*, **154** (2004), 437–446.

[29] C. A. K. Lovell and P. Schmidt, A comparison of alternative approaches to the measurement of productive efficiency, in A. Dogramaci and R. Färe, eds., *Applications of Modern Production Theory: Efficiency and Productivity*, Kluwer Academic Publishers, Boston, 1988.

[30] M. Luptáčik, Data envelopment analysis as a tool for measurement of eco-efficiency, in E. Dockner, R. F. Hartl, M. Luptáčik, and G. Sorger, eds., *Optimization, Dynamics and Economic Analysis*, Physica-Verlag, Würtzburg, Germany, 2000, 36–48.

[31] S. C. Ray, *Data Envelopment Analysis: Theory and Techniques for Economics and Operation Research*, Cambridge University Press, Cambridge, UK, 2004.

[32] R. R. Russell, Measures of technical efficiency, *J. Econ. Theory*, **35** (1985), 109–126.

[33] F. Scheel, Undesirable outputs in efficiency valuations, *Eur. J. Oper. Res.*, **132** (2001), 400–410.

[34] L. M. Seiford and R. M. Thrall, Recent developments in DEA: The mathematical programming approach to frontier analysis, *J. Econometrics*, **46** (1990), 7–38.

[35] J. K. Sengupta, Data envelopment analysis for efficiency measurement in the stochastic case, *Comput. Oper. Res.*, **14** (1987), 117–129.

[36] R. W. Shephard, *Cost and Production Functions*, Princeton University Press, Princeton, NJ, 1953.

[37] R. W. Shephard, *Theory of Cost and Production Functions*, Princeton University Press, Princeton, NJ, 1970.

[38] D. Tyteca, On the measurement of the environmental performance of firms: A literature review and a productivity efficiency perspective, *J. Environ. Manage.*, **46** (1996), 281–308.

[39] D. Tyteca, Linear programming models for the measurement of environmental performance of firms: Concepts and empirical results, *J. Productivity Anal.*, **8** (1997), 183–197.

6

Geometric Programming

> Which of you, intending to build a tower, sitteth not down first and counteth the cost, whether he have sufficient to finish it?
>
> *Luke* XIV,28,C.75

The open input–output model with continuous substitution between labor and capital, according to a Cobb–Douglas production function introduced in Section 1.2.8, leads to a mathematical programming problem in which the functions in the constraints are polynomials with positive coefficients (so-called *posynomials*).

Optimization problems with this class of functions emerged first in connection with engineering design problems. Zener [23] observed that many engineering design problems consisting of a sum of component costs could sometimes be maximized almost by inspection under suitable conditions. Although Zener discovered this remarkable fact through the process of engineering observation and inquiry, he soon realized that such an observation should have roots in a deeper mathematical theory. The mathematical foundation of Zener's discovery has been done through application of the arithmetic–geometric mean inequality relationship between sums and products of positive numbers, developed by Duffin [8]. Therefore, Duffin called this new branch of nonlinear optimization "geometric programming." In 1967, Duffin, Peterson, and Zener [9] published the classic well-known textbook that presented the first comprehensive description of the method, along with some example problems to illustrate this technique.

In a short time, geometric programming has been applied in a wide variety of fields (see [6] or the bibliographical note in the special issue of the *Journal of Optimization Theory and Applications* **26** (1978) devoted to geometric programming). Because also in economics and management science very often problems arise that can be formulated and solved by using this technique, we will now present the basic theory of geometric programming with some economic applications.

In Section 6.1, the basic principle of geometric programming—using the well-known economic lot size problem—will be explained. Section 6.2 deals with the theoretical fundamentals, and Section 6.3 deals with the models of geometric pro-

M. Luptáčik, *Mathematical Optimization and Economic Analysis*,
Springer Optimization and Its Applications 36, DOI 10.1007/978-0-387-89552-9_6,

gramming in economics. In Section 6.4, the transformation of some optimization problems into standard geometric programming models will be discussed.

6.1 The Principle of Geometric Programming

Let us consider the well-known economic lot size problem. For a given product, the manufacturer must decide how many pieces of this product he should put into stock periodically. The total variable cost consisting of manufacturing and storage cost is given by

$$y = \frac{lT}{2}x + rcx^{-1},$$

where x denotes the lot size (pieces per run), l the carrying cost per piece per month, r the annual requirements, c the setup cost (€ per run), and T the time period (1 year or 12 months). The first part in this objective function represents the total carrying cost, and the other part gives the total setup cost.

Let $a_1 = \frac{lT}{2}$ and $a_2 = rc$; then the function y to be minimized is of the form

$$y = a_1 x + a_2 x^{-1}. \tag{6.1}$$

The solution to this problem can easily be found through differential calculus,

$$\frac{dy}{dx} = a_1 - a_2 x^{-2} = 0,$$

from which it follows that

$$x^0 = \sqrt{\frac{a_2}{a_1}} = \sqrt{\frac{2rc}{lT}}.$$

The optimum (minimum) total cost is given by

$$y^0 = a_1 \left(\frac{2rc}{lT}\right)^{1/2} + a_2 \left(\frac{2rc}{lT}\right)^{-1/2},$$

or

$$y^0 = \underbrace{[a_1 a_2]^{1/2}}_{A} + \underbrace{[a_1 a_2]^{1/2}}_{B}. \tag{6.2}$$

We see that at the optimum the minimal cost is again composed of two component costs. If we look at the form (6.2), it is obvious that the ratio of these cost components is independent of the coefficients a_1 and a_2 because the term $a_1^{1/2} a_2^{1/2}$ is common to both cost components. Consequently, the optimal cost *distribution* is invariant under a change in the numerical values of a_1 and a_2. Algebraic analysis of cost components A and B reveals that A is *always* equal to the magnitude of B. That is, the *relative contribution* to the total optimal cost for carrying cost and for setup cost is invariant to changes in the economic coefficients a_1 and a_2. At optimality, one-half of the total cost will be attributed to the carrying cost and one-half to the setup cost. The key observation is that at an optimal solution, any other distribution of cost will yield a higher total cost.

Prior knowledge of this fact (later it will be shown how we can obtain this information) allows us to use this relationship in the following manner (from (6.1)):

$$a_1 x^0 = a_2 \frac{1}{x^0}.$$

Hence

$$x^0 = \sqrt{\frac{a_2}{a_1}} = \sqrt{\frac{2rc}{lT}},$$

which is the well-known *Harris–Anders form* from inventory theory (see, e.g., [24, p. 197]).

This result coincides with the optimal solution obtained previously via calculus and leads us to an entirely new way of solving mathematical programming problems. Instead of searching first for the optimal lot size and then determining the (resultant) optimal cost, we first search for the optimal *cost distribution* and then determine the resulting solution variables. This is the logic through which geometric programming was originally discovered by Duffin and Zener.

6.2 The Theory of Geometric Programming

The input–output model with substitution possibilities between primary factors from Section 1.2.8 is a special case of a general geometric programming problem that may be formulated as follows:

$$\min_{\mathbf{x}} \text{imize} \quad g_0(\mathbf{x}) \tag{6.3}$$

$$\text{subject to} \quad g_k(\mathbf{x}) \leqq 1 \qquad (k = 1, 2, \ldots, m), \tag{6.4}$$

$$x_j > 0 \qquad (j = 1, 2, \ldots, n), \tag{6.5}$$

where

$$g_k(\mathbf{x}) = \sum_{t=1}^{T_k} c_{kt} \prod_{j=1}^{n} x_j^{a_{ktj}} \quad (k = 0, 1, \ldots, m). \tag{6.6}$$

T_k denotes the number of terms in the polynomial k ($k = 0, 1, \ldots, m$), and the exponents a_{ktj} are arbitrary real numbers, but the coefficients c_{kt} are assumed to be positive. Thus the functions $g_k(\mathbf{x})$ are called posynomials.

The program (6.3)–(6.6) is termed the *primal problem*, the posynomial $g_o(\mathbf{x})$ is the *primal function*, and the variables $x_1, x_2, \ldots, x_n$ are the *primal variables*. The constraints imposed by (6.4) are called *forced constraints*, whereas those imposed by (6.5) are referred to as *natural constraints*. Collectively, the constraints (6.4)–(6.5) are termed *primal constraints*.

It should be mentioned that the posynomials in (1.22) from the input–output model of Section 1.2.8 are nonconvex.

Making a change of variables by letting

$$e^{z_j} = x_j \quad (z_j = \ln x_j) \quad (j = 1, 2, \ldots, n),$$

the primal problem (6.3)–(6.6) may be transformed into the following program:

$$\underset{\mathbf{z}}{\text{minimize}} \quad f_0(\mathbf{z}) \tag{6.7}$$

$$\text{subject to} \quad f_k(\mathbf{z}) \leqq 1 \qquad (k = 1, 2, \ldots, m), \tag{6.8}$$

where

$$f_k(\mathbf{z}) = \sum_{t=1}^{T_k} c_{kt} e^{\sum_{j=1}^{n} a_{ktj} z_j} \quad (k = 0, 1, \ldots, m) \tag{6.9}$$

is a positive exponential function.

The most important aspect of program (6.7)–(6.9) is brought out in the following.

Theorem 6.1 (see [9, p. 83]). *The transformed primal program* (6.7)–(6.9) *is a convex program. Each positive exponential function* (6.9) *is convex.*

Since the logarithmic function is monotone increasing, the solution of the transformed primal problem (6.7)–(6.9) can be found by solving the following program:

$$\underset{\mathbf{z}}{\text{minimize}} \quad \ln f_0(\mathbf{z}) \tag{6.10}$$

$$\text{subject to} \quad \ln f_k(\mathbf{z}) \leqq 0 \quad (k = 1, 2, \ldots, m). \tag{6.11}$$

The reader should have no trouble showing that the function $\ln f_k(\mathbf{z})$ $(k = 0, 1, 2, \ldots, m)$ is convex for arbitrary real numbers a_{ktj} and positive real numbers c_{kt}. Thus problem (6.10)–(6.11) is convex, and the duality theory developed in Chapter 3 can be applied.

The dual problem corresponding to the primal problem (6.10)–(6.11) is the following:

$$\underset{\mathbf{z},\boldsymbol{\lambda}}{\text{maximize}} \quad \Phi(\mathbf{z}, \boldsymbol{\lambda}) \tag{6.12}$$

$$\text{subject to} \quad \frac{\partial \Phi}{\partial \mathbf{z}} = \mathbf{0}, \tag{6.13}$$

$$\boldsymbol{\lambda} \geqq \mathbf{0}, \tag{6.14}$$

where

$$\Phi(\mathbf{z}, \boldsymbol{\lambda}) = \sum_{k=0}^{m} \lambda_k \ln f_k(\mathbf{z}) \quad \text{with } \lambda_0 = 1. \tag{6.15}$$

Differentiation of (6.15) with respect to z_j yields

$$\frac{\partial \Phi}{\partial z_j} = \sum_{k=0}^{m} \lambda_k \frac{\sum_{t=1}^{T_k} c_{kt} e^{\sum_{j=1}^{n} a_{ktj} z_j} a_{ktj}}{\sum_{t=1}^{T_k} c_{kt} e^{\sum_{j=1}^{n} a_{ktj} z_j}}. \tag{6.16}$$

Letting

$$\delta_{kt} = \lambda_k \frac{c_{kt} e^{\sum_{j=1}^{n} a_{ktj} z_j}}{\sum_{t=1}^{T_k} c_{kt} e^{\sum_{j=1}^{n} a_{ktj} z_j}} \quad (k = 0, 1, \ldots, m) \tag{6.17}$$

and using (6.16), the constraints (6.13) become

$$\sum_{k=0}^{m} \sum_{t=1}^{T_k} \delta_{kt} a_{ktj} = 0 \quad (j = 1, 2, \ldots, n). \tag{6.18}$$

Because $\lambda_0 = 1$, it follows from (6.17) for $k = 0$ that

$$\sum_{t=1}^{T_0} \delta_{0t} = 1, \tag{6.19}$$

where T_0 denotes the number of terms in the primal objective function (6.3) and

$$\sum_{t=1}^{T_k} \delta_{kt} = \lambda_k \quad (k = 1, 2, \ldots, m). \tag{6.20}$$

The nonnegativity constraints (6.14) and definition (6.17) imply that

$$\delta_{kt} \geqq 0 \quad (t = 1, 2, \ldots, T_k;\ k = 0, 1, \ldots, m). \tag{6.21}$$

Relation (6.17) can be rewritten as

$$\frac{\delta_{kt}}{c_{kt}} \sum_{t=1}^{T_k} c_{kt} e^{\sum_{j=1}^{n} a_{ktj} z_j} = \lambda_k e^{\sum_{j=1}^{n} a_{ktj} z_j} \quad (k = 0, 1, \ldots, m). \tag{6.22}$$

After raising (6.22) to a power of δ_{kt} and then taking a logarithm, we obtain

$$\delta_{kt} \ln \frac{\delta_{kt}}{c_{kt}} + \delta_{kt} \ln f_k(\mathbf{z}) = \delta_{kt} \ln \lambda_k + \delta_{kt} \sum_{j=1}^{n} a_{ktj} z_j \tag{6.23}$$
$$(k = 0, 1, \ldots, m;\ t = 1, 2, \ldots, T_k).$$

Summing (6.22) over k and t and taking into account (6.18), (6.19), and (6.20) gives

$$\sum_{k=0}^{m} \sum_{t=1}^{T_k} \delta_{kt} \ln \frac{\delta_{kt}}{c_{kt}} + \sum_{k=0}^{m} \lambda_k \ln f_k(\mathbf{z}) = \sum_{k=1}^{m} \lambda_k \ln \lambda_k,$$

and thus

$$\Phi(\mathbf{z},\boldsymbol{\lambda}) = \sum_{k=0}^{m}\sum_{t=1}^{T_k} \delta_{kt} \ln \frac{c_{kt}}{\delta_{kt}} + \sum_{k=1}^{m} \lambda_k \ln \lambda_k$$
$$= \ln \left\{ \prod_{k=0}^{m}\prod_{t=1}^{T_k} \left(\frac{c_{kt}}{\delta_{kt}}\right)^{\delta_{kt}} \prod_{k=1}^{m} \lambda_k^{\lambda_k} \right\} = \ln v(\boldsymbol{\delta}).$$

The dual objective function (6.12) is now expressed as a function of the new variables δ_{kt} (notice (6.20)). Moreover, the function $\ln v(\boldsymbol{\delta})$ is concave on its domain of definition, namely, the positive orthant $\delta_{kt} > 0$, $k = 0, 1, \ldots, m$; $t = 1, 2, \ldots, T_k$ (see [9, pp. 121–122]), such that the dual problem (6.12)–(6.14) becomes the convex programming problem with linear constraints

$$\begin{array}{ll} \text{maximize} & \ln v(\boldsymbol{\delta}) \\ \text{subject to} & (6.18)\text{–}(6.19) \text{ and } (6.21). \end{array}$$

Because of the monotonicity property of the logarithmic function, $v(\boldsymbol{\delta})$ and $\ln v(\boldsymbol{\delta})$ have the same set of maximizing points. Thus the dual program corresponding to the primal geometric programming problem (6.3)–(6.6) may be formulated as follows:

$$\text{maximize} \qquad v(\boldsymbol{\delta}) = \prod_{k=0}^{m}\prod_{t=1}^{T_k} \left(\frac{c_{kt}}{\delta_{kt}}\right)^{\delta_{kt}} \prod_{k=1}^{m} \lambda_k(\boldsymbol{\delta})^{\lambda_k(\boldsymbol{\delta})}, \tag{6.24}$$

where

$$\lambda_k(\boldsymbol{\delta}) = \sum_{t=1}^{T_k} \delta_{kt} \quad (k = 1, 2, \ldots, m),$$

$$\text{subject to} \quad \sum_{k=0}^{m}\sum_{t=1}^{T_k} a_{ktj}\delta_{kt} = 0 \qquad (j = 1, 2, \ldots, n), \tag{6.18}$$

$$\sum_{t=1}^{T_0} \delta_{0t} = 1, \tag{6.19}$$

$$\delta_{kt} \geqq 0 \qquad \begin{pmatrix} k = 0, 1, \ldots, m, \\ t = 1, 2, \ldots, T_k \end{pmatrix}, \tag{6.21}$$

where $T = \sum_{k=0}^{m} T_k$ is the total number of terms in the primal problem.

In evaluating the product function $v(\boldsymbol{\delta})$, it must be understood that $z^z = z^{-z} = 1$ for $z = 0$. This will make $v(\boldsymbol{\delta})$ continuous over its domain of definition.

The function $v(\boldsymbol{\delta})$ is termed the *dual function*, and the components of T-dimensional vector $\boldsymbol{\delta}$ are called *dual variables*. Relation (6.21) is termed the *non-negativity condition*, (6.19) is called the *normality condition*, and (6.18) constitutes the *orthogonality condition*. Collectively, these conditions are referred to as *dual constraints*.

Notice how the dual problem (6.24) subject to (6.18), (6.19), and (6.21) is obtained from its corresponding primal problem (6.3)–(6.6). The factors c_{kt} appearing in the dual function $v(\boldsymbol{\delta})$ are the coefficients of the posynomials $g_k(\mathbf{x})$ $(k = 0, 1, \ldots, m)$. Each term of $g_k(\mathbf{x})$ $(k = 0, 1, \ldots, m)$ is associated with one and only one of the dual variables δ_{kt} $(t = 1, 2, \ldots, T_k;\ k = 0, 1, \ldots, m)$. Each factor $\lambda_k(\boldsymbol{\delta})^{\lambda_k(\boldsymbol{\delta})}$ of $v(\boldsymbol{\delta})$ comes from a forced constraint $g_k(\mathbf{x}) \leqq 1$. Notice that none of these factors appears from the primal function because the normality condition forces $\lambda_0(\boldsymbol{\delta})$ to be 1. The normality condition is the only part of the dual problem (6.24) subject to (6.18), (6.19), and (6.21) that distinguishes between the primal function $g_0(\mathbf{x})$ and those posynomials $g_k(\mathbf{x})$ $(k = 1, 2, \ldots, m)$ that appear in the forced constraints. Finally, the coefficients a_{ktj} appearing in the orthogonality condition are simply the exponents of primal problem (6.3)–(6.6).

The difference between the number of variables and the number of independent linear equations is conventionally called the number of degrees of freedom. Note that there are n orthogonality conditions, one for each variable x_j, a single normality condition and T dual variables, one for each term. Hence (6.18)–(6.19) have $T-(n+1)$ degrees of freedom. Duffin, Peterson, and Zener [9] suggest calling this quantity the *degree of difficulty*. A geometric programming problem with zero degrees of difficulty (see the problem discussed in Section 6.1) is readily solved by first solving its dual. The solution vector $\boldsymbol{\delta}$ is easily determined because the dual constraints are linear. Since the vector $\boldsymbol{\delta}$ is the only solution to the dual constraints, it is also the maximizing vector for the dual problem. If the degree of difficulty is greater than zero, we have a nonlinear programming problem with a nonlinear objective function but with linear constraints.

The question that arises now is that of how to obtain the optimal solution of the primal problem $\mathbf{x}^0$ from the knowledge of the optimal solution $\boldsymbol{\delta}^0$ of the dual problem (6.24), (6.18)–(6.19), and (6.21).

Applying the weak duality theorem (Theorem 3.13), the so-called *main lemma of geometric programming* can be derived.

Lemma 6.1. *If* $\mathbf{x}$ *satisfies the constraints* (6.4)–(6.5) *and* $\boldsymbol{\delta}$ *satisfies the constraints* (6.18)–(6.19) *and* (6.21), *then*

$$g_0(\mathbf{x}) \geqq v(\boldsymbol{\delta}). \tag{6.25}$$

Inequality (6.25) follows immediately from (3.32).

Duffin, Peterson, and Zener [9, pp. 115–116] proved this lemma using the geometric inequality (a weighted geometric mean of positive numbers is always less than or equal to the corresponding weighted arithmetic mean). For this reason, they called this class of nonlinear programming problems *geometric* programming.

Relation (6.25) implies that the constrained minimum value of the primal function $g_0(\mathbf{x})$ cannot be lower than the constrained maximum value of the dual function $v(\boldsymbol{\delta})$. In this way the dual program provides a lower bound for the optimal value of the primal function $g_0(\mathbf{x})$.

Under the assumption of the Slater constraint qualification (Definition 3.6), which corresponds to the definition of superconsistency of the primal problem in [9], the application of Theorems 3.9 and 3.14 leads to the *first duality theorem* of geometric programming (originally proved by Duffin, Peterson, and Zener [9, pp. 117–119]).

Theorem 6.2. *Assume that the Slater constraint qualification for the primal problem* (6.3)–(6.6) *is met and that the primal function* $g_0(\mathbf{x})$ *attains its constrained minimum value at a feasible point* $\mathbf{x}^0$. *Then the following hold:*

(i) *The corresponding dual problem* (6.24) *subject to* (6.18), (6.19), *and* (6.21) *has an optimal solution.*

(ii) $g_0(\mathbf{x}^0) = v(\boldsymbol{\delta}^0)$, *where* $\boldsymbol{\delta}^0$ *denotes a feasible point at which the dual function* $v(\boldsymbol{\delta})$ *attains its constrained maximum value.*

(iii) *If* $\mathbf{x}^0$ *is an optimal solution of the primal problem* (6.3)–(6.6), *then there are nonnegative Lagrange multipliers* μ_k^0 $(k = 1, 2, \ldots, m)$ *such that the Lagrange function*

$$L(\mathbf{x}, \boldsymbol{\mu}) = g_0(\mathbf{x}) + \sum_{k=1}^{m} \mu_k (g_k(\mathbf{x}) - 1)$$

has the property

$$L(\mathbf{x}^0, \boldsymbol{\mu}) \leqq g_0(\mathbf{x}^0) = L(\mathbf{x}^0, \boldsymbol{\mu}^0) \leqq L(\mathbf{x}, \boldsymbol{\mu}^0)$$

for arbitrary $x_j > 0$ *and arbitrary* $\mu_k \geqq 0$. *Moreover, there is a maximizing vector* $\boldsymbol{\delta}^0$ *for the dual problem* (6.24) *subject to* (6.18), (6.19), *and* (6.21) *whose components are*

$$\delta_{kt}^0 = \begin{cases} \dfrac{c_{kt} x_1^{a_{kt1}} \cdots x_n^{a_{ktn}}}{g_0(\mathbf{x})} & (k = 0;\ t = 1, 2, \ldots, T_0), \\ \dfrac{\mu_k c_{kt} x_1^{a_{kt1}} \cdots x_n^{a_{ktn}}}{g_0(\mathbf{x})} & (k = 1, 2, \ldots, m;\ t = 1, 2, \ldots, T_k), \end{cases} \tag{6.26}$$

where $\mathbf{x} = \mathbf{x}^0$ *and* $\boldsymbol{\mu} = \boldsymbol{\mu}^0$. *Furthermore,*

$$\lambda_k(\boldsymbol{\delta}^0) = \frac{\mu_k}{g_0(\mathbf{x}^0)} \quad (k = 1, 2, \ldots, m). \tag{6.27}$$

(iv) *If* $\boldsymbol{\delta}^0$ *is an optimal solution for the dual problem* (6.24) *subject to* (6.18), (6.19), *and* (6.21), *each optimal solution* $\mathbf{x}^0$ *for the primal problem* (6.3)–(6.6) *satisfies the system of equations*

$$c_{kt} x_1^{a_{kt1}} \cdots x_n^{a_{ktn}} = \begin{cases} \delta_{kt}^0 v(\boldsymbol{\delta}^0) & (k = 0;\ t = 1, 2, \ldots, T_0), \\ \dfrac{\delta_{kt}^0}{\lambda_k(\boldsymbol{\delta}^0)} & (k = 1, 2, \ldots, m;\ t = 1, 2, \ldots, T_k), \end{cases} \tag{6.28}$$

where k *ranges over all positive integers for which* $\lambda_k(\boldsymbol{\delta}^0) > 0$.

Relation (6.26) allows us to compute an optimal solution vector $\boldsymbol{\delta}^0$ from the knowledge of a minimizing vector $\mathbf{x}^0$ and appropriate Lagrange multipliers μ_k^0 $(k = 1, 2, \ldots, m)$ of the primal problem (6.3)–(6.6). The relationship between the Lagrange multipliers λ_k of the problem (6.12)–(6.14) and the Lagrange multipliers μ_k^0 of the problem (6.3)–(6.6), described by formula (6.27), can be derived by comparing both problems and taking into account the transformation of the variables.

On the other hand, (6.28) provides a formula for computing an optimal solution $\mathbf{x}^0$ of the primal problem (6.3)–(6.6) from the knowledge of an optimal solution $\boldsymbol{\delta}^0$ of the dual problem (6.24) subject to (6.18), (6.19), and (6.21). System (6.28) can be solved easily because by taking the logarithm of both sides of each equation, we obtain a system of linear equations with variables $\ln x_j$ $(j = 1, 2, \ldots, n)$.

The first duality theorem of geometric programming supposes that an optimal solution for the primal problem exists. The following theorem provides a sufficient condition for this hypothesis to be satisfied [9, pp. 120–121].

Theorem 6.3. *If the primal problem* (6.3)–(6.6) *has a feasible solution, and if there is a feasible solution* $\boldsymbol{\delta}^*$ *of the dual problem with positive components, then there exists an optimal solution of the primal problem* (6.3)–(6.6).

The reader may prove that the following corollary is valid.

Corollary. *If the primal problem* (6.3)–(6.6) *has a feasible solution and if its dual problem* (6.24) *subject to* (6.18), (6.19), *and* (6.21) *has an optimal solution* $\boldsymbol{\delta}^0$ *with strictly positive components, then all the forced constraints for the primal problem are active at an optimal solution* $\mathbf{x}^0$, *that is,*

$$g_k(\mathbf{x}^0) = 1 \quad (k = 1, 2, \ldots, m).$$

6.3 Models of Geometric Programming in Economics

Most applications of geometric programming as a technique for solving nonlinear optimization problems can be found in engineering [6, 20]. Starting with the paper by Sengupta and Portillo-Campbell [21] and the book by Nijkamp [16], geometric programming has been used as an instrument of economic analysis. The well-known and very often-used Cobb–Douglas production function is a posynomial. Therefore, models of production with these types of production functions, allowing continuous substitution between inputs, are the typical field for economic applications of geometric programming. Regional economics [16, 7, 11], environmental economics [17, 1, 13], marketing mix problems [4], capital budgeting problems [18], and manpower planning [15] are other fields of geometric programming applications in economics and management science.

In this section, we present some economic models and show how geometric programming can support the decision-making process and contribute to explanation of economic phenomena.

6.3.1 The Economic Lot Size Problem

In order to demonstrate the fundamentals and the workability of geometric programming for economic analysis, we turn to the simplest model introduced in the first section. The problem contains the minimization of the total variable cost described by the primal function:

$$y = \frac{lT}{2}x + rcx^{-1},$$

or

$$y = a_1x + a_2x^{-1} \quad \text{where } a_1 = \frac{lT}{2} \text{ and } a_2 = rc.$$

The corresponding dual problem becomes

$$\begin{aligned} &\text{maximize} && v(\boldsymbol{\delta}) = \left(\frac{a_1}{\delta_{01}}\right)^{\delta_{01}} \left(\frac{a_2}{\delta_{02}}\right)^{\delta_{02}} \\ &\text{subject to} && \delta_{01} + \delta_{02} = 1, \\ & && \delta_{01} - \delta_{02} = 0. \end{aligned}$$

The dual variable δ_{01} is related to the first term of the primal objective function describing the total carrying cost, and the dual variable δ_{02} to the second part describing the total setup cost. The primal problem consists of two terms (thereupon two dual variables) and one primal variable; the degree of difficulty is zero. The dual problem yields the unique optimal solution

$$\delta_{01}^0 = \delta_{02}^0 = \frac{1}{2}.$$

According to (6.17), the dual variable δ_{01} (δ_{02}) gives us the proportion of the carrying cost (setup cost) on the total variable cost. The optimal solution of the dual model provides (as mentioned in Section 6.1) prior knowledge of the *optimal cost distribution*. Without knowing the optimal lot size—and independently of the coefficients a_1 and a_2—at the optimal solution, one-half of the total cost will be attributed to the carrying cost and half to the setup cost. Knowing this fact, we can easily determine the optimal lot size by setting

$$a_1x = a_2x^{-1},$$

from which the *Harris–Anders form* follows:

$$\mathbf{x}^0 = \sqrt{\frac{a_2}{a_1}} = \sqrt{\frac{2rc}{lT}}.$$

The same result yields the application of formula (6.26), where the optimal dual objective function value is

$$v(\delta^0) = \left(\frac{a_1}{1/2}\right)^{1/2} \left(\frac{a_2}{1/2}\right)^{1/2} = \sqrt{4a_1a_2} = \sqrt{2lTrc}.$$

At the optimum, the value of the dual objective function must coincide with the value of the primal function

$$y^0 = a_1x^0 + a_2x^{0^{-1}} = \frac{lT}{2}\sqrt{\frac{2rc}{lT}} + rc\sqrt{\frac{lT}{2rc}} = \sqrt{2lTrc}.$$

There are some extended versions of the basic economic lot size problem formulated as models of geometric programming in the literature. An economic order quantity model with obsolescence cost and an inventory model with shortage cost as geometric programming problems can be found in [10, pp. 688–689].

6.3.2 The Minimization of Cost

We again consider the producer's problem (1.7). It is to find, for a particular output level q^* and with a given structure of input prices, what input levels would constitute the cheapest way of producing this output and what would be the minimum cost. This question can be answered for all possible levels of output and the minimum cost would depend on the level of output to be produced.

Assuming that the technology is defined by the Cobb–Douglas production function (for simplicity, but without loss of generality, with two inputs only), the solution of model (1.9) will yield the cost function $C(\mathbf{r}, q)$, expressing minimum cost as a function of input prices $\mathbf{r}$ and output level q. A slight modification of the constraint in model (1.9) leads to the following geometric programming problem:

$$\begin{aligned} \underset{x_1,x_2}{\text{minimize}} \quad & M(\mathbf{x}) = r_1x_1 + r_2x_2 \\ \text{subject to} \quad & \frac{q^*}{a}x_1^{-\alpha}x_2^{-\beta} \leqq 1, \\ & x_1 > 0, \quad x_2 > 0. \end{aligned} \tag{6.29}$$

Problem (6.29) can be solved very easily because of the zero degree of difficulty ($d = T - n - 1 = 3 - 2 - 1 = 0$). It is obvious that this property is preserved also for $n > 2$ inputs in the Cobb–Douglas production function.

The normality and orthogonality conditions are

$$\begin{aligned} \delta_{01} + \delta_{02} &= 1, \\ \delta_{01} - \alpha\delta_{11} &= 0, \\ \delta_{02} - \beta\delta_{11} &= 0. \end{aligned}$$

We find that $\delta_{01}^0 = \frac{\alpha}{\alpha+\beta}$, $\delta_{02}^0 = \frac{\beta}{\alpha+\beta}$, and $\delta_{11}^0 = \frac{1}{\alpha+\beta}$. Because $\lambda_1^0 = \delta_{11}^0 > 0$, the constraint must be tight at optimality; this is, of course, quite obvious since production higher than the given level q^* increases the cost. The more interesting result provides the dual variables δ_{01} and δ_{02}. According to (6.26), they express the proportion of the cost of the first and second inputs, respectively, to the minimum cost of producing output level q^*. In other words, δ_{01} and δ_{02} determine the optimal cost structure. As in the economic lot size problem—again without knowing the optimal input levels—it follows from the above dual solution that the optimal cost structure is determined only by the elasticity coefficients α and β of the Cobb–Douglas production and is *independent* of the output level. It is a well-known result in microeconomic theory that the expansion path generated by the Cobb–Douglas production function is linear: Independently of the change of the output level, the structure of the input cost remains constant.

Using the dual function at $\boldsymbol{\delta}^0$,

$$\begin{aligned} v(\boldsymbol{\delta}^0) &= \left(\frac{r_1}{\delta_{01}^0}\right)^{\delta_{01}^0} \left(\frac{r_2}{\delta_{02}^0}\right)^{\delta_{02}^0} \left(\frac{q^*}{a\delta_{11}^0}\right)^{\delta_{11}^0} \delta_{11}^{0\ \delta_{11}^0} \\ &= (\alpha+\beta)\left(\frac{r_1}{\alpha}\right)^{\frac{\alpha}{\alpha+\beta}} \left(\frac{r_2}{\beta}\right)^{\frac{\beta}{\alpha+\beta}} \left(\frac{q^*}{a}\right)^{\frac{1}{\alpha+\beta}}, \end{aligned} \tag{6.30}$$

and formula (6.28),

$$r_1 x_1^0 = \delta_{01}^0 v(\boldsymbol{\delta}^0), \qquad r_2 x_2^0 = \delta_{02}^0 v(\boldsymbol{\delta}^0),$$

we obtain the optimal input quantities, x_i^0, as a function of input prices, and the output level:

$$x_1^0 = \left(\frac{\alpha r_2}{\beta r_1}\right)^{\frac{\beta}{\alpha+\beta}} \left(\frac{q^*}{a}\right)^{\frac{1}{\alpha+\beta}}$$

and

$$x_2^0 = \left(\frac{\beta r_1}{\alpha r_2}\right)^{\frac{\alpha}{\alpha+\beta}} \left(\frac{q^*}{a}\right)^{\frac{1}{\alpha+\beta}}.$$

These functions are the *conditional* (on the level of output) input demand functions and are homogeneous of degree zero in input prices. Thus given an output level, a proportional increase or decrease in all input prices will leave unchanged all input demands, and hence only relative prices matter.

The cost function is given by

$$\begin{aligned} C(\mathbf{r}, q) &= r_1 x_1^0 + r_2 x_2^0 \\ &= r_1^{\frac{\alpha}{\alpha+\beta}} r_2^{\frac{\beta}{\alpha+\beta}} \left(\frac{q^*}{a}\right)^{\frac{1}{\alpha+\beta}} \left[\left(\frac{\alpha}{\beta}\right)^{\frac{\beta}{\alpha+\beta}} + \left(\frac{\beta}{\alpha}\right)^{\frac{\alpha}{\alpha+\beta}}\right]. \end{aligned} \tag{6.31}$$

The reader can verify that at the optimum the values of the dual function (6.30) and the cost function (6.31) (due to Theorem 6.2) coincide. Under constant returns to scale, that is, $\alpha + \beta = 1$, the cost function (6.31) becomes

$$C(\mathbf{r}, q) = r_1^{\alpha} r_2^{1-\alpha} \frac{q^*}{a} \left[\left(\frac{\alpha}{1-\alpha} \right)^{1-\alpha} + \left(\frac{1-\alpha}{\alpha} \right)^{\alpha} \right] = r_1^{\alpha} r_2^{1-\alpha} \frac{q^*}{a} \alpha^{-\alpha} (1-\alpha)^{\alpha-1}.$$

In a similar way, problem (1.10) can be analyzed. Since the output q is positive, instead of maximization of q, we minimize q^{-1}, and model (1.10) can be rewritten as a geometric programming problem:

$$\begin{aligned} &\text{minimize} && q^{-1} = \frac{1}{a} x_1^{-\alpha} x_2^{-\beta} \\ &\text{subject to} && \frac{r_1}{M} x_1 + \frac{r_2}{M} x_2 \leqq 1, \\ &&& x_1 > 0, \quad x_2 > 0. \end{aligned} \tag{6.32}$$

It is again a problem with zero degree of difficulty. The dual problem is

$$\begin{aligned} &\text{maximize} && v(\boldsymbol{\delta}) = \left(\frac{1}{a\delta_0} \right)^{\delta_0} \left(\frac{r_1}{M\delta_{11}} \right)^{\delta_{11}} \left(\frac{r_2}{M\delta_{12}} \right)^{\delta_{12}} \lambda_1(\boldsymbol{\delta})^{\lambda_1(\boldsymbol{\delta})} \\ &\text{subject to} && \delta_0 = 1, \\ &&& -\alpha\delta_0 + \delta_{11} = 0, \\ &&& -\beta\delta_0 + \delta_{12} = 0, \\ &&& \delta_{11} \geqq 0, \quad \delta_{12} \geqq 0, \end{aligned}$$

where $\lambda_1(\boldsymbol{\delta}) = \delta_{11} + \delta_{12}$. The dual solution is easy to find: $\delta_0^0 = 1$ (there is only one term in the primal function), $\delta_{11}^0 = \alpha$, and $\delta_{12}^0 = \beta$.

Using the dual function at $\boldsymbol{\delta}^0$,

$$v(\boldsymbol{\delta}^0) = \frac{1}{a} \left(\frac{r_1}{\alpha M} \right)^{\alpha} \left(\frac{r_2}{\beta M} \right)^{\beta} (\alpha + \beta)^{(\alpha+\beta)},$$

and formula (6.28),

$$r_1 x_1^0 = \frac{\delta_{11}^0}{\lambda_1^0}, \qquad r_2 x_2^0 = \frac{\delta_{12}^0}{\lambda_1^0},$$

the optimal input quantities x_i^0 will be obtained:

$$x_1^0 = \frac{\alpha M}{(\alpha+\beta) r_1} \quad \text{and} \quad x_2^0 = \frac{\beta M}{(\alpha+\beta) r_2}.$$

The demand for input i is a decreasing function of its price r_i and an increasing function of the available budget M. The maximal output level is

$$q^0 = \frac{1}{a} \left(\frac{r_1}{\alpha M} \right)^{\alpha} \left(\frac{r_2}{\beta M} \right)^{\beta} (\alpha + \beta)^{(\alpha+\beta)},$$

which coincides with the value of the dual function $v(\boldsymbol{\delta}^0)$.

Looking at the dual variables δ_{11}^0 and δ_{12}^0, an interesting interpretation is offered. Because $\delta_{11}^0 = \alpha$ and $\delta_{12}^0 = \beta$, they indicate the percentage increase of production due to the increase of the first and second inputs, respectively, by 1%. Hence the Lagrange multiplier $\lambda_1^0 = \delta_{11}^0 + \delta_{12}^0 = \alpha + \beta$; it can be interpreted as the scale elasticity coefficient. Under constant returns to scale, it equals 1; under increasing (decreasing) returns to scale, it is greater (smaller) than 1.

6.3.3 The Economic Interpretation of Dual Variables as Elasticity Coefficients

In the production model (6.32), the dual variables corresponding to terms in the forced constraints were exactly equal to the elasticity coefficients of the underlying Cobb–Douglas production function. Dual variables as elasticity coefficients can be important indicators in economic analysis because they are without dimension and therefore easily comparable. The question that now arises is whether such an interpretation of dual variables is valid in general or specifically for the underlying model.

Let us consider the following mathematical programming problem:

$$\begin{aligned}
&\text{minimize} && g_0(\mathbf{x}) \\
&\text{subject to} && f_k(\mathbf{x}) = \sum_{t=1}^{T_k} f_{kt}(\mathbf{x}) \leqq b_k \quad (k = 1, 2, \ldots, m), \\
& && x_j > 0 \quad (j = 1, 2, \ldots, n),
\end{aligned} \tag{6.33}$$

where $g_0(\mathbf{x})$ and $f_k(\mathbf{x})$ $(k = 1, 2, \ldots, m)$ are posynomials and b_k $(k = 1, 2, \ldots, m)$ are given positive real numbers. Problem (6.33) written as a geometric programming problem has the form

$$\begin{aligned}
&\text{minimize} && g_0(\mathbf{x}) \\
&\text{subject to} && g_k(\mathbf{x}) = \frac{f_k(\mathbf{x})}{b_k} \leqq 1 \quad (k = 1, 2, \ldots, m), \\
& && x_j \geqq 0 \quad (j = 1, 2, \ldots, n).
\end{aligned} \tag{6.34}$$

For the Lagrange multipliers u_k^0 $(k = 1, 2, \ldots, m)$ of problem (6.33), the following holds (see (2.9)):

$$\frac{\partial g_0(\mathbf{x}^0(\mathbf{b}))}{\partial b_k} = -u_k^0 \quad (k = 1, 2, \ldots, m).$$

Under the same assumptions, the following property [12, pp. 67–68] can be proven for the Lagrange multipliers μ_k^0 $(k = 1, 2, \ldots, m)$ of problem (6.34).

Theorem 6.4. *If the derivation* $\frac{\partial g_0(\mathbf{x}^0(\mathbf{b}))}{\partial b_k}$ *exists, then it holds that*

$$\frac{\partial g_0(\mathbf{x}^0(\mathbf{b}))}{\frac{\partial b_k}{b_k}} = -\mu_k^0 \quad (k = 1, 2, \ldots, m). \tag{6.35}$$

Hence the Lagrange multipliers μ_k^0 $(k = 1, 2, \ldots, m)$ express the change of the value of the primal function due to a change of the constraint b_k $(k = 1, 2, \ldots, m)$ by 1%.

Substituting (6.35) for μ_k^0 in (6.27) yields

$$\lambda_k(\boldsymbol{\delta}^0) = \frac{\frac{\partial g_0(\mathbf{x}^0(\mathbf{b}))b_k}{\partial b_k}}{g_0(\mathbf{x}^0)} = \frac{\frac{\partial g_o(\mathbf{x}^0(\mathbf{b}))}{g_0(\mathbf{x}^0)}}{\frac{\partial b_k}{b_k}} \quad (k = 1, 2, \ldots, m), \tag{6.36}$$

which is exactly the definition of *elasticity*. Hence $\lambda_k(\boldsymbol{\delta}^0)$ $(k = 1, 2, \ldots, m)$ indicates the percentage change of the value of the primal function when the component b_k $(k = 1, 2, \ldots, m)$ changes by 1%.

It follows from (6.28) that

$$\delta_{kt}^0 = \lambda_k(\boldsymbol{\delta}^0) c_{kt} x_1^{a_{kt1}} x_2^{a_{kt2}} \cdots x_n^{a_{ktn}} \quad (k = 1, 2, \ldots, m;\ t = 1, 2, \ldots, T_k). \tag{6.37}$$

For abbreviation, we denote

$$w_{kt} = c_{kt} x_1^{a_{kt1}} x_2^{a_{kt2}} \cdots x_n^{a_{ktn}} \quad (k = 1, 2, \ldots, m;\ t = 1, 2, \ldots, T_k)$$

and substitute (6.36) for $\lambda_k(\boldsymbol{\delta}^0)$ in (6.37):

$$\delta_{kt}^0 = \frac{\frac{\partial g_0(\mathbf{x}^0)}{g_0(\mathbf{x}^0)}}{\frac{\partial b_k}{b_k}} w_{kt} = \frac{\frac{\partial g_0(\mathbf{x}^0)}{g_0(\mathbf{x}^0)}}{\frac{\partial b_k}{b_k}} \frac{f_{kt}}{b_k} = \frac{\frac{\partial g_0(\mathbf{x}^0)}{g_0(\mathbf{x}^0)}}{\frac{\partial f_{k1} + \cdots + \partial f_{kt} + \cdots + \partial f_{kT_k}}{b_k}} \frac{f_{kt}}{b_k}.$$

Suppose that only one term in f_k will change and all others remain constant. Then we obtain

$$\delta_{kt}^0 = \frac{\frac{\partial g_0(\mathbf{x}^0)}{g_0(\mathbf{x}^0)}}{\frac{\partial f_{kt}}{f_{kt}}} \quad (k = 1, 2, \ldots, m;\ t = 1, 2, \ldots, T_k),$$

or the *elasticity* of the value of the primal function with respect to the term t in the inequality constraint k. Hence the dual variables δ_{kt}^0 $(t = 1, 2, \ldots, T_k;\ k = 1, 2, \ldots, m)$ *indicate the percentage change of the value of the primal function when the corresponding term in the forced constraints changes by* 1% (and all other terms remain constant). Therefore, we interpret the dual variables δ_{kt}^0 $(t = 1, 2, \ldots, T_k;\ k = 1, 2, \ldots, m)$ in geometric programming as elasticity coefficients in comparison with the interpretation of dual variables as marginal coefficients in linear programming.

6.3.4 An Open Input–Output Model with Continuous Substitution between Primary Factors

In Section 1.2.8, the basic input–output was extended by substitution possibilities for labor and capital inputs according to a Cobb–Douglas production function for each sector of the economy. Under the objective of labor input minimization and perfect transferability of capital between the particular sectors of the economy, the following geometric programming problem arises:

$$\text{minimize} \quad L = \sum_{j=1}^{n} L_j \tag{1.25}$$

subject to the forced constraints

$$\sum_{j \neq i} d_{ij} L_j^{\alpha_j} K_j^{\beta_j} L_i^{-\alpha_i} K_i^{-\beta_i} + c_i L_i^{-\alpha_i} K_i^{-\beta_i} \leqq 1 \quad (i = 1, 2, \ldots, n), \tag{1.22}$$

where

$$d_{ij} = \frac{a_{ij}\varepsilon_j}{(1 - a_{ii})\varepsilon_i} \geqq 0 \quad \text{and} \quad c_i = \frac{y_i}{(1 - a_{ii})\varepsilon_i} > 0 \quad (i, j = 1, 2, \ldots, n),$$

due to the standard assumption $a_{ii} < 1$ and the positivity of the final demand,

$$\text{capital constraint} \quad \frac{1}{K} \sum_{j=1}^{n} K_j \leqq 1, \tag{1.23'}$$

and the natural constraints

$$L_j > 0, \quad K_j > 0 \quad (j = 1, 2, \ldots, n). \tag{1.26}$$

The conditions in (1.22) imply that the total sum of the proportions of the deliveries from sector i (into all other sectors and to the final demand) to the net production of sector i cannot be greater than one. In other words, the gross production of sector i must be sufficient to meet the deliveries of this sector to all other sectors of the economy and to the exogenously given final demand.

Condition (1.23′) expresses the constraint for the disposable capital input in the economy that can be allocated between the particular sectors.

The solution of this geometric programming model yields—for exogenously given final demand—the optimal allocation of labor and capital to the particular sectors of the economy and thereupon, due to the Cobb–Douglas production function (1.21), the optimal gross production for each sector of the economy.

To get deeper insight into the interdependencies described by model (1.25) subject to (1.22), (1.23′), and (1.26), we turn to the corresponding dual model. For simplicity, but without loss of generality, let us reduce the number of sectors to only two sectors ($n = 2$). In this case, the dual model takes the form

$$\begin{aligned} \text{maximize} \quad v(\boldsymbol{\delta}) = {} & \left(\frac{1}{\delta_{01}}\right)^{\delta_{01}} \left(\frac{1}{\delta_{02}}\right)^{\delta_{02}} \left(\frac{d_{12}}{\delta_{11}}\right)^{\delta_{11}} \left(\frac{c_1}{\delta_{12}}\right)^{\delta_{12}} \\ & \cdots \left(\frac{d_{21}}{\delta_{21}}\right)^{\delta_{21}} \left(\frac{c_2}{\delta_{22}}\right)^{\delta_{22}} \left(\frac{\frac{K_1}{K}}{\delta_{31}}\right)^{\delta_{31}} \left(\frac{\frac{K_2}{K}}{\delta_{32}}\right)^{\delta_{32}} \\ & \cdots \lambda_1(\boldsymbol{\delta})^{\lambda_1(\boldsymbol{\delta})} \lambda_2(\boldsymbol{\delta})^{\lambda_2(\boldsymbol{\delta})} \lambda_3(\boldsymbol{\delta})^{\lambda_3(\boldsymbol{\delta})}, \end{aligned}$$

where

$$\lambda_k(\boldsymbol{\delta}) = \sum_{t=1}^{T_k} \delta_{kt} \quad \text{for } k = 1, 2, 3,$$

subject to $\quad \delta_{01} \geqq 0, \quad \delta_{02} \geqq 0, \quad \ldots, \quad \delta_{32} \geqq 0,$

$$\begin{aligned}
\delta_{01} + \delta_{02} &= 1, \\
\delta_{01} - \alpha_1\delta_{11} - \alpha_1\delta_{12} + \alpha_1\delta_{21} &= 0, \\
\delta_{02} + \alpha_2\delta_{11} - \alpha_2\delta_{21} - \alpha_2\delta_{22} &= 0, \\
-\beta_1\delta_{11} - \beta_1\delta_{12} + \beta_1\delta_{21} + \delta_{31} &= 0, \\
\beta_2\delta_{11} - \beta_2\delta_{21} - \beta_2\delta_{22} + \delta_{32} &= 0,
\end{aligned}$$

where the first index at δ_{kt} relates to the constraint (or to the primal function) and the second to the term in this constraint (or in the primal function). The number of dual variables is equal to the number of terms in the primal problem (for two sectors, there are eight), and to every primal variable there corresponds one orthogonality condition. Thereupon the degree of difficulty is 3 ($d = T - (n + 1)$).

The economic interpretation of the dual variables δ_{01} and δ_{02} corresponding to the term in the primal function follows immediately from (6.26). They give us the proportion of labor input (employment) of sector one and two, respectively, to the total labor input (employment) in the economy.

According to the interpretation of dual variables δ_{kt} ($k = 1, 2, \ldots, m$; $t = 1, 2, \ldots, T_k$) as elasticity coefficients, the variable $\lambda_1(\boldsymbol{\delta})$ indicates the percentage increase of total employment due to increase of net production in sector one by 1%. This effect can be decomposed in two parts (in the general case of n sectors to n parts): δ^0_{11} describes the percentage increase of total employment due to increase of the deliveries of sector one to sector two by 1%; δ^0_{12} indicates the percentage increase of total employment due to increase of the final demand of sector one by 1%. In a similar way, the dual variables $\lambda_2(\boldsymbol{\delta}^0)$ and δ^0_{21}, δ^0_{22} can be interpreted.

The dual variable $\lambda_3(\boldsymbol{\delta}^0)$ describes the substitution effect; it indicates the percentage decrease of total employment when the disposable capital stock increases by 1%.

It is worthwhile to note that either all δ^0_{kt} for given k are equal to zero or all δ^0_{kt} are positive. If $\lambda_k(\boldsymbol{\delta}^0) = 0$ for a given k, then all δ^0_{kt} must be zero because of the nonnegativity condition for dual variables δ^0_{kt}. If $\delta^0_{kt} = 0$ for given k and one t, then it follows from (6.37) that $\lambda_k(\boldsymbol{\delta}^0) = 0$. Conversely, $\lambda_k(\boldsymbol{\delta}^0) > 0$ implies $\delta^0_{kt} > 0$ for all t. From the economic interpretation point of view, this result follows from the following property of the Cobb–Douglas production function:

$$F(L, 0) = 0, \qquad F(0, K) = 0.$$

In other words, both production factors are essential, which coincides with the positivity or natural constraints for primal variables in the model of geometric programming.

One of the most important questions in the framework of input–output analysis is that of how changes in the exogenously given final demand influence gross production and consequently the employment in the economy. (For example, what are the production and employment effects for the Austrian economy due to increasing international trade with the transforming Central European countries?) Under the assumption of the Leontief production function with fixed input coefficients, the answer is given by (1.15)—extended by labor input coefficients—and can be found in every textbook on input–output analysis (see, e.g., [14]). How do we estimate the employment effect under the assumption of substitution possibilities between labor and capital described by the Cobb–Douglas production function?

Applying the technique developed by Beightler and Phillips [6] called the *geometric programming inflationary rule*, model (1.25) subject to (1.22), (1.23′), and (1.26) provides an answer to the question above.

Suppose that the final demand $\mathbf{y}$ will be increased, and we have a new vector $\mathbf{y}^*$ (all other parameters remain unchanged). This implies new coefficients $c_1^*, c_2^*, \ldots, c_n^*$ in the constraints (1.22). How will the value of primal function or total employment in the economy be changed?

Due to Theorem 6.2, it holds that

$$\begin{aligned} L^0 = v(\boldsymbol{\delta}^0) = {} & \left(\frac{1}{\delta_{01}^0}\right)^{\delta_{01}^0} \left(\frac{1}{\delta_{02}^0}\right)^{\delta_{02}^0} \cdots \left(\frac{d_{12}}{\delta_{11}^0}\right)^{\delta_{11}^0} \left(\frac{d_{13}}{\delta_{12}^0}\right)^{\delta_{12}^0} \cdots \left(\frac{c_1}{\delta_{1n}^0}\right)^{\delta_{1n}^0} \\ & \cdots \left(\frac{d_{21}}{\delta_{21}^0}\right)^{\delta_{21}^0} \cdots \left(\frac{c_2}{\delta_{2n}^0}\right)^{\delta_{2n}^0} \cdots \left(\frac{c_n}{\delta_{nn}^0}\right)^{\delta_{nn}^0} \left(\frac{\frac{K_1}{K}}{\delta_{n+1,1}^0}\right)^{\delta_{n+1,1}^0} \\ & \cdots \left(\frac{\frac{K_n}{K}}{\delta_{n+1,n}^0}\right)^{\delta_{n+1,n}^0} \lambda_1(\boldsymbol{\delta}^0)^{\lambda_1(\boldsymbol{\delta}^0)} \lambda_2(\boldsymbol{\delta}^0)^{\lambda_2(\boldsymbol{\delta}^0)} \\ & \cdots \lambda_n(\boldsymbol{\delta}^0)^{\lambda_n(\boldsymbol{\delta}^0)} \lambda_{n+1}(\boldsymbol{\delta}^0)^{\lambda_{n+1}(\boldsymbol{\delta}^0)}, \end{aligned}$$

where L^0 is the optimal value of the primal function (the optimal level of total employment) and $v(\boldsymbol{\delta}^0)$ is the optimal value of the dual function corresponding to the original final demand $\mathbf{y}$. The vector of dual variables $\boldsymbol{\delta}$ remains feasible although the final demand, and consequently the coefficients c_i $(i = 1, 2, \ldots, n)$, will change.

From the main lemma of geometric programming, it follows that

$$L^* \geqq v^*(\boldsymbol{\delta}^0), \tag{6.38}$$

where L^* denotes the new level of total employment and $v^*(\boldsymbol{\delta}^0)$ is the value of dual function corresponding to the solution $\boldsymbol{\delta}^0$ of the dual problem, but with the new coefficients c_i^* $(i = 1, 2, \ldots, n)$ in the dual function.

Dividing both sides of (6.38) by L^0, we obtain

$$\frac{L^*}{L^0} \geqq \left(\frac{c_1^*}{c_1}\right)^{\delta_{1n}^0} \left(\frac{c_2^*}{c_2}\right)^{\delta_{2n}^0} \cdots \left(\frac{c_n^*}{c_n}\right)^{\delta_{nn}^0} = \left(\frac{y_1^*}{y_1}\right)^{\delta_{1n}^0} \left(\frac{y_2^*}{y_2}\right)^{\delta_{2n}^0} \cdots \left(\frac{y_n^*}{y_n}\right)^{\delta_{nn}^0}, \tag{6.39}$$

or

$$L^* \geqq L^0 \left(\frac{y_1^*}{y_1}\right)^{\delta_{1n}^0} \left(\frac{y_2^*}{y_2}\right)^{\delta_{2n}^0} \cdots \left(\frac{y_n^*}{y_n}\right)^{\delta_{nn}^0}. \tag{6.40}$$

The right-hand side of the form (6.39) provides a lower bound on the relative increase of employment due to the increase of the final demand, or in other words, a lower bound on *the nonlinear employment multiplier*.

Using the form (6.40)—without estimating the new solution for the primal and dual model—we get a lower bound for the new level of employment corresponding to the new vector of final demand.

Another interesting aspect of the model under consideration is related to the transferability of capital between the particular sectors of the economy. As mentioned in Section 1.2.8, the opposite case to the perfect transferability postulated in model (1.25) subject to (1.22), (1.23′), and (1.26) is a model with constraints for capital stock in each sector of the economy:

$$K_j \leqq \overline{K_j} \quad (j = 1, 2, \dots, n), \tag{1.24}$$

where $\overline{K_j}$ indicates the disposable capital stock in sector j.

The question that arises now is, under what conditions are the solution of the model with perfect transferability (constraint (1.23)) and the solution of the model with nontransferability of capital (constraints (1.24)) the same? To find an answer, we will compare the following geometric programming problems:

$$\begin{aligned} &\text{minimize} && g_0(\mathbf{x}) \\ &\text{subject to} && g_k(\mathbf{x}) \leqq 1 \quad (k = 1, 2, \dots, m), \\ &&& g_{m+i}(\mathbf{x}) \leqq 1 \quad (i = 1, 2, \dots, l), \\ &&& \mathbf{x} \geqq \mathbf{0} \end{aligned} \tag{A}$$

and

$$\begin{aligned} &\text{minimize} && g_0(\mathbf{x}) \\ &\text{subject to} && g_k(\mathbf{x}) \leqq 1 \quad (k = 1, 2, \dots, m), \\ &&& \sum_{i=1}^{l} \gamma_i g_{m+i}(\mathbf{x}) \leqq 1, \\ &&& \mathbf{x} \geqq \mathbf{0}, \end{aligned} \tag{B}$$

where $g_k(\mathbf{x}) = \sum_{t=1}^{T_k} c_{kt} \prod_{j=1}^{n} x_j^{a_{ktj}}$ $(k = 0, 1, \dots, m)$, $g_{m+1}(\mathbf{x}) = c_{m+i} \prod_{j=1}^{n} x_j^{a_{m+ij}}$ $(i = 1, 2, \dots, l)$ with $c_{kt} > 0$ $(t = 1, 2, \dots, T_k;\ k = 0, 1, 2, \dots, m, m+1, \dots, m+l)$, $\gamma_i \geqq 0$ $(i = 1, 2, \dots, l)$, and $\sum_{i=1}^{l} \gamma_i = 1$. The coefficients γ_i $(i = 1, 2, \dots, l)$ are called *surrogate multipliers* [6].

Let us denote the optimal primal solution of problem (A) by $\mathbf{x}_{\mathrm{A}}^0$ and the optimal dual solution by $\boldsymbol{\delta}_{\mathrm{A}}^0$, and denote the optimal primal solution of problem (B) by $\mathbf{x}_{\mathrm{B}}^0$ and the optimal dual solution by $\boldsymbol{\delta}_{\mathrm{B}}^0$. Obviously, problem (A) corresponds to the above-formulated input–output model under nontransferability of capital, and problem (B) to the model with perfect transferability of capital across the sectors of the economy. It is left to the reader to prove the following result.

Proposition 6.1. *If the surrogate multipliers* γ_i $(i = 1, 2, \ldots, l)$ *are proportional to the elasticity coefficients* $\delta^0_{m+i,\mathrm{A}}$*, i.e.,* $\gamma_i = b\delta^0_{m+i,\mathrm{A}}$ *for certain* $b > 0$ *and* $i = 1, 2, \ldots, l$*, then models* (A) *and* (B) *yield the same optimal allocation of labor and capital to the particular sectors of the economy.*

6.4 Transformation of Some Optimization Problems into Standard Geometric Programming Models

Many optimization problems can be transformed into standard geometric programming problems, even though they are not explicitly expressed in posynomial form.

Let us return to the input–output model that allows primary factor substitution, as described in Section 6.3.4. However, instead of the Cobb–Douglas production function, the more general-type so-called constant elasticity of substitution (CES) production function [3] is postulated:

$$x_j = \varepsilon_j\{(1-\beta_j)L_j^{-\rho_j} + \beta_j K_j^{-\rho_j}\}^{-\frac{1}{\rho_j}} \quad (j = 1, 2, \ldots, n), \tag{6.41}$$

where ε_j is an efficiency parameter, β_j $(0 < \beta_j < 1)$ is a distribution parameter, and ρ_j characterizes the elasticity of factor substitution.[1] The elasticity of substitution between the factors capital and labor is defined as

$$\sigma_j = \frac{d\log(K_j|L_j)}{d\log R_j} = \frac{L_j}{K_j}R_j\frac{d(K_j|L_j)}{dR_j} \quad (j = 1, 2, \ldots, n),$$

where $R_j = -\frac{dK_j}{dL_j}$ is the marginal rate of substitution. The parameter ρ_j is interpreted in terms of the constant elasticity of substitution σ_j:

$$\sigma_j = \frac{1}{1+\rho_j}, \quad \text{or} \quad \rho_j = \frac{1}{\sigma_j} - 1.$$

Two limiting cases of the CES function arise. As $\rho_j \to \infty$ (and $\sigma_j \to 0$), there ceases to be any substitution between factors and the production function becomes one of fixed coefficients. As $\rho_j \to 0$ (and $\sigma_j \to 1$), the function is replaced by the Cobb–Douglas form.

Taking the CES production function (6.41) instead of the Cobb–Douglas function (1.21) into the constraints (1.17), we get

$$\sum_{j=1}^{n}(\delta_{ij} - a_{ij})\varepsilon_j\{(1-\beta_j)L_j^{-\rho_j} + \beta_j K_j^{-\rho_j}\}^{-\frac{1}{\rho_j}} - y_i \geqq 0 \quad (i = 1, 2, \ldots, n).$$

This system of inequalities can be written as

[1] An input–output model with a CES production function can be found in [22, pp. 138–164].

$$\sum_{j \neq i}^{n} d_{ij}\{(1-\beta_j)L_j^{-\rho_j} + \beta_j K_j^{-\rho_j}\}^{-\frac{1}{\rho_j}}\{(1-\beta_i)L_i^{-\rho_i} + \beta_i K_i^{-\rho_i}\}^{\frac{1}{\rho_i}}$$
$$+ \frac{y_i}{(1-a_{ii})\varepsilon_i}\{(1-\beta_i)L_i^{-\rho_i} + \beta_i K_i^{-\rho_i}\}^{\frac{1}{\rho_i}} \leqq 1 \quad (i = 1, 2, \ldots, n), \tag{6.42}$$

where $d_{ij} = \frac{a_{ij}\varepsilon_j}{(1-a_{ii})\varepsilon_i} \geqq 0 \ (i, j = 1, 2, \ldots, n)$.

We introduce the constraint

$$\{(1-\beta_j)L_j^{-\rho_j} + \beta_j K_j^{-\rho_j}\}N_j^{-1} \leqq 1 \quad (j = 1, 2, \ldots, n), \tag{6.43}$$

where N_j is an additional independent variable. Then the constraints (6.42) are rewritten as

$$\begin{aligned} d_{12}N_2^{-\frac{1}{\rho_2}} N_1^{\frac{1}{\rho_1}} + \cdots + d_{1n}N_n^{-\frac{1}{\rho_n}} N_1^{\frac{1}{\rho_1}} &+ \frac{y_1}{(1-a_{11})\varepsilon_1} N_1^{\frac{1}{\rho_1}} \leqq 1, \\ \vdots \qquad\qquad \vdots \qquad &\qquad\qquad \vdots \\ d_{n1}N_1^{-\frac{1}{\rho_1}} N_n^{\frac{1}{\rho_n}} + \cdots + d_{nn-1}N_{n-1}^{\overline{\rho_{n-1}}^{1}} N_n^{\frac{1}{\rho_n}} &+ \frac{y_n}{(1-a_{nn})\varepsilon_n} N_n^{\frac{1}{\rho_n}} \leqq 1. \end{aligned} \tag{6.44}$$

Hence we have transformed the problem of minimizing the labor input (as described by (1.25)) subject to the constraints (6.42), (1.23′), and (1.26) into the standard geometric programming problem

$$\text{minimize} \quad L = \sum_{j=1}^{n} L_j$$

subject to the forced constraints (6.43)–(6.44) and (1.23′) and the natural constraints

$$L_j > 0, \qquad K_j > 0, \qquad N_j > 0 \quad (j = 1, 2, \ldots, n).$$

The reader can verify that for $\rho_j > 0$ (or $\sigma_j < 1$), the constraints (6.43) at the optimum are fulfilled as equalities (due to the objective function (1.25)), and thereupon the equivalence of the constraints (6.42) and (6.44) is ensured. From an economic interpretation point of view, the elasticity of substitution $\sigma_j > 1$ $(-1 < \rho_j < 0)$ implies that output per capita increases indefinitely as capital used (relative to labor) increases. On the other hand, as the use of labor increases ($\frac{K_j}{L_j} \to 0$), output per capita declines to the limit $(1-\beta_j)^{-1/\rho_j}$ [2, p. 54]. There is also some empirical evidence (see, e.g., [3, 5, 19]) confirming the assumption of elasticity of substitution below one.

As a second problem that can be transformed into standard geometric programming model, let us consider the marketing mix problem for a large brewery described by Balachandran and Gensch [4]. In this paper, the authors analyze the question of how the different marketing decision variables (price, advertising expenditure, the number of salesmen, etc.) should be combined in order to maximize sales. For this purpose, Balachandran and Gensch [4, pp. 163–168] estimated the function for sales, denoted by x_0, as

$$
\begin{aligned}
\text{sales}(x_0) = 16.212 &+ 3.937 A2_{t-1}^{0.91} B_t^{1.31} \\
&- 0.0021 A1_{t-1}^{-0.95} P_t^{-0.68} T_t^{-0.84} C_t^{-0.28} - 0.00305 A_{c_t}^{-0.18} Q_t^{1.76} \\
&- 0.0046 I_t^{-0.9} S_{t-1}^{-1.1} - 0.0053 P_t^{-0.76} D_t^{-1.12},
\end{aligned} \tag{6.45}
$$

with the following interpretation of the marketing decision variables:

- The relative advertising expenditure A_t is defined as the ratio of dollars spent on advertising by the particular firm during a given time period to the total industry advertising effort in dollars spent. $A1_t$ represents advertising emphasizing price, and $A2_t$ represents "mood" or image advertising.
- The relative in-store promotion I_t is defined as the ratio of dollars spent on in-store promotion by the particular firm during a given time period (e.g., displays, signs, and small customer gifts) to the total industries' in-store effort in dollars spent.
- The relative price P_t is the retail price of the firm's product divided by the average price charged by competing firms.
- The relative price differential C_t is the dollar change in retail price from one time period to the next divided by the average retail price change in the industry.
- T_t denotes special discounts a firm allows its wholesalers and retailers, and S_t denotes the total salary and commissions of salesmen.
- D_t represents the availability of the firm's product and is defined as the percentage of weighted (by the quantity of the product sold) retail outlets that carried the firm's brand.
- The last three marketing variables are relative packaging (B_t), relative quality (Q_t), and the age composition of the population (A_{c_t}).

The brewery will maximize total beer sales (x_0) under some budget constraints that are imposed by the management on the marketing mix variables. Without going into details (see [4, p. 169]), we can summarize these constraints as

$$
g_k(\mathbf{x}) \leqq 1 \quad (k = 1, 2, \ldots, m), \tag{6.46}
$$

where $g_k(\mathbf{x})$ are polynomials.

Instead of maximizing the sales function (6.45), we minimize

$$
-x_0 = f(\mathbf{x}) - u(\mathbf{x}),
$$

where

$$
\begin{aligned}
f(\mathbf{x}) = 0.0021 A1_{t-1}^{-0.95} P_t^{-0.68} T_t^{-0.84} C_t^{-0.28} &+ 0.0035 A_{c_t}^{-0.18} Q_t^{1.76} \\
+ 0.0046 I_t^{-0.9} S_{t-1}^{-1,1} &+ 0.0053 P_t^{-0.76} D_t^{-1.12}
\end{aligned}
$$

and

$$
u(\mathbf{x}) = 3.937 A2_{t-1}^{0.91} B_t^{1.31}.
$$

Suppose that the maximum value of x_0 is known to be positive. In this case, the constraint

$$u(\mathbf{x}) - f(\mathbf{x}) \geqq x_0,$$

or

$$x_0 + f(\mathbf{x}) - u(\mathbf{x}) \leqq 0,$$

is feasible. Then the maximization of function (6.45) subject to the constraints (6.46) is equivalent to the following geometric programming problem:

$$\begin{array}{ll} \text{minimize} & x_0^{-1} \\ \text{subject to} & x_0 u(\mathbf{x})^{-1} + f(\mathbf{x}) u(\mathbf{x})^{-1} \leqq 1, \\ & g_k(\mathbf{x}) \leqq 1 \quad (k = 1, 2, \ldots, m), \\ & x_0 > 0, \quad \mathbf{x} > \mathbf{0}, \end{array}$$

where $\mathbf{x}$ denotes the vector of the marketing decision variables described above.

In this way, a profit maximization problem with $f(\mathbf{x})$ as a cost function and $u(\mathbf{x})$ as a revenue function (supposed with only one term) can be reformulated into a geometric programming model.

As shown by Duffin, Peterson, and Zener [9, pp. 94–97], optimization problems with functions of the form

$$G(\mathbf{x}) = \sum_i \prod_j \frac{[g_{ij}(\mathbf{x})]^{a_{ij}}}{[1 - p_{ij}(\mathbf{x})]^{b_{ij}}},$$

where both the functions $g_{ij}(\mathbf{x})$ and $p_{ij}(\mathbf{x})$ are posynomials and the constants a_{ij} and b_{ij} are positive, can be reduced to geometric programming. These generalized posynomials G can appear both as the function to be minimized and as functions in the forced constraints. For the minimization problem, it is assumed that the functions $1 - p_{ij}(\mathbf{x})$ are positive.

In concluding this section, we should mention that the additional primal variables (N_j in the input–output model and x_0 in the marketing mix problem) used to transform these problems to geometric programming models have a meaningful economic interpretation.

References and Further Reading

[1] R. Abrams, Optimization models for regional air pollution control, in A. Charnes and W. R. Synn, eds., *Mathematical Analysis of Decision Problems in Ecology*, Springer-Verlag, Berlin, 1975.

[2] R. G. D. Allen, *Macro-Economic Theory: Mathematical Treatment*, Macmillan, London, 1968.

[3] K. J. Arrow, H. B. Chenery, B. S. Minhas, and R. M. Solow, Capital–labor substitution and economic efficiency, *Rev. Econ. Statist.*, **43** (1961), 225–250.

[4] V. Balachandran and D. Gensch, Solving the marketing-mix problem using geometric programming, *Manage. Sci.*, **21** (1974), 160–171.

[5] J. R. Behrman, Sectoral elasticities of substitution between capital and labor in a developing economy: Time series analysis in the case of post-war Chile, *Econometrica*, **40**-2 (1972), 311–326.

[6] Ch. S. Beightler and D. T. Phillips, *Applied Geometric Programming*, Wiley, New York, 1976.

[7] J. J. Dinkel, G. A. Kochenberger, and Y. Seppala, On the solution of regional planning models via geometric programming, *Environ. Planning*, **5** (1973), 397–408.

[8] R. J. Duffin, Cost minimization problems treated by geometric means, *Oper. Res.*, **10** (1962), 668–675.

[9] R. J. Duffin, E. L. Peterson, and C. Zener, *Geometric Programming: Theory and Application*, Wiley, New York, 1967.

[10] H. A. Eiselt, G. Pederzoli, and C. L. Sandblom, *Continuous Optimization Models*, de Gruyter, Berlin, 1987.

[11] A. S. Kadas, An application of geometric programming to regional economics, *Rep. Regional Sci. Assoc.*, **34** (1975), 95–106.

[12] M. Luptáčik, Geometrische Programmierung und ökonomische Analyse, Mathematical Systems in Economics vol. 32, Verlag Anton Hain, Meisenheim am Glau, Germany, 1977.

[13] A. P. Mastenbroek and P. Nijkamp, A spatial environmental model for an optimal allocation of investments, in P. Nijkamp, ed., *Environmental Economics, Vol.* 2: *Methods*, Nijhoff Social Sciences, Leiden, The Netherlands, 1976.

[14] R. E. Miller and P. D. Blair, *Input–Output Analysis: Foundations and Extensions*, Prentice–Hall, Englewood Cliffs, NJ, 1985.

[15] B. Nicolleti and L. Mariani, Generalized polynomial programming: Its approach to the solution of some management science problems, *Eur. J. Oper. Res.*, **1** (1977), 239–246.

[16] P. Nijkamp, *Planning of Industrial Complexes by Means of Geometric Programming*, University Press, Rotterdam, 1972.

[17] P. Nijkamp and J. H. P. Paelinck, An interregional model of environmental choice, *Papers Regional Sci. Assoc.*, **31** (1977), 51–71.

[18] W. Pinney and J. Modi, An application of geometric programming to the capital budgeting problem, in *ORSA Meeting, San Juan, Puerto Rico*, ORSA (INFORMS), Hanover, MD, 1974, 46.

[19] A. Rainer, *Produktionsfunktionen der österreichischen Industrie*, Research Memorandum 88, Institute for Advanced Studies, Vienna, 1974.

[20] M. J. Rijckaert, Engineering applications of geometric programming, in M. Avriel, M. J. Rijckaert, and D. J. Wilde, eds., *Optimization and Design*, Prentice–Hall, Englewood Cliffs, NJ, 1973.

[21] J. K. Sengupta and J. H. Portillo-Campbell, The approach of geometric programming with economic applications, *Z. Staatswirtschaft*, **128** (1972), 437–455.

[22] J. Schumann, *Input–Output Analyse*, Springer-Verlag, Berlin, 1968.

[23] C. Zener, Mathematical aid in optimizing engineering designs, *Proc. Math. Acad. Sci.*, **47** (1961), 537–539.

[24] H. J. Zimmermann, *Einführung in die Grundlagen des Operations Research*, Verlag Moderne Industrie, Munich, 1971.

Part II

Multiobjective Optimization

Principeremo col definire un termine di ciu è comodo fare uso per scansare lungaggini. Diremo che i componenti di una collettività godono, in una certa posizione, del massimo di ofelimità, quanto è impossibile allontanarsi pochissimo da quella posizione giovando, o nuocendo, a tutti i componenti la collettività; ogni piccolissimo spostamento da quella posizione avendo necessariamente per effetto di giovare a parte dei componenti la collettivitàe di nuocere ad altri.

Wilfredo Pareto
1906

7

Fundamentals of Multiobjective Optimization

To manage a business is to balance a variety of needs and goals.

Milan Zelený (1982)

In all of the mathematical programming problems considered thus far, we have assumed that one particular objective function, such as the maximization of profit or minimization of cost, was prespecified by some decision maker. In general, however, there exist a large variety of objectives—including maximizing profit, revenue, and market share; increasing environmental quality; etc. Zeleny [89, p. 1] states that "multiple objectives are all around us." There are some empirical studies supporting this hypothesis. Smith, Boyes, and Peseau [70] found that for 557 large U.S. firms, sales revenue and profits were objectives followed by the firms. Beedles [6] used time series data for the time period 1929–1973 for three large firms and showed that the firms pursued sales revenue, profits, and stock price as their objectives. The relevance of different objectives in the relatively wide set of a firm's objectives was analyzed by Fritz [26]. The 1994 Nobel Prize winner for economics, Reinhard Selten, referring to the new developments in this field, told the Austrian newspaper *Die Presse* that "At the firms it is still more necessary to take into account that they are faced with multiple objectives" (July 28, 2001; translated from German).

A wide array of theoretical and empirical contributions to multiple-objective decision analysis can be found in the literature of the past three decades. (For a very good survey of the state of art, see Gal, Stewart, and Hanne [27]). The book by Ballestero and Romero [3] makes reference to a bibliographical survey by Steuer, Gardiner, and Gray [73] that reveals more than 1,200 reviewed journal articles published on multiple-criteria decision making just between 1987 and 1992. The majority of applications are on a microeconomic level (project and plan evaluation, capital budgeting, financial and investment planning, marketing policy, etc.). But macroeconomic policy analysis also has a long history of treating multiple goals.

An excellent survey on the mathematical development of multicriteria optimization, or the vector maximum problem, is provided by Stadler [71]. This review describes the development of the subject from 1776 to 1960.

M. Luptáčik, *Mathematical Optimization and Economic Analysis*,
Springer Optimization and Its Applications 36, DOI 10.1007/978-0-387-89552-9_7,

That starting point of 1776 coincides exactly with the development of utility theory as given by Stigler [74]. In Stadler's view, "The mathematical foundation of the consideration of N different desiderata in an optimization process have their origin in economics, in particular in the development of welfare theory and utility theory" [71, p. 16].

This chapter is organized as follows: In Section 7.1, we present some examples of multiobjective programming models in economics. The extension of the Kuhn–Tucker conditions to a vector minimization problem is given in Section 7.2, and Section 7.3 deals with duality theory in multiobjective optimization. The behavior of the firm facing a bicriteria objective under regulatory constraint is analyzed in Section 7.4.

7.1 Examples of Multiobjective Programming Models in Economics

In the past three decades, multiple-objective decision analysis has become one of the most promising methodologies; it enhances the quality of decision making by providing deeper insight into the structure of the modeled system and the trade-offs that have to be made with respect to the objectives—an operational framework for actual decision making.

Let us start with some examples from different fields of economics.

7.1.1 Welfare Economics

Welfare economics is concerned with the conditions that determine the total economic welfare of a society. In the broadest sense, the welfare of a society depends on the levels of satisfaction of all of its individuals or social groups. But almost every alternative economic state to be judged by welfare economists will have favorable effects on some people (or social groups) and unfavorable effects on others. This implies interpersonal comparability of utility. But there is no obvious way to determine whether individual A or individual B derives more satisfaction from the consumption of a given bundle of goods. In order to dispense with interpersonal comparability of utility, Lange [44] proposed defining the social welfare not as the sum of the utilities of the individuals (a scalar quantity) but as a *vector*. The utilities of the individuals are the components of this vector. Let there be n individuals or social groups in the community, and let u_i be the utility of the ith individual. Total welfare is then the vector

$$\mathbf{u} = (u_1, u_2, \dots, u_n).$$

Each component of $\mathbf{u}$ measures the utility or welfare of the corresponding individual or social group.

Let the utility of each individual depend on the amount of commodities in his possession. Denoting by $x_{i1}, x_{i2}, \dots, x_{im}$ the quantities of m commodities in the possession of the ith individual, his utility (function) is

$$u_i = u_i(x_{i1}, x_{i2}, \dots, x_{im}).$$

The total amount of the jth commodity is then given by

$$x_j = \sum_{i=1}^{n} x_{ij} \quad (j = 1, 2, \dots, m).$$

These amounts depend on technological transformation, described by a function $f(x_1, x_2, \dots, x_m) = 0$. The problem is to maximize total welfare subject to the constraint of the transformation function.

Without reference to Pareto, Lange [44, p. 216] defines an optimum as follows: "A *maximum* of total welfare occurs when conditions cannot be changed so as to increase the vector **u**, i.e., when it is impossible to increase the utility of any person without decreasing that of others."

Thereupon he formulated the following optimization problem for the ith individual $(i = 1, 2, \dots, n)$:

$$\text{maximize} \quad u_i(x_{i1}, x_{i2}, \dots, x_{im}) \tag{7.1}$$

$$\text{subject to} \quad u_k(x_{k1}, x_{k2}, \dots, x_{kn}) = \text{const} \quad (k = 1, \dots, i-1, i+1, \dots, n), \tag{7.2}$$

$$x_j = \sum_{i=1}^{n} x_{ij} \quad (j = 1, 2, \dots, m), \tag{7.3}$$

$$f(x_1, x_2, \dots, x_m) = 0. \tag{7.4}$$

Lange [44, pp. 216–217] notes that this is equivalent to maximizing the Lagrange function

$$L(\mathbf{x}, \boldsymbol{\mu}, \nu) = \sum_{i=1}^{n} \alpha_i u_i(\mathbf{x}_i) + \sum_{j=1}^{m} \mu_j \left(\sum_{i=1}^{n} x_{ij} - x_j \right) + \nu f(x_1, x_2, \dots, x_m).$$

Besides the technological transformation (7.4), an optimal solution of the problem (7.1)–(7.4) is determined by the constants on the right-hand side of (7.2). As indicated by Lange [44], it follows from the maximization of the individuals' utilities that these constants are uniquely related to the money incomes of the respective individuals:

$$\text{maximize} \quad u_i(x_{i1}, x_{i2}, \dots, x_{im})$$

$$\text{subject to} \quad \sum_{j=1}^{m} p_j x_{ij} = M_i,$$

where M_i is the income of the ith individual and p_j is the price of the jth commodity. The commodity prices can be determined from the equalities of demand and supply of each commodity, but the income M_i remains arbitrary. Therefore, the problem of determining the constants on the right-hand side of (7.2) is reduced to that of determining the distribution of incomes.

One way of dealing with this problem, as indicated by Lange [44], is to establish a social valuation of the distribution of commodities or incomes between the individuals, which can be expressed in the form of a scalar function of the vector $\mathbf{u}$, i.e., $W(\mathbf{u})$. The new problem is now

$$\text{maximize} \quad W(u_1, u_2, \ldots, u_n)$$

subject to (7.3) and (7.4).

This model of an exchange economy can easily be extended to a general equilibrium model for a competitive economy. Denote by $\bar{\mathbf{x}}_i$ an initial endowment of the ith consumer ($\bar{x}_{ij} > 0$ for $j = 1, 2, \ldots, m$) and by $\mathbf{y}_k$ a production vector of the kth firm ($k = 1, 2, \ldots, r$), whose element $\mathbf{y}_{kj} > 0\ (< 0)$ is the output (input) of the jth good. Let Y_k be the possible set of $\mathbf{y}_k$, i.e., the set of $\mathbf{y}_k$ that satisfies the restriction on production $F_k(\mathbf{y}_k) \geqq 0$. The distribution of incomes is described by λ_{ik}, indicating the proportion of profit of the kth firm distributed to the ith consumer.

An equilibrium point under perfect competition is defined as follows [56, pp. 92–93].

Definition 7.1. The following are the conditions of an *equilibrium point* $(\mathbf{x}_i, \mathbf{y}_k, \mathbf{p})$:

(a) Equalities of demand and supply for nonfree goods:

$$\sum_{i=1}^{n} x_{ij} - \sum_{k=1}^{r} y_{kj} - \sum_{i=1}^{n} \bar{x}_{ij} \leqq 0,$$

$$p_j \left(\sum_{i=1}^{n} x_{ij} - \sum_{k=1}^{r} y_{kj} - \sum_{i=1}^{n} \bar{x}_{ij} \right) = 0 \quad (j = 1, 2, \ldots, m).$$

(b) The equilibrium of consumers: $\mathbf{x}_i$ is a maximum point of $u_i(\mathbf{x}_i)$ subject to

$$\sum_{j=1}^{m} p_j x_{ij} \leqq \sum_{j=1}^{m} p_j \bar{x}_{ij} + \max \left[0, \sum_{k=1}^{r} \lambda_{kj} \sum_{j=1}^{m} p_j y_{kj} \right] \equiv M_i \quad (i = 1, 2, \ldots, n).$$

(c) The equilibrium of firms: $\mathbf{y}_k$ is a maximum of $\sum_{j=1}^{m} p_j y_{kj}$ subject to

$$F_k(\mathbf{y}_k) \geqq 0 \quad (\mathbf{y}_k \in Y_k) \quad (k = 1, 2, \ldots, r).$$

A welfare maximum is defined as follows.

Definition 7.2. Consider a social welfare function as the weighted sum of utility functions $\sum_{i=1}^{n} \alpha_i u_i(\mathbf{x}_i)$ with weights $\alpha_i > 0$ $(i = 1, 2, \ldots, n)$. We call a point $(\mathbf{x}_i, \mathbf{y}_k)$ that maximizes it, subject to the condition of no excess of demand over supply, $\sum_{i=1}^{n} x_{ij} \leqq \sum_{i=1}^{n} \bar{x}_{ij} + \sum_{k=1}^{r} y_{kj}$ for $j = 1, 2, \ldots, m$, and production subject to the restriction on $F_k(\mathbf{y}_k) \geqq 0$ $(k = 1, 2, \ldots, r)$, a *welfare maximum point*.

Then under specific—and, from the economic interpretation viewpoint, widely used—assumptions on the utility function $u_i(\mathbf{x}_i)$ and production restrictions, Negishi [56] proved that a competitive equilibrium is a welfare maximum point with the weight of a consumer that is equal to the reciprocal value of equilibrium marginal utility of income.

The main problem in the above-described model of welfare economics is the formulation of a social welfare function (SWF). The discussion of the welfare criteria for choosing among efficient allocations of resources, in other words between different vectors $\mathbf{u}$, moves between the goals of efficiency and equity.

The first pole—*maximum efficiency*—is represented by *utilitarian* SWF, for which the society's welfare is equal to the sum of utilities of the different individuals or social groups. This concept implies that, according to Lange [44], the so-called *marginal social significance* of the ith individual, $W_i = \frac{\partial W}{\partial u_i}$, is the same for each individual. In the other words, an increase in the welfare of a rich person by one unit has the same social value as an increase of the welfare of a poor person by one unit. This type of SWF can provide very unequal allocations of wealth between the individuals.

The second type of SWF was first posed by the philosopher John Rawls [61]. He asserts that members of society would choose to depart from perfect equality only on the condition that the worst-off person under an unequal distribution of utilities would actually be better off than under equality. In other words, for a *Rawlsian* SWF, the welfare of the society depends on the utility of only the poorest or worst-off individual or social group. The use of this kind of SWF will favor the *maximum equity*, but it can provide poor aggregate performance in terms of overall social welfare. The Rawlsian criterion suggests that many efficient allocations may not be socially desirable and that societies may choose equality even at considerable efficiency cost.

In order to provide a compromise between efficiency and equity, Romero [64] proposed a general model in which the views underlying both the utilitarian and the Rawlsian criteria are taken into account simultaneously. The following notation is used: $\mathbf{x} = (x_1, x_2, \ldots, x_m)$ denotes a vector of policy instruments, $\mathbf{b} = (b_1, b_2, \ldots, b_z)$ is a vector of model parameters, and $\mathbf{u} = (u_1, u_2, \ldots, u_m) = (h_1(\mathbf{x}, \mathbf{b}), h_2(\mathbf{x}, \mathbf{b}), \ldots, h_n(\mathbf{x}, \mathbf{b})) = h(\mathbf{x}, \mathbf{b})$ measures the policy outcome for the corresponding individual or social group. The utility possibility frontier, or the feasible domain of Pareto-efficient policies, is described by $T(u_1, u_2, \ldots, u_n) = k$, and the social welfare function by $W(h(\mathbf{x}, \mathbf{b}))$. The utilitarian SWF then takes the form

$$W_U(h(\mathbf{x}, \mathbf{b})) = \sum_{i=1}^{n} h_i(\mathbf{x}, \mathbf{b}),$$

and the Rawlsian SWF is represented by

$$W_R(h(\mathbf{x}, \mathbf{b})) = \operatorname*{Min}_i [h_i(\mathbf{x}, \mathbf{b})].$$

The model proposed by Romero [64] is the following multiobjective optimization problem:

$$\begin{array}{ll} \text{maximize} & W_U(h(\mathbf{x}, \mathbf{b})) \\ \text{maximize} & W_R(h(\mathbf{x}, \mathbf{b})) \\ \text{subject to} & T(u_1, u_2, \ldots, u_n) = k. \end{array}$$

7.1.2 Quantitative Economic Policy

As already mentioned, the treatment of multiple goals has been used in macroeconomic policy for a long time. Starting with Frisch [21, 22, 23], Tinbergen [81, 82] (both Nobel laureates for economics in 1969), Klein [40] (Nobel laureate for economics in 1980), and Theil [79], quantitative economic policy analysis, in which multiple objectives are increasingly integrated, was developed. By "economic policy," certain acts of economic behavior are indicated. In its broadest sense, therefore, the phrase includes the entire subject matter of economic theory. This is particularly true with regard to the "economic policy" of individuals or firms. This economic policy is directed toward the maximization of the ordinary ophelimity functions. In a narrower sense, we may restrict the meaning of "economic policy" to the behavior of organized groups, such as trade unions, agricultural or industrial organizations, etc. Here some *collective ophelimity function* will be the object to be maximized. Despite this broad definition, Tinbergen himself and also the majority of later authors restricted themselves to a "discussion of government economic policy." In quantitative economic policy, which leaves the structure and the organization of the economy unchanged, four types of variables are distinguished:

- *Data* are variables in some sense exogenous to the economic system considered. Their values or changes are given.
- *Target variables* represent the state of the economic system, and they are relevant to the general well-being (e.g., real national income, level of employment, or balance of payments). The decision maker is primarily interested in the values of these variables.
- *Instruments*, or *political parameters*, are variables that the decision maker can determine, and they are the tools through which the government can influence the economy.
- *Irrelevant variables* are all the variables "that, though indispensable in a true picture of the economy considered, are not considered interesting for the economic policy studied" [83, pp. 7–8].

In the so-called *fixed-target approach* [81, 83], the decision maker has to specify a priori the desired values for the targets, and then the values for the instruments can be derived from the economic model solving the system of linear equations. The mathematical conditions for the existence of solution are the well-known "law" of Tinbergen: In a linear system of independent equations the number of target variables should be equal to the number of instruments. In the general case in which there are more instruments than targets, the system has, in principle, infinite solutions, and then we can find one solution maximizing (however defined) welfare function. "And it is only the consequence of our method to replace the maximum problem by fixed targets" [83, p. 38].

Starting with the papers by Frisch [23] and van Eijk and Sandee [16], a real preference function that should measure social welfare was introduced. The preference function by Theil [80] is defined as

$$(\mathbf{y} - \mathbf{y}^*)' G_1 (\mathbf{y} - \mathbf{y}^*) + (\mathbf{z} - \mathbf{z}^*)' G_2 (\mathbf{z} - \mathbf{z}^*), \tag{7.5}$$

where $\mathbf{y}$ is an m-dimensional vector of the target variables, $\mathbf{y}^*$ is an m-dimensional vector of the desired (or reference) values for the target variables, $\mathbf{z}$ is an n-dimensional vector of the instruments, $\mathbf{z}^*$ is an n-dimensional vector of the desired values for the instruments, and G_1 and G_2 are $(m \times m)$ and $(n \times n)$ matrices of preference weights. In most cases, G_1 and G_2 are supposed to be diagonal so that there are no cross-effects among deviations from desired values. The cost function (7.5) is then minimized subject to the restrictions implied by the econometric model. The particular version of this type of model is the linear-quadratic model. The explicitly specified preference function to be minimized is quadratic in the target and instrument variables, and the constraints are linear in these variables. If the econometric model contains dynamic restrictions, the economic policy model leads to the optimal control model, which is very often used, especially for deriving optimal stabilization problems (see [18, 13, 20, 31, 15, 60, 30, 55] and numerous papers, especially those published in the *Journal of Economic Dynamics and Control*).

The main source of criticism of the cost function (7.5) is related to its symmetry, in which positive and negative deviations are equally penalized.

Despite rapid development in this field, the main difficulty of the aforementioned approach to practical decision making is the specification of a scalar-valued preference function. Therefore, several authors suggested the use of modern methods of interactive multiobjective programming (see [86, 85, 9, 10]). For this approach to econometric decision models no—ex ante–explicitly specified scalar-valued objective function is needed. Instead, the preference structure of the decision maker is revealed by an interactive question–answer procedure. Gruber [30, p. 1] expressed the main implication of the interactive vector optimization procedures by a variation on a theme by Leontief [47] as follows: "In econometric decision models, use observed preferences of the decision maker instead of theoretically assumed (hypothetical, 'plausible') preferences."

7.1.3 Optimal Monetary Policy

In the discussion of a monetary policy strategy lies the focus on flexible inflation targeting [75, 76]. This objective has been adopted by the central banks of New Zealand, the UK, Sweden, and other countries in the last decade [45]. Among the academic researchers, an alternative objective to flexible inflation targeting, namely, nominal income targeting, is intensely discussed (see, for instance, [19, 33, 52]).

According to Frisch and Staudinger [25], two new arguments provide support for further research into nominal income targeting or for an optimal monetary policy that takes into account movements in inflation and output, the two strategic variables of the central bank. First, the European Central Bank (ECB) announced a reference money supply growth of 4.5%, calculated as a sum of an inflation target, and a forecasted real output growth rate of 2.5%. The growth rate of nominal income is then the sum of the money supply growth rate and the change in the velocity of money. It is also given as the sum of the inflation rate and the growth rate of the real GDP. Second, due to the apparent overprediction of inflation and underprediction of real output growth in the U.S. economy, McCallum [51] suggested, instead of uncertain estimates of the output gap, using a nominal income growth as a strategic variable for a monetary policy.

Relying on Frisch and Staudinger [25] and on contributions by Calvo [12], Roberts [63], and McCallum [51], the so-called *new Keynesian model* will be used to analyze the policy of the central bank taking aim at two objectives, hitting the inflation and output gap targets. The model consists of a forward-looking Phillips curve in the form

$$\Pi_t = E_t \Pi_{t+1} + a x_t + \varepsilon_t \tag{7.6}$$

and a forward-looking "investment = saving" (IS) curve,

$$x_t = E_t x_{t+1} - b(i_t - E_t \Pi_{t+1}) + \eta_t . \tag{7.7}$$

The Phillips curve relates the inflation rate of the current period, denoted by Π_t, positively to the output gap $x_t = y_t - y^n$ of the current period. $E_t \Pi_{t+1}$ is the expected inflation rate of the next period based on the information available in period t; y_t denotes the growth rate of nominal income; y^n is the natural growth rate of nominal income; and the term ε_t denotes a supply shock.

The IS curve relates output negatively to the real interest rate. $E_t x_{t+1}$ denotes the expected output gap of the next period based on the information available in period t, and η_t denotes a demand shock, not correlated with the supply shock. In the paper by Frisch and Staudinger [25], the objective function of the central bank is the quadratic loss function involving the deviations from natural output and the deviations from an exogenously given inflation target Π^*. Therefore, the central bank minimizes $\Phi = E_t \sum_{j=0}^{\infty} \frac{1}{2} \delta^j [(\Pi_{t+j} - \Pi^*)^2 + \alpha x_{t+j}^2]$ subject to (7.6) and (7.7), where α measures the weight policy attached to output stabilization relative to inflation stabilization such that $0 < \alpha < \infty$. For $\alpha = 0$, a regime of strict inflation targeting [75] is obtained, whereas $\alpha > 0$ describes flexible inflation targeting. For a given α, the central bank controls the nominal interest rate i_t to affect output and inflation such that it minimizes the deviations from the inflation and output gap targets.

Because of the difficult task of estimating the parameter α, the following formulation of the above model as a vector optimization problem with two objectives is proposed:

$$\text{minimize} \quad \Phi_1 = E_t \sum_{j=0}^{\infty} \frac{1}{2} \delta^j (\Pi_{t+j} - \Pi^*)^2$$

$$\text{minimize} \quad \Phi_2 = E_t \sum_{j=0}^{\infty} \frac{1}{2} \delta^j x_{t+j}^2$$

subject to (7.6) and (7.7).

7.1.4 Optimal Behavior of a Monopolist Facing a Bicriteria Objective Function

Profit-maximizing behavior, one of the basic assumptions in the neoclassical theory of the monopolistic firm, has already been criticized by Hicks [34] and Scitovsky [68]. Following the contributions by Baumol [4, 5] and Williamson [87], different goals have been proposed as a description of the firm's managers' behavior. In Section 2.5.2, we discussed a model suggested by Baumol [4, 5] in which a firm maximizes its total revenue subject to a minimum profit constraint. It has been shown that at the constrained revenue-maximizing output, the marginal revenue is lower than the marginal cost and the profit is equal to the prescribed level. However, Baumol's approach implies [65] that a firm orders various outcomes (each outcome is a combination of a certain level of profit and a certain level of sales revenue) in a lexicographic manner. In other words, the firm is assumed to have a marginal rate of substitution of sales revenue for profit (the amount of profit it is willing to give up in order to increase the revenue by one unit) that is infinite as long as profit exceeds the minimum level acceptable to shareholders and that is always equal to zero as long as profit is below the minimum level.

Fisher [17] suggested a symmetrical alternative with profit maximization subject to a revenue constraint. Osborne [58] and Hall [32] investigated the implications of these two alternatives, and Williamson [87] and Marris [53] proposed a multicriteria objective function. Brown and Revankar [11] formalized this approach into a generalized theory of the firm that specifies a firm's utility function that includes revenue as well as profit:

$$U = U(R, \Pi),$$

where U is a twice-differentiable function with

$$U_\Pi = \frac{\partial U}{\partial \Pi} > 0 \quad \text{and} \quad U_R = \frac{\partial U}{\partial R} \geqq 0.$$

R is revenue, a function of the output, $q : R(q)$; and $\Pi(q) = R(q) - C(q)$ represents profit, where $C(q)$ denotes cost. The firm maximizes U, but it is constrained by

- its production function, $q = f(\mathbf{x})$, where $\mathbf{x}$ denotes a vector of inputs;
- the demand for its product, $q = h(p)$, where p is product price, or the product price is a function of the output $q : p = h^{-1}(q)$;
- the supply functions of the production factors, $x_i = g_i(r_i)$, for $i = 1, 2, \ldots, n$, where x_i is the ith input and r_i is the price of that input.

Then Brown and Revankar [11] show that if the marginal utility of revenue is positive,

- the product price is not likely to be less than the competitive price under profit maximization; the extent to which the product price exceeds marginal cost depends on the elasticity of product demand and on the marginal utility of profit and revenue, respectively;
- the marginal products of the factors are forced down below the neoclassical level in the same proportion, for given degrees of market imperfections;
- the marginal rates of substitution equilibrium conditions are identical to those derived under profit maximization;
- labor and capital incomes are benefited at the expense of the owners of the firm.

Although the main results are derived regardless of the form of the utility function, from the operational point of view and in order to reveal the pattern of "weight structure" reflected in the firm's utility function ("weights" referring to Π and R), the formulation of this model as a multiobjective optimization problem is proposed:

$$\begin{aligned} &\text{maximize} \quad \Pi(q) \\ &\text{maximize} \quad R(q) \\ &\text{subject to} \quad q = f(\mathbf{x}), \\ &\qquad\qquad\quad p = h^{-1}(q), \\ &\qquad\qquad\quad x_i = g_i(r_i) \quad \text{for } i = 1, 2, \ldots, n. \end{aligned}$$

7.1.5 Leontief Pollution Model with Multiple Objectives

In Section 4.6.3, we introduced the augmented Leontief model describing the relationships between the economic system and the environment. In this model, the levels of gross industrial outputs and the abatement activity levels depend on the exogenously given pollution standards. Thereupon the model allows us to analyze the impact of changes in the level or structure of final demand and/or of tolerated net pollution (or environmental standards) on the gross industrial production and abatement activities. Early in the discussion of environmental problems, it was often claimed that there is a trade-off between the goals of economic policy (like growth rate of GDP) and the improvement of environmental quality. Sometimes this is expressed as a requirement to find a reasonable compromise between the "destruction" of jobs by imposing cost-intensive constraints and the "destruction" of the environment.

However, in his seminal paper [46], Leontief concluded that the imposition of effective limits $\mathbf{y}_2$ on net pollution results in higher industrial production and value-added cost. Moreover, the notion of environmental quality is multidimensional, too. Decreasing emission of one pollutant during a given period can be accompanied by increasing emissions of another pollutant. For example, emission of SO_2 decreased in Austria from 400,000 tons in 1980 to 50,000 tons in 1997, and emission of CO from nearly 1,700,000 tons in 1980 to 1,000,000 tons in 1997. But emission of N_2O (a by-product of NO_x reduction using a catalyst) increased between 1980 and 1997 from nearly 6,000,000 tons to more than 7,000,000 tons [88].

For these reasons, the Leontief pollution model from Section 4.6.3 will now be formulated as an optimization model with multiple objectives. Luptáčik and Böhm [48] proposed two versions of the model according to different sets of criteria. The essential difference from model (4.52) is that the pollution standards $\mathbf{y}_2$ are not treated as exogenously given (it is no simple task to estimate their levels), but an objective to keep the levels of net emissions (the residual amount of emissions after abatement) as low as possible is considered. In other words, "minimization" of the emissions of all pollutants is postulated.

Among the economic objectives, in the first version of their model, the minimization of primary inputs to produce gross national output for the exogenously given vector of final demand is considered.

Using the same notation of variables as in model (4.52), the following multicriteria Leontief pollution model can be formulated:

$$\begin{aligned} &\text{minimize} && V(\mathbf{x}_1,\mathbf{x}_2) = \mathbf{v}_1{}'\mathbf{x}_1 + \mathbf{v}_2{}'\mathbf{x}_2 \\ &\text{minimize} && W(\mathbf{x}_1,\mathbf{x}_2) = A_{21}\mathbf{x}_1 - (E - A_{22})\mathbf{x}_2 \\ &\text{subject to} && (E - A_{11})\mathbf{x}_1 - A_{12}\mathbf{x}_2 \geqq \mathbf{y}_1, \\ & && \mathbf{x}_1 \geqq \mathbf{0}, \quad \mathbf{x}_2 \geqq \mathbf{0}, \end{aligned}$$

where $\mathbf{v}_1{}'$ is the n-dimensional row vector of primary inputs per unit of industrial production and $\mathbf{v}_2{}'$ is the k-dimensional row vector of primary inputs per unit of antipollution activities.

In the second version of the model, the final demand is not exogenously given, but the objective is now to maximize the value of final demand for given price vector $\mathbf{c}$.

The constraints of this model are described by the following set of inequalities:

$$-(E - A_{11})\mathbf{x}_1 + A_{12}\mathbf{x}_2 + \mathbf{y}_1 \leqq \mathbf{0}, \tag{7.8}$$

$$A_{21}\mathbf{x}_1 - (E - A_{22})\mathbf{x}_2 - \mathbf{y}_2 \leqq \mathbf{0}, \tag{7.9}$$

$$\mathbf{v}_1{}'\mathbf{x}_1 + \mathbf{v}_2{}'\mathbf{x}_2 \leqq \overline{V}, \tag{7.10}$$

$$\mathbf{x}_1 \geqq \mathbf{0}, \quad \mathbf{x}_2 \geqq \mathbf{0}, \quad \mathbf{y}_1 \geqq \mathbf{0}, \quad \mathbf{y}_2 \geqq \mathbf{0}, \tag{7.11}$$

where $\overline{V}$ denotes the disposable amount of the primary input (e.g., labor). The inequalities (7.8) are the balance conditions for the production of goods, (7.9) are those for net pollution, and (7.10) describes the constraint for the primary input.

The objective of environmental policy is the minimization of net pollution $\mathbf{y}_2$, and the multiobjective optimization problem becomes

$$\begin{aligned} &\text{maximize} && f_1(\mathbf{y}_1) = \mathbf{c}'\mathbf{y}_1 \\ &\text{minimize} && f_2(\mathbf{y}_2) = \mathbf{y}_2 \end{aligned}$$

subject to the constraints (7.8)–(7.11).[1]

[1] For the quantitative implications resulting from different economic objectives, see Luptáčik and Böhm [48], and for an empirical demonstration based on an interactive approach, see Böhm and Luptáčik [10].

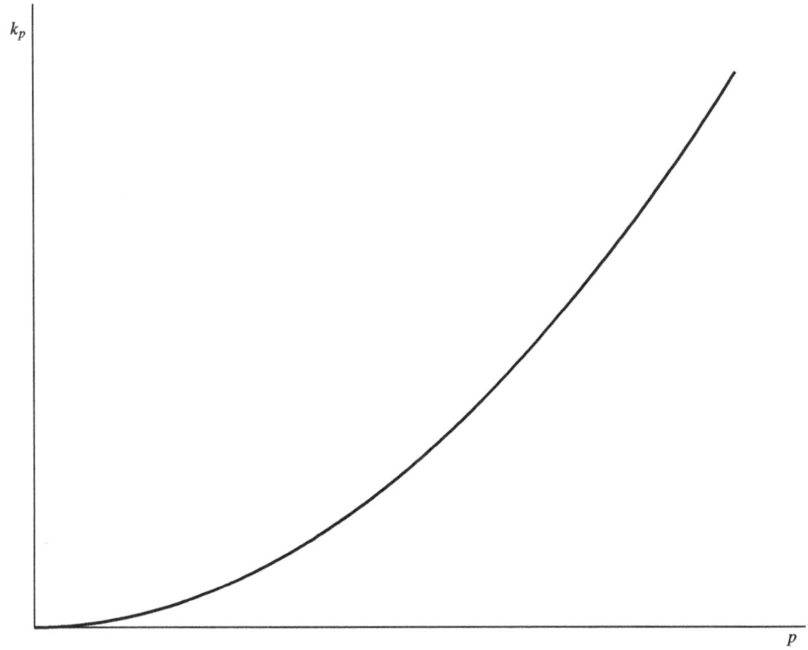

Fig. 7.1. A cost curve of environmental pollution.

7.1.6 A Nonlinear Model of Environmental Control

In the paper by Mastenbroek and Nijkamp [54], the following nonlinear model of environmental control was presented. The costs of environmental pollution consist of two components, the damage costs caused to the environment (the pollution costs) and the costs caused by abatement activities (the costs of environmental control). It is assumed that pollution costs, denoted by k_p, rise progressively with an increase in pollution p. In mathematical form a relation of this kind could be represented by the following nonlinear function (see Figure 7.1):

$$k_p = ap^{\mathcal{K}} \quad \text{with } \mathcal{K} > 1 \quad \text{and} \quad a > 0.$$

The second cost component is made up by the costs of abatement measures. These costs of environmental control (i.e., expenditures on environmental investments) will probably increase progressively, as the required reduction in pollution increases. This can be modeled as follows.

It is assumed that the total amount of pollution p consists of pollution generated by production of goods (y) and consumption (c):

$$p = by + dc,$$

where b and d are the corresponding average emission coefficients.

In contrast to the Leontief pollution model from Section 4.6.3, the emission coefficient b is now allowed to change by increasing the environmental investments (e.g., purification plants and filters, new technology, etc.). According to Mastenbroek and Nijkamp [54, p. 34], we can assume that

$$b = b_0 \left(\frac{i_0}{i}\right)^{\mu} \quad \text{with } 0 < \mu < 1,$$

where b_0 represents the average emission coefficient estimated at a certain current level of pollution p_0 and at a certain current volume i_0 of environmental investments. The parameter b is the new emission coefficient after the abatement measures have been put into effect. It is clear that $b = b_0$ if $i_0 = i$ and b decreases with increasing environmental investments i.

The output y can be allocated between environmental investments and consumption:

$$i + c \leqq y.$$

The question now arises of how to regulate production and the allocation of output between the cost of environmental control and consumption in order to keep simultaneously a high level of consumption and high environmental quality. In other words, the consumption should be maximized and the damage cost to the environment should be minimized. This leads to the following nonlinear multiobjective programming model:

$$\begin{aligned}
\text{minimize} \quad & f_1(p) = ap^{\mathcal{K}} \\
\text{maximize} \quad & f_2(c) = c \\
\text{subject to} \quad & b_0 \left(\frac{i_0}{i}\right)^{\mu} y + dc = p, \\
& i + c \leqq y, \\
& c > 0, \quad i > 0, \quad p > 0, \quad y > 0.
\end{aligned}$$

After a simple transformation, the multiobjective geometric programming model can be obtained:

$$\begin{aligned}
\text{minimize} \quad & f_1(p) = ap^{\mathcal{K}} \\
\text{minimize} \quad & (f_2(c))^{-1} = c^{-1} \\
\text{subject to} \quad & li^{-\mu} yp^{-1} + dcp^{-1} \leqq 1, \\
& iy^{-1} + cy^{-1} \leqq 1, \\
& c > 0, \quad i > 0, \quad p > 0, \quad y > 0,
\end{aligned}$$

where $l = b_0 i_0^{\mu} > 0$.

7.2 Kuhn–Tucker Conditions for the Multiobjective Programming Problem

The multiobjective mathematical programming problems described in the previous section can be generally written as

$$\begin{aligned} \text{Minimize} \quad & F(\mathbf{x}) = (f_1(\mathbf{x}), f_2(\mathbf{x}), \ldots, f_s(\mathbf{x})) \\ \text{subject to} \quad & G(\mathbf{x}) \leqq \mathbf{0}, \\ & \mathbf{x} \in R^n, \end{aligned} \tag{7.12}$$

where F and G are vector-valued functions from $R^n \to R^s$ and $R^n \to R^m$. We assume that all the functions are differentiable at a point that is an optimal point or a candidate for an optimal point of our problem (7.12). Without loss of generality, we consider a vector minimization problem with s objectives, n variables, and m constraints. Minimization (or Min) and minimization (or min) indicate that the problem under consideration is a minimization multiple-objective problem and minimization single-objective problem, respectively.

In the seminal paper by Kuhn and Tucker [43] and subsequently in the papers by Geoffrion [29], Bitran [8], Marusciac [50], and Singh [69]—to name only a few—the extension of necessary (and under the convexity assumption of sufficient) optimality conditions of Kuhn–Tucker (or Karush–John) type to a problem with multiple objectives is already provided.

Let $M = \{1, 2, \ldots, m\}$, $K = \{\mathbf{x} | \mathbf{x} \in R^n, G(\mathbf{x}) \leqq \mathbf{0}\}$, and $I = \{i \in M | g_i(\mathbf{x}^0) = 0$ for some fixed $\mathbf{x}^0 \in K\}$.

The term *vector minimum (maximum) problem* was introduced by Kuhn and Tucker [43, p. 488] as follows.

Definition 7.3. To find an $\mathbf{x}^0$ that minimizes (maximizes) the vector function $F(\mathbf{x})$ constrained by $G(\mathbf{x}) \leqq \mathbf{0}$, that is, to find an $\mathbf{x}^0$ satisfying the constraints and such that $F(\mathbf{x}) \leq F(\mathbf{x}^0)$ $[F(\mathbf{x}) \geq F(\mathbf{x}^0)]$ for no $\mathbf{x}$ satisfying the constraints.

This notion is equivalent to the notion of a Pareto minimal (maximal) point as given by Marusciac [50].

Definition 7.4. A point $\mathbf{x}^0 \in K$ is a Pareto minimal (maximal) point of F on K if and only if there exists no $\mathbf{x} \in K$ such that $F(\mathbf{x}) \leq F(\mathbf{x}^0)$ $[F(\mathbf{x}) \geq F(\mathbf{x}^0)]$.

Definition 7.5. A point $\mathbf{x}^0 \in K$ is a weak Pareto minimal (maximal) point of F on K if and only if there exists no $\mathbf{x} \in K$ such that $F(\mathbf{x}) < F(\mathbf{x}^0)$ $[F(\mathbf{x}) > F(\mathbf{x}^0)]$.

Another equivalent notion is Koopmans's efficient point type in production theory [41, p. 48].

Definition 7.6. A possible point in the commodity space is called efficient whenever an increase in one of its coordinates (the net output of one good) can be achieved only at the cost of a decrease in some other coordinate (the net output of another good).

The reader can verify that all above definitions correspond to Definition 5.10, and thereupon in what follows, we will use the term *Pareto–Koopmans efficiency*.

The following theorem provides the necessary conditions for a Pareto–Koopmans minimal point of problem (7.12).

Theorem 7.1. *Let*

(i) $\mathbf{x}^0$ *be a Pareto–Koopmans minimal point (weak Pareto–Koopmans minimal point) of problem* (7.12);
(ii) $F(\mathbf{x})$, $G(\mathbf{x})$ *be differentiable at* $\mathbf{x}^0$;
(iii) $g_i(\mathbf{x})$, *for* $i \in I$, *satisfy the constraint qualification at* $\mathbf{x}^0$.

Then there exist $\boldsymbol{\alpha}^0 \in R^s_+$, $\boldsymbol{\alpha}^0 \neq \mathbf{0}$, *and a vector of Lagrange multipliers* $\mathbf{u}^0 \in R^m$ *such that*

$$\boldsymbol{\alpha}^{0'} \nabla F(\mathbf{x}^0) + \mathbf{u}^{0'} \nabla G(\mathbf{x}^0) = \mathbf{0}, \tag{7.13}$$

$$\mathbf{u}^{0'} G(\mathbf{x}^0) = 0, \tag{7.14}$$

$$G(\mathbf{x}^0) \leqq \mathbf{0}, \tag{7.15}$$

$$\mathbf{u}^0 \geqq \mathbf{0}, \tag{7.16}$$

where $\nabla F(\mathbf{x}^0)$ *denotes the* $s \times n$ *matrix whose rows are the gradients at* $\mathbf{x}^0$ *of the components of F and* $\nabla G(\mathbf{x}^0)$ *denotes the* $m \times n$ *matrix whose rows are the gradients at* $\mathbf{x}^0$ *of the components G (Jacobians of F and G at the point* $\mathbf{x}^0$*).*

The proof of this theorem can be found in [69, p. 118].

Assuming the convexity of the functions $f_1(\mathbf{x}), f_2(\mathbf{x}), \ldots, f_s(\mathbf{x})$ and $g_1(\mathbf{x}), g_2(\mathbf{x}), \ldots, g_m(\mathbf{x})$, the conditions (7.13)–(7.16) are also sufficient for $\mathbf{x}^0$ to be a Pareto–Koopmans minimal point of problem (7.12) as proved by Singh [69, pp. 119–120].

Theorem 7.2. *Suppose the following:*

(i) $F(\mathbf{x})$, $G(\mathbf{x})$ *are differentiable at* $\mathbf{x}^0$;
(ii) $F(\mathbf{x})$, $G(\mathbf{x})$ *are convex;*
(iii) *there exist* $\boldsymbol{\alpha}^0 > \mathbf{0}$ *and a vector* $\mathbf{u}^0 \in R^m$ *such that conditions* (7.13)–(7.16) *are fulfilled.*

Then $\mathbf{x}^0$ *is a (weak) Pareto–Koopmans minimal point of* $F(\mathbf{x})$ *over* K.

The reader may verify that conditions (7.13)–(7.16) are exactly the Kuhn–Tucker conditions for the so-called *parametric* problem

$$\begin{aligned} &\text{minimize} \quad \sum_{k=1}^{s} \alpha_k f_k(\mathbf{x}) \\ &\text{subject to} \quad \mathbf{x} \in K, \end{aligned} \tag{7.17}$$

where $\boldsymbol{\alpha} = (\alpha_1, \ldots, \alpha_s)$ is a vector of nonnegative components.

The interest in problem (7.17) arises from the fact that if $\boldsymbol{\alpha} > \mathbf{0}$, then due to Theorem 7.2 every solution to problem (7.17) is Pareto–Koopmans efficient in problem (7.12).

Geoffrion [29] attempted to rule out efficient points with the property that the marginal gain in one criterion can be made arbitrarily large relative to each of the marginal losses incurred in other criteria. Therefore, he introduced the concept of *properly efficient* solution, defined as follows.

Definition 7.7. A point $\mathbf{x}^0$ is a properly efficient solution of problem (7.12) if it is (Pareto–Koopmans) efficient and if there is a scalar $\xi > 0$ such that $\mathbf{x} \in K$, $f_k(\mathbf{x}) < f_k(\mathbf{x}^0)$ implies

$$\frac{[f_k(\mathbf{x}^0) - f_i(\mathbf{x})]}{[f_l(\mathbf{x}) - f_l(\mathbf{x}^0)]} \leqq \xi$$

for some l with $f_l(\mathbf{x}) > f_l(\mathbf{x}^0)$.

These properties of the parametric problem (7.17) permit the application of nonlinear programming theory to problem (7.17) in order to study and characterize efficient points. In this way, the relationship between the saddle point of the Lagrange function and a properly efficient solution of the vector minimum problem (7.12) can be obtained [43, p. 489].

Theorem 7.3 (equivalence theorem). *Let the functions* $f_1(\mathbf{x}), f_2(\mathbf{x}), \ldots, f_s(\mathbf{x})$, $g_1(\mathbf{x}), g_2(\mathbf{x}), \ldots, g_m(\mathbf{x})$ *be convex as well as differentiable for* $\mathbf{x} \geqq \mathbf{0}$. *Then* $\mathbf{x}^0$ *is a properly efficient solution of problem* (7.12) *if and only if there is some* $\boldsymbol{\alpha}^0 > \mathbf{0}$ *such that* $\mathbf{x}^0$ *and some* $\mathbf{u}^0$ *give a solution of the saddle value problem for* $\Phi(\mathbf{x}, \mathbf{u}) = \boldsymbol{\alpha}^{0\prime} F(\mathbf{x}) + \mathbf{u}' G(\mathbf{x})$.

For economic applications, the approach based on a vector-valued Lagrange function, where an $(s \times m)$ matrix U of dual variables is associated with the constraints in K, is interesting. It seems reasonable to assume that the matrices of dual variables contain more information than the pairs $(\boldsymbol{\alpha}, \mathbf{u})$. The dual variables u_{ki} can be interpreted, under certain conditions [8, pp. 380–383], as the partial derivatives of the objectives with respect to the components of vector $\mathbf{b}'$ when $g_i(\mathbf{x})$ is written as

$$g_i(\mathbf{x}) = h_i(\mathbf{x}) - b_i \quad (i = 1, 2, \ldots, m).$$

They represent the rate of change of $(f_1, f_2, \ldots, f_s)$ at $\mathbf{x}^0$ with respect to small changes in the right-hand side of the constraints. In other words, the elements of matrix U are interpreted as shadow prices, similarly to the interpretation of dual variables in the single-objective optimization models (however, extended with respect to different objectives).

In the parametric approach mentioned above, a vector $\mathbf{u}$ of dual variables associated to the constraints can be seen as an aggregation of the matrix U. In fact, for fixed $\boldsymbol{\alpha}$, problem (7.17) can be interpreted as an aggregate version of problem (7.12).

The idea of a matrix of dual variables, rather than a vector, has been used by Ritter [62], Zowe [91, 92], Craven [14], and Bitran [8] in connection with linear multiobjective optimization problems by Gale, Kuhn, and Tucker [28], Isermann [36, 37, 38], and Rödder [67], and for a lexicographic linear programming by Isermann [39] and Turnovec [84]. In these papers, extensions of the Kuhn–Tucker conditions for a pair $(\mathbf{x}^0, U^0)$ are provided and a duality theory for multiobjective optimization problems developed.

According to Bitran [8, pp. 389–390], Kuhn–Tucker conditions for problem (7.12) using a matrix of dual variables U can be stated as follows.

Theorem 7.4. *Suppose the following:*

(i) $\mathbf{x}^0$ *is a Pareto–Koopmans minimal point of problem* (7.12);
(ii) $F(\mathbf{x})$, $G(\mathbf{x})$ *are differentiable at* $\mathbf{x}^0$.

Then there is a pair $(\mathbf{x}^0, U^0)$ *and* $\boldsymbol{\alpha}^0 > \mathbf{0}$ *such that*

$$\boldsymbol{\alpha}^0[\nabla F(\mathbf{x}^0) + U^0 \nabla G(\mathbf{x}^0)] = \mathbf{0}, \tag{7.18}$$

$$\boldsymbol{\alpha}^{0\prime} U^0 \geqq \mathbf{0}, \tag{7.19}$$

$$U^0 G(\mathbf{x}^0) = \mathbf{0}, \tag{7.20}$$

$$G(\mathbf{x}^0) \leqq \mathbf{0}. \tag{7.21}$$

If $F(\mathbf{x})$ *and* $G(\mathbf{x})$ *are convex and the Kuhn–Tucker conditions* (7.18)–(7.21) *hold at a pair* $(\mathbf{x}^0, U^0)$, *then* $\mathbf{x}^0$ *is Pareto–Koopmans efficient in problem* (7.12).

As already mentioned, the parametric approach to multiobjective optimization consists of determining efficient points in problem (7.12) by solving mathematical programming problems of type (7.17). In order to relate the approach using the matrix of Lagrange multipliers U to the parametric approach, Bitran [8, pp. 394–395] proved the following.

Theorem 7.5. *Let* $F(\mathbf{x})$ *and* $G(\mathbf{x})$ *be differentiable and convex on* R^n. *Then there is a pair* $(\mathbf{x}^0, U^0)$ *satisfying the Kuhn–Tucker conditions* (7.18)–(7.21) *if and only if* $\mathbf{x}^0$ *solves problem* (7.17) *for some* $\overline{\boldsymbol{\alpha}} > \mathbf{0}$, *i.e.,* $\mathbf{x}^0$ *is a Pareto–Koopmans minimal point and* $\mathbf{x}^0$ *solves the linear approximation to problem* (7.17) *at* $\mathbf{x}^0$.

Furthermore, Bitran [8] can show that $\boldsymbol{\alpha}$, $\mathbf{u}$, and U are related such that

$$\boldsymbol{\alpha}' U = \mathbf{u}', \tag{7.22}$$

where the α' can be interpreted as the weights corresponding to the objectives. In this way, as already mentioned, a vector $\mathbf{u}$ of dual variables is expressed as an aggregation of the matrix U.

As a special case of the multiobjective minimization problem (7.12), let us consider the following *lexicographic minimization problem*:

$$\text{lex min}\{F(\mathbf{x}) | \mathbf{x} \in R^n,\ G(\mathbf{x}) \leqq \mathbf{0}\}. \tag{7.23}$$

A vector $\mathbf{x} \in R^n$ is said to be *lexicographically nonnegative*, denoted by

$$\mathbf{x} \,\text{lex} \geqq \mathbf{0},$$

if either $\mathbf{x} = \mathbf{0}$ or its first nonzero component is positive.

An $m \times n$ matrix A is called lexicographically nonnegative, denoted by

$$A \,\text{lex} \geqq 0,$$

if all its columns are lexicographically nonnegative.

Let F be an s-dimensional vector-valued function defined on R^n and $K \subset R^n$. A vector $\mathbf{x}^* \in K$ is said to be a *lexicographically minimal point* of F with respect to K if for any $\mathbf{x} \in K$,

$$F(\mathbf{x}^*) \text{ lex} \leqq F(\mathbf{x}).$$

The problem of finding a lexicographically minimal point of F with respect to K is called a *lexicographic minimization problem*, denoted by

$$\text{lex} \min\{F(\mathbf{x})|\mathbf{x} \in K\}.$$

Defining $K = \{\mathbf{x}|G(\mathbf{x}) \leqq \mathbf{0}\}$, we obtain problem (7.23).

In the literature on multiple-criteria decision making, some examples of lexicographic optimization models can be found. Behringer [7] reported on several mathematical and game-theoretic applications. Nijkamp [57] described an application of a lexicographic optimization model to a land-use problem for industrial activities in an area of the Rhine delta region near Rotterdam (the so-called Meuse flat). These models reflect a structure that can be represented by a ranking of the objectives instead of an ordinary optimization problem with a scalar-valued objective function.

A properly efficient solution for the lexicographic optimization problems is defined as follows.

Definition 7.8. A lexicographically minimal point $\mathbf{x}^0$ is a properly optimal solution to (7.23) if there exists no $\mathbf{y} \in R^n$ such that

$$\nabla F(\mathbf{x}^0)\mathbf{y} \text{ lex} < \mathbf{0},$$
$$\nabla G_I(\mathbf{x}^0)\mathbf{y} \leqq \mathbf{0},$$

where G_I is a subvector of G consisting of all components of G corresponding to the constraints active at $\mathbf{x}^0$, i.e., $g_i(\mathbf{x}^0) = 0$.

Now the Kuhn–Tucker conditions for problem (7.23), as provided by Luptáčik and Turnovec [49, p. 261], can be written as follows.

Theorem 7.6.

(i) *If* $\mathbf{x}^0$ *is a properly optimal solution to problem* (7.23)*, then there exists a matrix* U^0 *of dimension* $s \times m$ *such that*

$$\nabla F(\mathbf{x}^0) + U^0 \nabla G(\mathbf{x}^0) = 0, \tag{7.24}$$
$$G(\mathbf{x}^0) \leqq \mathbf{0}, \tag{7.25}$$
$$U^0 G(\mathbf{x}^0) = \mathbf{0}, \tag{7.26}$$
$$U^0 \text{ lex} \geqq 0. \tag{7.27}$$

(ii) *If problem* (7.23) *is convex, then conditions* (7.24)–(7.27) *are also sufficient for* $\mathbf{x}^0$ *to be a properly optimal solution to* (7.23).

The reader can see that the form of conditions (7.24)–(7.27), except for a matrix of Lagrange multipliers U^0 instead of the vector $\mathbf{u}^0$ and the lexicographic relation (7.27) instead of nonnegativity in the usual sense, is the same as the Kuhn–Tucker conditions for a single-objective optimization problem described in Chapter 2.

To illustrate, let us consider the following lexicographic minimization problem:

$$\text{lex min } F(\mathbf{x}) = \begin{pmatrix} f_1(\mathbf{x}) \\ f_2(\mathbf{x}) \end{pmatrix} = \begin{pmatrix} x_1^2 - 2x_1x_2 + x_2^2 - 2x_1 + 2x_2 \\ -x_1 + x_2^2 \end{pmatrix}$$

$$\text{subject to } \quad g_1(\mathbf{x}) = x_1 + x_2 - 8 \leqq 0,$$
$$g_2(\mathbf{x}) = -x_1 + x_2 + 2 \leqq 0.$$

We compute

$$\nabla F(\mathbf{x}) = \begin{pmatrix} 2x_1 - 2x_2 - 2 & -2x_1 + 2x_2 + 2 \\ -1 & 2x_2 \end{pmatrix}$$

and

$$\nabla G(\mathbf{x}) = \begin{pmatrix} 1 & 1 \\ -1 & 1 \end{pmatrix}.$$

The Kuhn–Tucker conditions (7.24) become

$$\begin{pmatrix} 2x_1 - 2x_2 - 2 & -2x_1 + 2x_2 + 2 \\ -1 & 2x_2 \end{pmatrix} + \begin{pmatrix} u_{11} & u_{12} \\ u_{21} & u_{22} \end{pmatrix} \begin{pmatrix} 1 & 1 \\ -1 & 1 \end{pmatrix} = \begin{pmatrix} 0 & 0 \\ 0 & 0 \end{pmatrix},$$

or

$$2x_1 - 2x_2 - 2 + u_{11} - u_{12} = 0, \tag{7.28}$$
$$-2x_1 + 2x_2 + 2 + u_{11} + u_{12} = 0, \tag{7.29}$$
$$-1 + u_{21} - u_{22} = 0, \tag{7.30}$$
$$2x_2 + u_{21} + u_{22} = 0. \tag{7.31}$$

The conditions (7.25) are the feasibility conditions

$$x_1 + x_2 - 8 \leqq 0, \tag{7.32}$$
$$-x_1 + x_2 + 2 \leqq 0. \tag{7.33}$$

From (7.26), we obtain

$$\begin{pmatrix} u_{11} & u_{12} \\ u_{21} & u_{22} \end{pmatrix} \begin{pmatrix} x_1 + x_2 - 8 \\ -x_1 + x_2 + 2 \end{pmatrix} = \begin{pmatrix} 0 \\ 0 \end{pmatrix},$$

or

$$u_{11}(x_1 + x_2 - 8) + u_{12}(-x_1 + x_2 + 2) = 0, \tag{7.34}$$

$$u_{21}(x_1 + x_2 - 8) + u_{22}(-x_1 + x_2 + 2) = 0. \tag{7.35}$$

According to condition (7.27), the matrix of Lagrange multipliers U^0 must be lexicographically nonnegative:

$$\begin{pmatrix} u_{11} & u_{12} \\ u_{21} & u_{22} \end{pmatrix} \text{lex} \geqq \begin{pmatrix} 0 & 0 \\ 0 & 0 \end{pmatrix}. \tag{7.36}$$

As already mentioned, the dual variables u_{ki} $(k = 1, 2;\ i = 1, 2)$ can be interpreted as shadow prices of the resource i with respect to the objective k. It follows from (7.34) and (7.35) that if at least one u_{ki} $(k = 1, 2)$ is different from zero, the corresponding ith constraint must be fulfilled as equality; the ith resource is scarce.

Let us assume that $u_{11} = 0, u_{21} = 0$, and $u_{12} \neq 0$. Then it follows from (7.34) that condition (7.33) is satisfied as equality and $x_2 = x_1 - 2$. Solving (7.28) and (7.29), we will obtain the solution $u_{12} = 2$. Then (7.30) yields $u_{22} = -1$, and consequently (7.31) yields $x_2 = \frac{1}{2}$. Finally, $x_1 = 2 + x_2 = \frac{5}{2}$. The reader may verify that both objective functions $f_1(\mathbf{x})$ and $f_2(\mathbf{x})$ are convex, and due to Theorem 7.5 $\mathbf{x}^0 = (\frac{5}{2}, \frac{1}{2})$ is a properly optimal solution to our example.

7.3 Duality for Multiobjective Optimization Problems

As has been shown in the first part of this book, duality is an attractive and very useful concept in single-objective mathematical programming as well as in economics. The primal model concerns allocations of commodities, while the dual concerns prices. The two concepts intersect in an equilibrium problem involving both allocations and prices.

There seems to be no unified approach to dualization in multiobjective optimization. One of the difficulties is the fact that an efficient solution to a multiobjective problem is not unique but in general becomes a set. The definition of infimum (or supremum) of a set with a partial order plays a key role in development of duality theory in multiobjective optimization.

The first results on duality theory in multiobjective optimization can be found in [77, 62, 91, 92, 66, 78]. For the special case of linear multiobjective problems, the first papers are [28, 42, 35, 36, 37, 38].

Because of the different approaches to duality in multiobjective optimization—which require a deeper mathematical background—and our emphasis on economic applications, we will restrict our description to the parametric problem (7.17) and to the lexicographic minimization problem (7.23) only.

7.3.1 Duality for Multiobjective Optimization Problems in Parametric Form

As mentioned in the previous section, the parametric approach to multiobjective optimization consists of determining efficient points in problem (7.12) by solving nonlinear programming problems of the following type:

$$\text{minimize} \quad \{\boldsymbol{\alpha}' F(\mathbf{x}) | \mathbf{x} \in R^n, \ G(\mathbf{x}) \leqq \mathbf{0}\}, \tag{7.37}$$

where $F(\mathbf{x})$ and $G(\mathbf{x})$ are assumed to be differentiable and convex on R^n.

According to duality theory for convex programming from Section 3.3, the dual problem to problem (7.37) becomes

$$\begin{aligned} &\text{maximize} && \Phi(\mathbf{x}, \mathbf{u}) = \boldsymbol{\alpha} F(\mathbf{x}) + \mathbf{u} G(\mathbf{x}) \\ &\text{subject to} && \boldsymbol{\alpha}'[\nabla F(\mathbf{x})] + \mathbf{u}'[\nabla G(\mathbf{x})] = \mathbf{0}, \\ & && \mathbf{u} \geqq \mathbf{0}. \end{aligned} \tag{7.38}$$

In Theorem 7.5, the equivalence between the Kuhn–Tucker conditions (7.18)–(7.21) and the properly efficient solution $\mathbf{x}^0$ of problem (7.37) for some $\bar{\boldsymbol{\alpha}} > \mathbf{0}$ has been shown. The following corollary [8, p. 395] extends this result toward a strong duality for the parametric approach:

Corollary 7.1. *Let* $\mathbf{x}^0$ *and* U *satisfy the Kuhn–Tucker conditions* (7.18)–(7.21). *Then* $\mathbf{x}^0$ *solves problem* (7.37) *for any* $\boldsymbol{\alpha} > \mathbf{0}$, *such that* (7.18)–(7.19) *hold. Moreover,* $\boldsymbol{\alpha} U$ *solves the nonlinear programming dual problem* (7.38) *to problem* (7.37).

These properties of the parametric problem permit the application of nonlinear programming duality theory to problem (7.37) to study and characterize efficient points. The Kuhn–Tucker conditions provide a relation among $\boldsymbol{\alpha}$, $\mathbf{u}$, U that can be useful for practical purposes, for example, to generate weights $\boldsymbol{\alpha}$ in order to take into account the preferred choice of a decision maker with respect to a given efficient alternative. Several interactive methods of multiobjective optimization, like those of Zionts and Wallenius [90], Steuer [72], and others, are based on the solution of problem (7.37).

Another application of this approach will be presented in Chapter 9 for multiobjective geometric programming.

7.3.2 Duality Theory for Convex Lexicographic Programming

We consider the lexicographic minimization problem (7.23),

$$\text{lex}\min\{F(\mathbf{x}) | \mathbf{x} \in R^n, \ G(\mathbf{x}) \leqq \mathbf{0}\},$$

using the same notation as in Section 7.2. In what follows, problem (7.23) will be called the *primal problem*. The Kuhn–Tucker conditions for problem (7.23) are given by (7.24)–(7.27).

By $L(\mathbf{x}, U) = F(\mathbf{x}) + UG(\mathbf{x})$, we shall denote a vector-valued Lagrange function. Assuming convexity of all functions $f_k(\mathbf{x})$ $(k = 1, 2, \ldots, s)$ and $g_i(\mathbf{x})$ $(i = 1, 2, \ldots, m)$, the lexicographic maximization problem

$$\text{lex}\max\{L(\mathbf{x}, U) | \nabla_{\mathbf{x}} L(\mathbf{x}, U) = 0, \ U \text{ lex} \geqq 0\} \tag{7.39}$$

will be called the *dual problem*. $\nabla_x L(\mathbf{x}, U)$ denotes the $s \times n$ matrix whose rows are the gradients at $\mathbf{x}$ of the components of L.

For the pair of lexicographic programming problems (7.23) and (7.39), the weak duality properties are valid.

Theorem 7.7 (see [49, p. 263]). *Let F, G be convex functions with first partial derivatives, $\mathbf{x}$ a feasible solution to (7.23), $\mathbf{x}^0$ a properly optimal solution to (7.23), and $(\mathbf{y}, U)$ a feasible solution to the corresponding dual problem (7.39). Then*

(i) $F(\mathbf{x})\ \text{lex} \geqq L(\mathbf{y}, U)$;
(ii) *if $F(\mathbf{x}) = L(\mathbf{y}, U)$, then $\mathbf{x}$ is a properly optimal solution to (7.23) and $(\mathbf{y}, U)$ is an optimal solution to (7.39);*
(iii) *there exists a matrix U^0 such that $(\mathbf{x}^0, U^0)$ is an optimal solution to (7.39) and $F(\mathbf{x}^0) = L(\mathbf{x}^0, U^0)$.*

The proof can be found in[49, p. 264].

To illustrate, let us consider the numerical example from Section 7.2:

$$\begin{aligned} \text{lex min } F(\mathbf{x}) &= \begin{pmatrix} f_1(\mathbf{x}) \\ f_2(\mathbf{x}) \end{pmatrix} = \begin{pmatrix} x_1^2 - 2x_1x_2 + x_2^2 - 2x_1 + 2x_2 \\ -x_1 + x_2^2 \end{pmatrix} \\ \text{subject to} \quad & g_1(\mathbf{x}) = x_1 + x_2 - 8 \leqq 0, \\ & g_2(\mathbf{x}) = -x_1 + x_2 + 2 \leqq 0. \end{aligned}$$

Because of the convexity of the functions $f_1(\mathbf{x})$, $f_2(\mathbf{x})$, $g_1(\mathbf{x})$, and $g_2(\mathbf{x})$, the dual problem (7.39) can be formulated.

The vector-valued Lagrange function $L(\mathbf{x}, U)$ is given by

$$\begin{aligned} &L(\mathbf{x}, U) \\ &= \begin{pmatrix} x_1^2 - 2x_1x_2 + x_2^2 - 2x_1 + 2x_2 \\ -x_1 + x_2^2 \end{pmatrix} + \begin{pmatrix} u_{11} & u_{12} \\ u_{21} & u_{22} \end{pmatrix} \begin{pmatrix} x_1 + x_2 - 8 \\ -x_1 + x_2 + 2 \end{pmatrix} \\ &= \begin{pmatrix} x_1^2 - 2x_1x_2 + x_2^2 - 2x_1 + 2x_2 + u_{11}(x_1 + x_2 - 8) + u_{12}(-x_1 + x_2 + 2) \\ -x_1 + x_2^2 + u_{21}(x_1 + x_2 - 8) + u_{22}(-x_1 + x_2 + 2) \end{pmatrix}. \end{aligned}$$

According to (7.39), we can write the dual problem as

$$\begin{aligned} &\text{lex max } L(\mathbf{x}, U) \\ &\text{subject to} \quad \begin{pmatrix} 2x_1 - 2x_2 - 2 + u_{11} - u_{12} & -2x_1 + 2x_2 + 2 + u_{11} + u_{12} \\ -1 + u_{21} - u_{22} & 2x_2 + u_{21} + u_{22} \end{pmatrix} = \begin{pmatrix} 0 & 0 \\ 0 & 0 \end{pmatrix} \end{aligned}$$

and

$$\begin{pmatrix} u_{11} & u_{12} \\ u_{21} & u_{22} \end{pmatrix} \text{lex} \geqq \begin{pmatrix} 0 & 0 \\ 0 & 0 \end{pmatrix}.$$

As already mentioned in Section 7.2, the element u_{ki} of matrix U is the dual variable related to the kth objective and ith constraint.

Let $\mathbf{x}^* = (3, 0)$ be a feasible solution of the primal problem and $\mathbf{y}^* = (2, 0)$ and

$$U^* = \begin{pmatrix} 0 & 2 \\ \frac{1}{2} & -\frac{1}{2} \end{pmatrix}$$

be a feasible solution of the dual problem. The reader can easily verify that

$$\left[F(\mathbf{x}^*) = \begin{pmatrix} 3 \\ -3 \end{pmatrix}\right] \text{lex} > L(\mathbf{y}^*, U^*) = \begin{pmatrix} 0 \\ -5 \end{pmatrix}.$$

This result illustrates the first part of Theorem 7.7. As shown in the previous section, using the Kuhn–Tucker conditions (7.24)–(7.27), the properly optimal solution to the primal problem $\mathbf{x}^0$ and the matrix of Lagrange multipliers U^0 can be estimated:

$$\mathbf{x}^0 = \left(\frac{5}{2}, \frac{1}{2}\right), \qquad U^0 = \begin{pmatrix} 0 & 2 \\ 0 & -1 \end{pmatrix}.$$

Because $F(\mathbf{x}^0) = L(\mathbf{x}^0, U^0)$, $\mathbf{x}^0$ is the properly optimal solution of the primal problem and $(\mathbf{x}^0, U^0)$ the optimal solution of the dual problem.

7.4 Behavior of the Firm Facing a Bicriteria Objective Function under Regulatory Constraint

In Section 2.5.3, the behavior of the firm under the "fair rate of return" regulation as proposed by Averch and Johnson [1] has been analyzed.

The main result was a "misallocation of economic resources" in which the firm "has an incentive to substitute between the factors in an uneconomic fashion" [1, p. 1068]). The firm will substitute capital for labor (overcapitalization effect).

Essential assumption in the Averch–Johnson model is that the firm maximizes profit. Bailey and Malone [2] argue that if the firm maximizes either revenue or output, then it will tend to undercapitalize.

Taking into account both results, the following question arises: What is the impact of the rate of return regulation for the firm maximizing revenue as well as profit as described in Section 7.1.4?

In answering this question, let us accept the same assumption as postulated by Averch and Johnson. The model is static as before, depreciation and regulatory lags are ignored, the allowed rate of return set by the regulator (s) is assumed to be greater than the firm's cost of capital, and the firm is assumed to produce only a simple product.

Hence the problem for the firm facing a bicriteria objective function is to maximize profit and revenue subject to the regulatory constraint. Using the notation as in Section 2.5.4, the following multiobjective optimization problem arises:

$$\text{Maximize} \qquad F(\mathbf{x}) = \begin{Bmatrix} \Pi(\mathbf{x}) \\ R(\mathbf{x}) \end{Bmatrix}$$

$$\text{subject to} \quad pq - sx_1 - c_2x_2 \leqq 0,$$

where $q = f(x_1, x_2)$, $f(0, x_2) = f(x_1, 0) = 0$, and $p = p(q)$ with

$$p'(q) = \frac{dp}{dq} < 0.$$

The revenue is given by

$$R = p(q)q$$

and the profit by

$$\Pi = p(q)q - c_1 x_1 - c_2 x_2,$$

where x_1 and x_2 denote the quantities of capital and labor and c_1, c_2 the prices of capital and labor, respectively. Application of the Kuhn–Tucker conditions (7.13)–(7.16) for the multiobjective programming problem yields

$$\alpha_1[(p+p'q)f_1 - c_1] + \alpha_2(p+p'q)f_1 - u([(p+p'q)f_1 - s] = 0, \quad (7.40)$$
$$\alpha_1[(p+p'q)f_2 - c_2] + \alpha_2(p+p'q)f_2 - u[(p+p'q)f_2 - c_2] = 0, \quad (7.41)$$
$$u(pq - sx_1 - c_2x_2) = 0,$$
$$pq - sx_1 - c_2x_2 \leqq 0,$$
$$u \geqq 0.$$

Denoting by $(p+p'q)f_1$ the marginal revenue of capital by MR_1 and by $(p+p'q)f_2$ the marginal revenue of labor by MR_2, (7.40)–(7.41) can be rewritten as

$$(\alpha_1 + \alpha_2 - u)\,\mathrm{MR}_1 + us = \alpha_1 c_1, \quad (7.42)$$
$$(\alpha_1 + \alpha_2 - u)\,\mathrm{MR}_2 + uc_2 = \alpha_1 c_2. \quad (7.43)$$

Furthermore, it follows from (7.42) and (7.43) that

$$\frac{\mathrm{MR}_1}{\mathrm{MR}_2} = \frac{f_1}{f_2} = \frac{\alpha_1 c_1 - us}{(\alpha_1 - u)c_2}. \quad (7.44)$$

For the unregulated monopoly maximizing profit ($u = 0$ and $\alpha_1 = 1$, $\alpha_2 = 0$), the marginal rate of substitution of capital for labor is equal to the ratio of their prices.

Under the conditions of effective regulatory constraint ($u > 0$) and $s > c_1$ ($u \neq 1$), (7.44) discloses that the equality of the marginal rate of substitution to the ratio of the input prices is not fulfilled. For the revenue-maximizing firm under regulatory constraint ($\alpha_1 = 0$; $u > 0$), the form (7.44) yields

$$\frac{f_1}{f_2} = \frac{s}{c_2} > \frac{c_1}{c_2},$$

i.e., the undercapitalization effect shown by Bailey and Malone [2]. What kind of result will be obtained if the firm maximizes profit as well as revenue: over- or undercapitalization?

The answer depends on the relation between α_1 and u. If $\alpha_1 > u$ (the preference for profit maximization is relatively high or the regulatory constraint is not very tight), then under the basic assumption $s > c_1$ it can be shown that

$$\frac{f_1}{f_2} = \frac{\alpha_1 c_1 - us}{(\alpha_1 - u)c_2} < \frac{c_1}{c_2},$$

i.e., Averch–Johnson effect or overcapitalization occurs. The profit-maximizing firm ($\alpha_1 = 1, \alpha_2 = 0$) with $0 < u < 1$ as a special case of our model confirms this result.

In the opposite case, $\alpha_1 < u$ implies

$$\frac{f_1}{f_2} = \frac{\alpha_1 c_1 - us}{(\alpha_1 - u)c_2} > \frac{c_1}{c_2}.$$

The result is undercapitalization, and the firm has an incentive to substitute labor for capital. Consequently, from (7.44) we have the following proposition.

Proposition 7.1. *In the firm maximizing revenue as well as profit and underlying regulatory constraint ($u > 0$), the cost-minimizing allocation of production factors in the sense*

$$\frac{f_1}{f_2} = \frac{c_1}{c_2}$$

cannot be achieved independently of the firm's objective preferences. The overcapitalization effect of the profit maximization cannot be compensated by the undercapitalization effect of the revenue maximization.

It follows from the above proposition that regardless of the firm's objectives, the rate of return regulation leads to suboptimal allocation of production factors.

References and Further Reading

[1] H. Averch and L. L. Johnson, Behavior of the firm under regulatory constraint, *Amer. Econ. Rev.*, **52** (1962), 1052–1069.

[2] E. E. Bailey and J. C. Malone, Resource allocation and the regulated firm, *Bell J. Econ. Manage. Sci.*, **1**-1 (1970), 129–142.

[3] E. Ballestero and C. Romero, *Multiple Criteria Decision Making and Its Applications to Economic Problems*, Kluwer Academic Publishers, Dordrecht, The Netherlands, 1998.

[4] W. Baumol, *Business Behavior Value and Growth*, Macmillan, New York, 1959.

[5] W. Baumol, The theory of expansion of the firm, *Amer. Econ. Rev.*, **52** (1962), 1078–1087.

[6] W. L. Beedles, A micro-economic investigation of multi-objective firms, *J. Finance*, **32** (1977), 1217–1233.

[7] F. A. Behringer, Lexicographic quasiconcave multiobjective programming, *Z. Oper. Res.*, **21** (1977), 103–116.

[8] G. R. Bitran, Duality for nonlinear multiple-criteria optimization problems, *J. Oper. Theory Appl.*, **38** (1981), 367–401.

[9] B. Böhm and P. Brandner, Optimal economic policy making with an econometric model using interactive methods of vector optimization, in J. Gruber, ed., *Econometric Decision Models: New Methods of Modeling and Applications*, Lecture Notes in Economics and Mathematical Systems vol. 366, Springer-Verlag, Berlin, 1991, 149–163.

[10] B. Böhm and M. Luptáčik, Optimisation with an environmental input–output model, in G. Boero and A. Silberston, eds., *Environmental Economics*, St. Martin's Press, New York, 1995, 21–33.

[11] M. Brown and N. Revankar, A generalized theory of the firm: An integration of the sales and profit maximization hypothesis, *Kyklos*, **24**-3 (1971), 427–443.

[12] G. A. Calvo, Staggered contracts in a utility-maximizing framework, *J. Monetary Econ.*, **12** (1983), 383–398.

[13] G. C. Chow, *Analysis and Control of Dynamic Systems*, Wiley, New York, 1975.

[14] B. D. Craven, Strong vector minimization and duality, *Z. Angew. Math. Mech.*, **60** (1980), 1–5.

[15] M. Despontin, P. Nijkamp, and J. Spronk, eds., *Macro-Economic Planning with Conflicting Goals*, Lecture Notes in Economics and Mathematical Systems vol. 230, Springer-Verlag, Berlin, 1984.

[16] C. J. van Eijk and J. Sandee, Quantitative determination of an optimum economic policy, *Econometrica*, **27** (1959), 1–33.

[17] F. M. Fisher, Book review of W. Baumol, *Business Behavior Value and Growth*, *J. Political Econ.*, **68** (1960), 314–315.

[18] K. A. Fox, J. K. Sengupta, and E. Thorbecke, *The Theory of Quantitative Economic Policy with Applications to Economic Growth and Stabilization*, North-Holland, Amsterdam, 1966.

[19] J. A. Frankel and M. Funke, A two-country analysis of international targeting of nominal GNP, *Rev. Political Econ.*, **83** (1993), 69–106.

[20] B. M. Friedman, *Economic Stabilization Policy: Methods in Optimization*, North-Holland, Amsterdam, 1975.

[21] R. Frisch, Circulation planning: Proposal for a national organisation of a commodity and service exchange, *Econometrica*, **3** (1934), 68–82.

[22] R. Frisch, *A Memorandum on Price-Wage-Tax-Subsidy Policies as Instruments in Maintaining Optimal Envelopment*, Document E (CN1/Dub 2), United Nations, New York, 1949; reprinted as Memorandum, Institute of Economics, University of Oslo, Oslo, 1953.

[23] R. Frisch, Macroeconomics and linear programming, in *Twenty-Five Economic Essays in Honour of Eric Lindahl*, Ekonomisk Tidskrift, Stockholm, 1956.

[24] R. Frisch, *Numerical Determination of a Quadratic Preference Function for Use in Macroeconomic Programming* Memorandum 14, Institute of Economics, University of Oslo, Oslo, 1957; reprinted in "Studies in Honour of Gustavo del Vecchio," *G. Econ. Ann. Econ.*, **1** (1961), 43–83.

[25] H. Frisch and S. Staudinger, Inflation targeting versus nominal income targeting, *J. Econ.*, **78** (2003), 113–137.

[26] W. Fritz, Empirische Zielforschung und Operations Research: Eine Untersuchung unternehmerischer Ziele und ihre Konsequenzen für das Operations Research, *Oper. Res. Spektrum*, **8** (1986) 99–108.

[27] T. Gal, Th. J. Stewart, and Th. Hanne, eds., *Multicriteria Decision Making: Advances in MCDM Models, Algorithms, Theory, and Applications*, Kluwer Academic Publishers, Dordrecht, The Netherlands, 1999.

[28] D. Gale, H. W. Kuhn, and A. W. Tucker, Linear programming and the theory of games, in T. C. Koopmans, ed., *Activity Analysis of Production and Allocation*, Cowles Commission Monograph 13, Wiley, New York, 1951, Chapter XIX, 317–329.

[29] A. M. Geoffrion, Proper efficiency and the theory of vector maximization, *J. Math. Anal. Appl.*, **22** (1968), 618–630.

[30] J. Gruber, ed., *Econometric Decision Models*, Lecture Notes in Economics and Mathematical Systems vol. 208, Springer-Verlag, Berlin, 1983.

[31] J. Gruber, ed., *Econometric Decision Models: New Methods of Modeling and Applications*, Lecture Notes in Economics and Mathematical Systems vol. 366, Springer-Verlag, Berlin, 1991.

[32] M. Hall, On the goals on the firm: Comment, *Q. J. Econ.*, **80** (1966) 154–158.

[33] R. Hall and N. G. Mankiw, Nominal income targeting, in N. G. Mankiw, ed., *Monetary Policy*, National Bureau of Economic Research, Cambridge, MA, 1994.
[34] J. R. Hicks, Annual survey of economic theory: The theory of monopoly, *Econometrica*, **34** (1935), 1–20.
[35] H. Isermann, Existence and duality in multiple objective linear programming, in H. Thiriez and S. Zionts, eds., *Multiple Criteria Decision Making*, Lecture Notes in Economics and Mathematical Systems vol. 130, Springer-Verlag, Berlin, 1976, 64–75.
[36] H. Isermann, The relevance of duality in multiple objective linear programming, *TIMS Stud. Manage. Sci.*, **6** (1977), 241–262.
[37] H. Isermann, Duality in multiple objective linear programming, in S. Zionts, ed., *Multiple Criteria Problem Solving: Proceedings of a Conference, Buffalo*, 1977, Lecture Notes in Economics and Mathematical Systems vol. 155, Springer-Verlag, Berlin, 1978, 274–285.
[38] H. Isermann, On some relations between a dual pair of multiple objective linear programs, *Z. Oper. Res.*, **22** (1978), 33–41.
[39] H. Isermann, Linear lexicographic optimization, *Oper. Res. Spektrum*, **4** (1982), 223–228.
[40] L. R. Klein, The use of econometric models as a guide to economic policy, *Econometrica*, **15** (1947), 33–47.
[41] T. C. Koopmans, Analysis of production as an efficient combination of activities, in T. C. Koopmans, ed., *Activity Analysis of Production and Allocation*, Cowles Commission Monograph 13, Wiley, New York, 1951, 33–97.
[42] J. S. H. Kornbluth, Duality indifference and sensitivity analysis in multiple objective linear programming, *Oper. Res. Q.*, **25** (1974), 599–614.
[43] H. W. Kuhn and A. W. Tucker, Nonlinear programming, in J. Neyman, ed., *Proceedings of the 2nd Berkeley Symposium on Mathematical Statistics and Probability*, University of California Press, Berkeley, CA, 1951, 481–492.
[44] O. Lange, The foundations of welfare economics, *Econometrica*, **10** (1942), 215–228.
[45] L. Leiderman and L. E. O. Svensson, *Inflation Targets*, CEPR, London, 1995.
[46] W. W. Leontief, Environmental repercussions and the economic structure: An input–output approach, *Rev. Econ. Statist.*, **52** (1970), 262–271.
[47] W. Leontief, Theoretical assumptions and nonobserved facts, *Amer. Econ. Rev.*, **61** (1971), 1–7.
[48] M. Luptáčik and B. Böhm (1994), An environmental input–output model with multiple criteria, *Ann. Oper. Res.*, **54**, 119–127.
[49] M. Luptáčik and F. Turnovec, Lexicographic geometric programming, *Eur. J. Oper. Res.*, **51** (1991), 259–269.
[50] I. Marusciac, On Fritz John type optimality criterion in multiobjective optimization, *Anal. Numér. Théorie Approx.*, **11** (1982), 109–114.
[51] B. T. McCallum, *The Alleged Instability of Nominal Income Targeting*, Paper G97/6, Reserve Bank of New Zealand, Wellington, 1997.
[52] B. T. McCallum and E. Nelson, Nominal income targeting in an open economics optimizing model, *J. Monetary Econ.*, **11** (1999), 139–168.
[53] R. Marris, *The Economic Theory of "Managerial" Capitalism*, Free Press of Glencoe, Glencoe, IL, 1964.
[54] A. P. Mastenbroek and P. Nijkamp, A spatial environmental model for an optimal allocation of investments, in P. Nijkamp, ed., *Environmental Economics, Vol. 2: Methods*, Nijhoff Social Sciences, Leiden, The Netherlands, 1976.
[55] R. Neck, Ch. Richter, and P. Mooslechner, eds., *Quantitative Economic Policy: Essays in Honour of Andrew Hughes Hallet*, Springer, Berlin, 2008.
[56] T. Negishi, Welfare economics and existence of an equilibrium for a competitive economy, *Metroeconomica*, **12** (1960), 92–97.

[57] P. Nijkamp, *Environmental Policy Analysis*, Wiley, Chichester, UK, 1980.
[58] D. K. Osborne, On the goals of the firm, *Q. J. Econ.*, **77** (1964), 592–603.
[59] V. Pareto, *Manuale di Economia Politica*, Societa Editrice Libraria, Milan, 1906 (in Italian); *Manual of Political Economy*, Macmillan, New York, 1971 (in English; A. S. Schwier, transl.).
[60] M. L. Petit, *Control Theory and Dynamic Games in Economic Policy Analysis*, Cambridge University Press, Cambridge, UK, 1990.
[61] J. Rawls, *A Theory of Justice*, Harvard University Press, Cambridge, MA, 1971.
[62] K. Ritter, Optimization theory in linear spaces III, *Math. Ann.*, **184** (1970), 133–154.
[63] J. M. Roberts, New Keynesian economics and the Phillips curve, *J. Money Credit Banking*, **27** (1995), 975–984.
[64] C. Romero, A note on distributive equity and social efficiency, *J. Agric. Econ.*, **52**-2 (2001), 110–112.
[65] R. Rosenberg, Profit constraint revenue maximization: A note, *Amer. Econ. Rev.*, **61** (1971), 208–209.
[66] E. E. Rosinger, Duality and alternative in multiobjective optimization, *Proc. Amer. Math. Soc.*, **64** (1977), 307–312.
[67] W. A. Rödder, A generalized saddle-point theory, *Eur. J. Oper. Res.*, **1** (1977), 55–59.
[68] T. Scitovsky, A note on profit maximization and implications, *Rev. Econ. Stud.*, **11** (1943), 57–60.
[69] C. Singh, Optimality conditions in multiobjective differentiable programming, *J. Oper. Theory Appl.*, **53** (1987), 115–123.
[70] D. J. Smith, W. J. Boyes, and D. E. Peseau, *Size, Growth, Profits, and Executive Compensation in the Large Corporation*, Macmillan, London, 1975.
[71] W. Stadler, A survey of multicriteria optimization or the vector maximum problem, Part I: 1776–1960, *J. Oper. Theory Appl.*, **29** (1979), 1–51.
[72] R. E. Steuer, An interactive multiple objective linear programming procedure, *TIMS Stud. Manage. Sci.*, **6** (1977), 225–239.
[73] R. E. Steuer, L. R. Gardiner, and J. Gray, A bibliographic survey of the activities and international nature of multiple criteria decision making, *J. Multicriteria Decision Anal.*, **5** (1996), 195–217.
[74] G. Stigler, The development of utility theory, *J. Political Econ.*, **58** (1950), 307–327.
[75] L. E. O. Svensson, Inflation forecast targeting: Implementing and monitoring inflation targets, *Eur. Econ. Rev.*, **41** (1997), 1111–1146.
[76] L. E. O. Svensson, Optimal inflation targets, conservative central banks, and linear inflation contracts, *Amer. Econ. Rev.*, **87** (1997), 98–114.
[77] K. P. Schönfeld, Some duality theorems for the nonlinear vector maximum problem, *Unternehmensforschung*, **14** (1970), 51–63.
[78] T. Tanino and Y. Sawaragi, Duality theory in multiobjective programming, *J. Optim. Theory Appl.*, **27** (1979), 509–529.
[79] H. Theil, Econometric models and welfare maximization, *Weltwirtschaftliches Archiv*, **72** (1954), 60–83.
[80] H. Theil, *Optimal Decision Rules for Government and Industry*, North-Holland, Amsterdam, 1964.
[81] J. Tinbergen, *On the Theory of Economic Policy*, 1st printing, North-Holland, Amsterdam, 1952.
[82] J. Tinbergen, *Economic Policy: Principles and Design*, 1st ed., North-Holland, Amsterdam, 1956.
[83] J. Tinbergen, *On the Theory of Economic Policy*, 4th printing, North-Holland, Amsterdam, 1966.

[84] F. Turnovec, Lexicografická maximalizace (Lexicographic maximization), *Ekon. Mat. Obz.*, **3** (1983), 275–297.

[85] H. Wallenius, Optimizing macroeconomic policy: A review of approaches and applications, *Eur. J. Oper. Res.*, **10** (1982), 221–228.

[86] H. Wallenius, I. Wallenius, and P. Vartia, An experimental investigation of an interactive programming approach to solving macroeconomic policy problems, in K. B. Haley, ed., *Operational Research* '78, North-Holland, Amsterdam, 1979, 878–894.

[87] O. Williamson, *The Economics of Discretionary Behaviour: Managerial Objectives in a Theory of the Firm*, Prentice–Hall, Englewood Cliffs, NJ, 1964.

[88] M. E. Wolf and J. Hanauer, NAMEA der Luftschadstoffe Österreich 1980–1997, *Statist. Nachr.*, **5** (2000), 401–418.

[89] M. Zeleny, *Multiple Criteria Decision Making*, McGraw–Hill, New York, 1982.

[90] S. Zionts and J. Wallenius, An interactive programming method for solving the multicriteria problem, *Manage. Sci.*, **22** (1976), 652–663.

[91] J. Zowe, A duality theory for a convex programming in order-complete vector lattices, *J. Math. Anal. Appl.*, **50** (1975), 273–287.

[92] J. Zowe, The saddle-point theorem of Kuhn and Tucker in ordered vector spaces, *J. Math. Anal. Appl.*, **57** (1977), 41–55.

8

Multiobjective Linear Programming

As in single-objective mathematical programming, the most developed part of multiobjective optimization—from the theoretical as well as the applications point of view—is multiobjective linear programming. In 1951, Gale, Kuhn, and Tucker [7] considered a pair of general matrix linear programming problems, i.e., a linear programming problem with a matrix-valued linear objective function, and established some theorems of existence and duality. With matrix linear programming problems containing linear programming problems with a vector-valued as well as a scalar-valued objective function as special cases, the developed theory comprises the respective theoretical framework for vector linear programming problems as well as for ordinary linear programming problems.

The chapter consists of four sections. In Section 8.1, some existence results for multiple-objective linear programming are presented. A duality concept based on Isermann [13, 15, 16, 17] and the multiple-objective simplex method are the subjects of Section 8.2. In Section 8.3, interactive procedures represented by the Zionts–Wallenius method are described. Section 8.4 is devoted to the analysis of the Leontief pollution model consisting of multiple objectives.

8.1 Linear Vector Optimization Problems

The Leontief pollution model from Section 7.1.5 provides an example of a linear multiobjective programming problem. In both versions of this model, the objective functions as well as the constraints are linear.

For linear functions $f_k(\mathbf{x})$ $(k = 1, 2, \ldots, s)$ and $g_i(\mathbf{x})$ $(i = 1, 2, \ldots, m)$, the multiobjective programming problem (7.12) can be written as

$$\text{Minimize} \quad \{F(\mathbf{x}) = C\mathbf{x} | A\mathbf{x} = \mathbf{b},\ \mathbf{x} \geqq \mathbf{0}\}, \tag{8.1}$$

with A being an $m \times n$ matrix with rank $(A) = m$, $\mathbf{b} \in R^m$, and C is the $s \times n$ criterion matrix. Adding the slack variables, we replaced the inequality constraints in (7.12) by equalities and explicitly introduced the nonnegativity constraints for the

M. Luptáčik, *Mathematical Optimization and Economic Analysis*,
Springer Optimization and Its Applications 36, DOI 10.1007/978-0-387-89552-9_8,

variables $x_j (j = 1, 2, \ldots, n)$. The linear minimum problem (8.1) can be considered as the problem of finding all $\mathbf{x}^0 \in K = \{\mathbf{x} | A\mathbf{x} = \mathbf{b},\ \mathbf{x} \geqq \mathbf{0}\}$, which are efficient per Definition 7.4, and the respective values $F(\mathbf{x}^0)$. The set of all efficient solutions $\mathbf{x}^0$ for (8.1) will be denoted by K^0. It can be shown that for linear vector optimization problems, efficient solutions and properly efficient solutions coincide [11, Theorem 2, p. 620]. Note that for $s = 1$, problem (8.1) reduces to the linear programming problem

$$\text{minimize} \quad \{f(\mathbf{x}) = \mathbf{c}'\mathbf{x} | A\mathbf{x} = \mathbf{b},\ \mathbf{x} \geqq \mathbf{0}\},$$

where $\mathbf{c} \in R^n$ and the notion of an efficient solution or of a Pareto minimal point in Definition 7.4 is equivalent to the notion of optimality in ordinary linear programming problems.

For simplicity and the possibility of a graphical presentation, let us return to the production problem used in Section 4.5. Under the constraints for land, capital, and labor, the farmer has to decide how many acres of wheat and how many acres of potatoes he should plant in order to maximize his return revenue. Suppose that we consider an additional objective of the farmer is, for example, maximizing the utilization of the production factor land. Using the same notation as in Section 4.5, the following multiple-objective linear programming (MOLP) problem is obtained:

$$\text{Maximize} \qquad F(\mathbf{x}) = \begin{Bmatrix} f_1(\mathbf{x}) = 40x_1 + 120x_2, \\ f_2(\mathbf{x}) = x_1 + x_2 \end{Bmatrix} \tag{8.2}$$

$$\begin{aligned} \text{subject to} \qquad x_1 + x_2 &\leqq 100, \\ 10x_1 + 20x_2 &\leqq 1100, \qquad (8.3) \\ x_1 + 4x_2 &\leqq 160, \\ x_1 &\geqq 0, \\ x_2 &\geqq 0. \qquad (8.4) \end{aligned}$$

The graphical representation including the set of feasible solutions K with all its extreme points is given in Figure 8.1. There are five extreme points, denoted by $\mathbf{x}^{(1)}$, $\mathbf{x}^{(2)}$, $\mathbf{x}^{(3)}$, $\mathbf{x}^{(4)}$, and $\mathbf{x}^{(5)}$. The corresponding values of the objective functions $f_1(\mathbf{x})$ and $f_2(\mathbf{x})$ respectively can be found in Table 8.1.

Single optimization with respect to the first objective $f_1(\mathbf{x})$ yields the unique optimal solution $\mathbf{x}^{(3)} = (60, 25)$ and $f_1(\mathbf{x}^{(3)}) = 5{,}400$. Maximization of the second objective function $f_2(\mathbf{x})$ leads to infinite many optimal solutions—represented by the linear convex combination of points $\mathbf{x}^{(4)}$ and $\mathbf{x}^{(5)}$—with the objective function value equal to 100. Moving from point $\mathbf{x}^{(5)}$ toward point $\mathbf{x}^{(4)}$, the value of $f_2(\mathbf{x})$ remains constant, but the value of the first objective function is increasing (from 4,000 to 4,800). Therefore, point $\mathbf{x}^{(5)}$ cannot be a Pareto-efficient solution of the MOLP (8.2)–(8.4). Similarly, the extreme points $\mathbf{x}^{(1)}$ and $\mathbf{x}^{(2)}$ are dominated by points $\mathbf{x}^{(3)}$ and $\mathbf{x}^{(4)}$ and therefore cannot be candidates for an efficient solution of problem (8.2)–(8.4). The graphical representation in criterion space is provided in Figure 8.2.

Table 8.1. The values of the objective functions for the extreme points.

	$\mathbf{x}^1 = \begin{pmatrix} 0 \\ 0 \end{pmatrix}$	$\mathbf{x}^2 = \begin{pmatrix} 0 \\ 40 \end{pmatrix}$	$\mathbf{x}^3 = \begin{pmatrix} 60 \\ 25 \end{pmatrix}$	$\mathbf{x}^4 = \begin{pmatrix} 90 \\ 10 \end{pmatrix}$	$\mathbf{x}^5 = \begin{pmatrix} 100 \\ 0 \end{pmatrix}$
$f_1(\mathbf{x}^{(k)})$	0	4,800	5,400 (max)	4,800	4,000
$f_2(\mathbf{x}^{(k)})$	0	40	85	100 (max)	100 (max)

The ideal solution, consisting of the optimal values from the single optimization represented by the point $\hat{F}$, is not possible. Consequently, the solution of the linear vector optimization problem (8.2)–(8.4) consists in finding Pareto-efficient points of F on the set of feasible solution K. The set of efficient solutions for the MOLP (8.2)–(8.4) is then described as the set of all linear convex combinations of the extreme points $\mathbf{x}^{(3)}$ and $\mathbf{x}^{(4)}$:

$$K^0 = \{\mathbf{x} | \mathbf{x} = \lambda \mathbf{x}^{(3)} + (1 - \lambda)\mathbf{x}^{(4)}, \ 0 \leqq \lambda \leqq 1\}. \tag{8.5}$$

Before we try to do it, we will present some existence results for multiple-objective linear programming.

With reference to the presentation of duality for single-objective linear programming in Chapter 4, problem (8.1) will be rewritten as

$$\text{Maximize} \quad \{F(\mathbf{x}) = C(\mathbf{x}) | \mathbf{x} \in K\}. \tag{8.6}$$

Let us consider the linear system

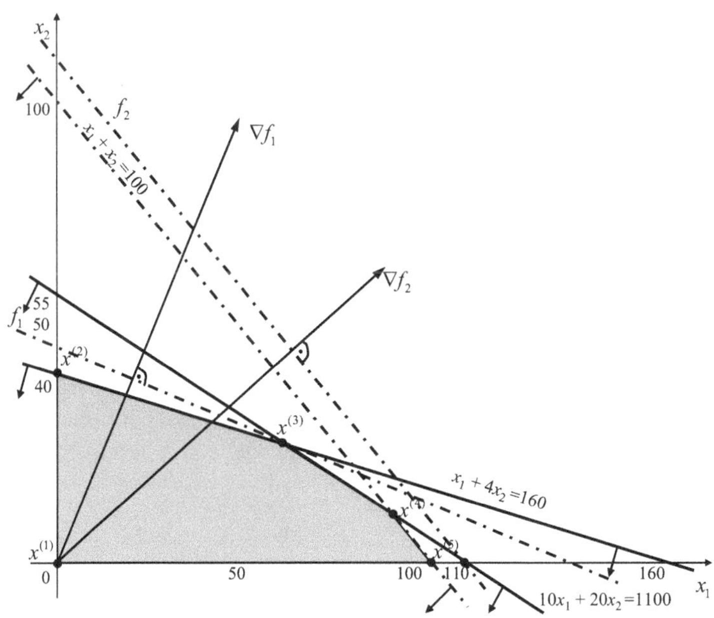

Fig. 8.1. Graph of example (8.2)–(8.4) in decision space.

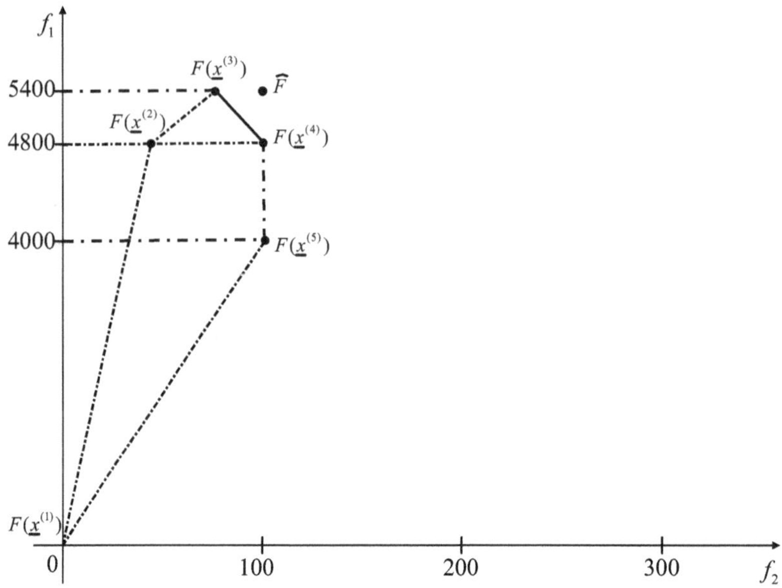

Fig. 8.2. Graph of example (8.2)–(8.4) in criterion space.

$$A\mathbf{x} \leqq \mathbf{b} \quad \mathbf{b} \geqq \mathbf{0}, \quad \mathbf{x} \geqq \mathbf{0} \tag{8.7}$$

and the homogeneous linear system

$$A\mathbf{x} \leqq \mathbf{0}, \quad \mathbf{x} \geqq \mathbf{0}. \tag{8.8}$$

Kreko [20, pp. 184–185] proved the following.

Theorem 8.1. *The set of feasible solutions of system* (8.7) *is not bounded if and only if system* (8.8) *has a nontrivial solution* ($\mathbf{x} \geq \mathbf{0}$).

Assuming K is not empty, Isermann [17] distinguished two exhaustive cases:

(i) K is bounded; i.e., (8.7) has a solution $\mathbf{x}'$, but (8.8) has no semipositive solution ($\mathbf{x} \geq \mathbf{0}$).

(ii) K is not bounded; i.e., (8.7) as well as (8.8) have solutions $\mathbf{x}' \geqq \mathbf{0}$ and $\mathbf{x}'' \geq \mathbf{0}$, respectively.

Following the respective definition for single-objective linear programming, Isermann [17, p. 34] defined $F(\mathbf{x})$ to be bounded from above in K if and only if

$$C\mathbf{x} \geq \mathbf{0}, \quad A\mathbf{x} = \mathbf{0}, \quad \mathbf{x} \geqq \mathbf{0} \tag{8.9}$$

have no solution. According to Theorem 8.1, $F(\mathbf{x})$ is not bounded from above in K if and only if (8.9) has a solution $\mathbf{x}'' \geq \mathbf{0}$. If $F(\mathbf{x})$ is not bounded from above in K, then for each $\mathbf{x} \in K$ there exists some $\mathbf{x}' \in K$ such that $C\mathbf{x}' \geq C\mathbf{x}$.

Then Isermann [17, p. 35] proved the following two lemmas.

Lemma 8.1. *Let $K \neq \varnothing$. Problem* (8.6) *has no efficient solution if and only if $F(\mathbf{x})$ is not bounded from above in K.*

Lemma 8.2. *If problem* (8.6) *has an efficient solution, at least one feasible basic solution for* (8.6) *is efficient.*

Summarizing, Isermann [17, p. 35] formulated the existence properties of a linear vector maximum problem (8.6) as follows.

Proposition 8.1. *Problem* (8.6) *has no efficient solution if and only if*

(i) $K = \varnothing$ *or*
(ii) $F(\mathbf{x})$ *is not bounded from above in $K \neq \varnothing$.*

The reader can easily verify that system (8.9) with the data from example (8.2)–(8.4) has no solution, and thereupon $F(\mathbf{x})$ is bounded from above in K.

To illustrate Lemma 8.1, let us consider a small numerical example of the multi-criteria Leontief pollution model described in Section 7.1.5. Only two commodities or sectors and one pollutant are considered with the matrix of input coefficients

$$A_{11} = \begin{pmatrix} 0.25 & 0.4 \\ 0.4 & 0.12 \end{pmatrix}$$

and with the following vector of primary inputs per unit of output: $\mathbf{v}_1' = (0.8;\ 3.6)$. The amounts of pollutant per unit of good i $(i = 1, 2)$ are given by the vector $(0.5;\ 0.2)$. For elimination of the pollutant by one unit, 0.2 unit of input from sector 2 and one unit of primary input are needed. The disposable amount of the primary input is restricted to 362 units.

Maximization of the final demand and the minimization of net pollution under the constraints (7.8)–(7.11) lead to

$$\text{Maximize} \quad F(\mathbf{x}, \mathbf{y}) = \begin{pmatrix} 0 & 0 & 0 & 1 & 1 & 0 \\ 0 & 0 & 0 & 0 & 0 & -1 \end{pmatrix} \begin{pmatrix} \mathbf{x} \\ \mathbf{y} \end{pmatrix} \tag{8.10}$$

$$\text{with} \quad \mathbf{x}' = (x_1, x_2, x_3) \quad \text{and} \quad \mathbf{y}' = (y_1, y_2, y_3)$$

$$\text{subject to} \quad -0.75x_1 + 0.4x_2 + y_1 \leqq 0,$$

$$0.14x_1 - 0.88x_2 + 0.2x_3 + y_2 \leqq 0, \tag{8.11a}$$

$$0.5x_1 + 0.2x_2 - x_3 - y_3 \leqq 0,$$

$$0.8x_1 + 3.6x_2 + x_3 \leqq 362, \tag{8.11b}$$

$$x_1, x_2, x_3, y_1, y_2, y_3 \geqq 0, \tag{8.12}$$

where y_1 and y_2 denote the final demand for the commodities (which are now endogenous variables) and y_3 is the residual amount of emissions after abatement, which should be minimized.

The reader can easily verify that the homogeneous system corresponding to system (8.11) has the trivial solution $x_1 = x_2 = x_3 = y_1 = y_2 = y_3 = 0$ only; in other words, system (8.9) has no solution.

The set of feasible solutions described by (8.11)–(8.12) is bounded. The single maximization of the final demand (the first objective function) yields the following optimal solution: $x_1^{(1)} = 264$, $x_2^{(1)} = 42$, $x_3^{(1)} = 0$; $y_1^{(1)} = 181$, $y_2^{(1)} = 0$, and $y_3^{(1)} = 140$. The corresponding value of the first objective function $f_1(\mathbf{x}^{(1)}, \mathbf{y}^{(1)}) = 181$ and of the second objective function—the level of net pollution—$f_2(\mathbf{x}^{(1)}, \mathbf{y}^{(1)}) = 140$.

The single minimization of the net pollution leads to the trivial solution $x_1^{(2)} = x_2^{(2)} = x_3^{(2)} = y_1^{(2)} = y_2^{(2)} = y_3^{(2)} = 0$ and $f_1(\mathbf{x}^{(2)}, \mathbf{y}^{(2)}) = f_2(\mathbf{x}^{(2)}, \mathbf{y}^{(2)}) = 0$.

Deleting the constraint (8.11b) for the primary factor from system (8.11) implies the existence of a nontrivial solution for the corresponding homogeneous system (and for system (8.9)), and therefore the vector-valued function (8.10) is not bounded from above with respect to (8.11a)–(8.12) and the set of efficient solutions is empty.

8.2 Duality in Multiple-Objective Linear Programming

As already mentioned in Section 7.3. there are different duality concepts for multiple-objective optimization. In the paper by Isermann [16], three duality concepts in multiple-objective linear programming—developed by Gale, Kuhn, and Tucker [7], Kornbluth [19], and Isermann [13, 15, 17]—are described and related to each other. A new concept of duality based on a set-expansion process for the computation of optimal solutions without scalarization is proposed by Galperin and Guerra Jimenez [8]. With reference to the theory of linear programming presented in Chapter 4, we will now introduce a duality concept elaborated by Isermann [13, 15, 17].

Let us call problem (8.6) the *primal problem*. As in Section 7.2, we define an $(s \times m)$-dimensional matrix $U = \{u_{ki}\}$ of dual variables, where u_{ki} is the dual variable assigned to the ith primal constraint and to the kth criteria. The multiple-objective program

$$\text{Minimize} \quad \{H(U) = U\mathbf{b} | U \in T\}, \tag{8.13}$$

with $T = \{U | UA\mathbf{w} \leq C\mathbf{w} \text{ for no } \mathbf{w} \geqq \mathbf{0}\}$, is called the *dual problem* of (8.6). $H(U) = (h_1(U), h_2(U), \ldots, h_s(U))'$ is the vector-valued objective function with $h_k(U) = \sum_{i=1}^{m} u_{ki} b_i$ for all $k = 1, 2, \ldots, s$, and "Minimize" is the notation for finding all efficient solutions in a minimizing sense, i.e., all $U^0 \in T$ for which there exists no $U^* \in T$ such that $H(U^*) \leq H(U^0)$. For $s = 1$, problem (8.13) reduces to a linear program because the problem

$$\text{minimize} \quad \{h(\mathbf{u}) = \mathbf{u}'\mathbf{b} | \mathbf{u}'A\mathbf{w} < \mathbf{c}'\mathbf{w} \text{ for no } \mathbf{w} \geqq \mathbf{0}\},$$

with $\mathbf{u} \in R^m$ is equivalent to

$$\text{minimize} \quad \{h(\mathbf{u}) = \mathbf{u}'\mathbf{b} | \mathbf{u}'A \geqq \mathbf{c}'\}. \tag{8.14}$$

Problem (8.14) is the dual problem to the problem

$$\text{maximize} \quad \{f(\mathbf{x}) = \mathbf{c}'\mathbf{x} | A\mathbf{x} = \mathbf{b},\ \mathbf{x} \geqq \mathbf{0}\},$$

obtained as a single-objective case ($s = 1$) of problem (8.6).

As for the primal problem (8.6), we can state similar necessary and sufficient conditions that ensure the existence of at least one efficient solution for the dual problem (8.13). We consider the system

$$U A\mathbf{w} \leq C\mathbf{w} \quad \text{for no } \mathbf{w} \geqq \mathbf{0}$$

as well as the homogeneous system

$$U A\mathbf{w} \leq \mathbf{0} \qquad \text{for no } \mathbf{w} \geqq \mathbf{0}.$$

Assuming that T is not empty, $H(U)$ is bounded from below in T if and only if

$$U\mathbf{b} \leq \mathbf{0}, \quad U A\mathbf{w} \leq \mathbf{0} \qquad \text{for no } \mathbf{w} \geqq \mathbf{0} \tag{8.15}$$

have no solution U. $H(U)$ is not bounded from below in T if and only if (8.15) has a solution U''.

Similarly to Proposition 8.1 above, Isermann [17, pp. 35–36] stated the following.

Proposition 8.2. *Problem* (8.13) *has no efficient solution if and only if*

(i) $T = \varnothing$ *or*
(ii) $H(U)$ *is not bounded from below in* $T \neq \varnothing$.

Isermann [16, p. 276] summarized the existence and duality properties of (8.6) and (8.13) in the following.

Theorem 8.2. *Consider problems* (8.6) *and* (8.13).

(i) *The following statements are equivalent:*
- *Both* (8.6) *and* (8.13) *have a feasible solution.*
- *Both* (8.6) *and* (8.13) *have an efficient solution and there exists at least one pair* $(\mathbf{x}^0, U^0)$ *of efficient solutions such that* $C\mathbf{x}^0 = U^0\mathbf{b}$.
- *The linear program*

$$\begin{aligned} \textit{minimize} \quad & \mathbf{u}'\mathbf{b} \\ \textit{subject to} \quad & \mathbf{u}'A - \boldsymbol{\alpha}'C \geqq \mathbf{0}', \\ & \boldsymbol{\alpha}' \geqq \mathbf{e}', \end{aligned} \tag{8.16}$$

where $\mathbf{e}' = (1, 1, \ldots, 1)$, *has an optimal solution* $(\mathbf{u}^*, \boldsymbol{\alpha}^*)$.

(ii) $\mathbf{x}^0$ *is an efficient solution for* (8.6) *if and only if there exists a feasible solution* U^0 *for* (8.13) *such that* $C\mathbf{x}^0 = U^0\mathbf{b}$. U^0 *is then itself an efficient solution for* (8.13).

(iii) U^0 *is an efficient solution for* (8.13) *if and only if there exists a feasible solution* $\mathbf{x}^0$ *for* (8.6) *such that* $C\mathbf{x}^0 = U^0\mathbf{b}$. $\mathbf{x}^0$ *is then itself an efficient solution for* (8.6).

The dual pair of multiple-objective linear programming problems under consideration can provide interesting and useful insights into the multiple-criteria decision-making process. For this purpose, let us return to the production problem (8.2)–(8.4). The respective dual problem can be interpreted as finding two (generally s) price vectors with the components u_{ki} ($k = 1, 2$; $i = 1, 2, 3$; generally $i = 1, 2, \ldots, m$) describing the marginal contributions of the ith scarce resource with respect to the kth objective such that the s-dimensional (in our example, two-dimensional) total value of the resources is minimal. An efficient allocation of the scarce resources implies a set of shadow prices such that the total value of the resources equals each of the primal objective function values considered. In the same way, a properly selected set of shadow prices implies an efficient production plan. How do we find the matrix of shadow prices U^0?

In connection with Theorem 8.2, efficient solutions for both the primal problem (8.6) and the dual problem (8.13) can be extracted directly from the multiple-objective simplex tableau of the primal problem (8.6). The multiple-objective simplex method (see, e.g., [4, 14]), as an extension of the single-objective simplex method described in Section 4.5, at first seeks feasibility for the dual problem while maintaining feasibility in the primal problem. As soon as feasible solutions for both problems are determined, an initial pair ($\mathbf{x}^0$, U^0) of efficient solutions is found such that $C\mathbf{x}^0 = U^0\mathbf{b}$. The further procedure of the multiple-objective simplex method consists in finding all feasible solutions for the primal problem while maintaining feasibility in the dual problem.

The initial simplex tableau in Table 4.4 is now extended by replacing a vector $\mathbf{c}$ by a matrix C (see Table 8.2). Just as in single-objective linear programming, multiple-objective simplex tableaux can be transformed by multiplication of their columns by the inverse basis B^{-1} (see Table 8.3).

Table 8.2. Initial simplex tableau for the multiple-objective linear program.

$\mathbf{C}$		
A	E	$\mathbf{b}$
$-C$	$\mathbf{0}$	0

Table 8.3. Transformed simplex tableau.

$B^{-1}A$	B^{-1}	$B^{-1}\mathbf{b}$
$C_B B^{-1}A - C$	$C_B B^{-1}$	$C_B B^{-1}\mathbf{b}$

A feasible basic solution for the primal problem (8.6), $\mathbf{x}^0$, is then described by the vector of basic variables $\mathbf{x}_B^0 = B^{-1}\mathbf{b}$ and by $\mathbf{x}_N^0 = \mathbf{0}$, where $\mathbf{x}_N^0$ denotes the vector of nonbasic variables. The values of the objective functions are given by $F(\mathbf{x}^0) = C_B B^{-1}\mathbf{b}$. Moreover, the feasibility for the dual problem

$$(C_B B^{-1}A - C)\mathbf{w} \leq 0 \quad \text{for no } \mathbf{w} \geqq \mathbf{0}$$

Table 8.4.

			f_2	1	1	0	0	0	
		f_1		40	120	0	0	0	
i	B	c_B	c_B	P_1	P_2	P_3	P_4	P_5	P_0
1	P_3	0	0	1	1	1	0	0	100
2	P_4	0	0	10	20	0	1	1	1100
3	P_5	0	0	1	$\boxed{4}$	0	0	0	160
4		$z_j^{(1)} - c_j^{(1)}$		−40	−120	0	0	0	0
5			$z_j^{(2)} - c_j^{(2)}$	−1	−1	0	0	0	0

implies $C_B B^{-1} A\mathbf{x} \leq C\mathbf{x}$ for no $\mathbf{x} \in K$ and because of $C_B B^{-1} A\mathbf{x} = C\mathbf{x}^0$. Finally, $C\mathbf{x}^0 \leq C\mathbf{x}$ for no $\mathbf{x} \in K$. Summarizing, if a feasible basic solution $\mathbf{x}^0$ is dual feasible, i.e., $U^0 = C_B B^{-1} \in T$, then $\mathbf{x}^0$ is an efficient solution for the primal problem (8.6). Moreover, $U^0 = C_B B^{-1}$ is an efficient solution for the dual problem (8.13) according to Theorem 8.2 because the expression $C_B B^{-1}\mathbf{b}$ in Table 8.3 shows the corresponding values of primal *and* dual objective functions. The expression $C_B B^{-1}\mathbf{b}$ can be written as $H(U^0) = U^0\mathbf{b}$ and as $F(\mathbf{x}^0) = C_B\mathbf{x}^0$, respectively. Analogously to single-objective linear programming, the multiple-objective simplex tableau for the primal problem (8.6) yields, in connection with each efficient basic solution $\mathbf{x}^0$ for (8.6), an efficient solution U^0 for (8.13) such that $F(\mathbf{x}^0) = H(U^0)$.

To illustrate, let us return again to the farmer problem as described in Section 8.1. The initial multiple-objective simplex tableau for the MOLP (8.2)–(8.4) concerning the allocation of the scarce production factors—land, capital, and labor—for the production of wheat and potatoes has the form given in Table 8.4.

Looking at the graphical presentation in Figure 8.1, the starting basic solution is described by the point $\mathbf{x}^{(1)}$. According to the rules of the simplex method, with column P_2 being the pivot column and row 3 being the pivot row (see Section 4.5), we will obtain the new basic solution given in Table 8.5 and described by the point $\mathbf{x}^{(2)}$ in Figure 8.1.

The interpretation of the elements in the criterion rows 4 (for the first objective) and 5 (for the second objective) is analogous to the interpretation of the elements $z_j - c_j$ in the single-objective simplex tableau. They indicate the change in the first and second objective function by introducing a particular nonbasic variable into the basis. Because the elements in the first column in both criterion rows are negative, the basic solution $\mathbf{x}^{(2)}$ cannot be efficient. The planting of potatoes can increase the values of both objective functions, the revenue of the farmer as well as the utilization of the production factor land. The result is given in Table 8.6.

As the reader can see in Figure 8.2, the received solution $\mathbf{x}^{(3)}$ is an efficient solution for the primal problem (8.2)–(8.4). Planting 60 acres of potatoes and 25 acres of wheat yields return revenue in the amount of €5,400 and 85 acres of land are used. As indicated in the criterion rows 4 and 5 of Table 8.6, the value of the second objective can be increased by introducing the nonbasic variable x_5 into the basis (the

Table 8.5.

			f_2	1	1	0	0	0	
		f_1		40	120	0	0	0	
i	B	c_B	c_B	P_1	P_2	P_3	P_4	P_5	P_0
1	P_3	0	0	$\frac{3}{4}$	0	1	0	$-\frac{1}{4}$	60
2	P_4	0	0	$\boxed{5}$	0	0	1	-5	300
3	P_2	120	1	$\frac{1}{4}$	1	0	0	$\frac{1}{4}$	40
4		$z_j^{(1)} - c_j^{(1)}$		-10	0	0	0	300	4800
5			$z_j^{(2)} - c_j^{(2)}$	$-\frac{3}{4}$	0	0	0	$\frac{1}{4}$	40

Table 8.6.

			f_2	1	1	0	0	0	
		f_1		40	120	0	0	0	
i	B	c_B	c_B	P_1	P_2	P_3	P_4	P_5	P_0
1	P_3	0	0	0	0	1	$-\frac{3}{20}$	$\frac{1}{2}$	15
2	P_1	40	1	1	0	0	$\frac{1}{5}$	-1	60
3	P_2	120	1	0	1	0	$-\frac{1}{20}$	$\frac{1}{2}$	25
4		$z_j^{(1)} - c_j^{(1)}$		0	0	0	2	20	5400
5			$z_j^{(2)} - c_j^{(2)}$	0	0	0	$\frac{3}{20}$	$-\frac{1}{2}$	85

element in criterion row 5 and column P_5 is negative), however, only with decreasing value of the first objective function (the element in criterion row 4 and column P_5 is positive). In that sense, the elements in the criterion rows of the multiple-objective simplex tableau characterize the trade-off between the different objectives, caused by taking up of the particular product (e.g., planting potatoes) into the production plan. By introducing the nonbasic variable x_4 into the basis, the values of both objective functions will decrease.

Due to Theorem 8.2, if $\mathbf{x}^{(3)}$ is an efficient solution for (8.2)–(8.4), Table 8.6 must provide an efficient solution U^0 for the dual problem. According to the interpretation of the elements in Table 8.3, the dual solution U^0 will be found in the criterion rows (in our example, rows 4 and 5) and in the columns of the basic variables from the initial simplex tableau (in our case, in columns P_3, P_4, and P_5). Consequently, the efficient solution U_3^0 connected with efficient solution $\mathbf{x}^{(3)}$ is given by

$$U_3^0 = \begin{pmatrix} 0 & 2 & 20 \\ 0 & \frac{3}{20} & -\frac{1}{2} \end{pmatrix}.$$

The elements of the matrix U_3^0 are the shadow prices of the production factors land, capital, and labor (the columns of the matrix U_3^0) with respect to the first objective function (the first row of the matrix U^0) and to the second objective function (the second row of the matrix U^0), respectively. The increase of capital by €1 would increase the return revenue by €2 and the utilization of land by $\frac{3}{20}$ acre. One additional man-day of labor would increase the return revenue by €20 and decrease the utilization of land by $\frac{1}{2}$ acre. One additional acre of land will not change the values of the objective functions. For the efficient solution $\mathbf{x}^{(3)}$, the marginal contribution of the production factor land to both objectives equals zero.

Due to Theorem 8.2, the efficient solution $\mathbf{x}^{(3)}$ for the MOLP (8.2)–(8.4) and the properly selected shadow prices for the production factors imply that the total value of all three resources equals the value of the first and second objective function, respectively. Indeed,

$$C\mathbf{x}^{(3)} = \begin{pmatrix} 40 & 120 \\ 1 & 1 \end{pmatrix} \begin{pmatrix} 60 \\ 25 \end{pmatrix} = \begin{pmatrix} 5400 \\ 85 \end{pmatrix},$$

$$U_3^0\mathbf{b} = \begin{pmatrix} 0 & 2 & 20 \\ 0 & \frac{3}{20} & -\frac{1}{2} \end{pmatrix} \begin{pmatrix} 100 \\ 1100 \\ 160 \end{pmatrix} = \begin{pmatrix} 5400 \\ 85 \end{pmatrix}.$$

The next efficient solution for the primal problem (8.2)–(8.4), represented by point $\mathbf{x}^{(4)}$ in Figure 8.1 and point $F(\mathbf{x}^{(4)})$ in Figure 8.2, can be obtained by entering the nonbasic variable x_5 into the basis. The resulting simplex tableau is given in Table 8.7.

The farmer should plant 90 acres of potatoes and 10 acres of wheat. The return revenue has been decreased to €4,800, but the utilization of land has been raised. All 100 acres of land will be used. The efficient dual solution U_4^0 connected with the efficient solution $\mathbf{x}^{(4)}$ is given in Table 8.7 as

$$U_4^0 = \begin{pmatrix} -40 & 8 & 0 \\ 1 & 0 & 0 \end{pmatrix}.$$

Table 8.7.

			f_2	1	1	0	0	0	
		f_1		40	120	0	0	0	
i	B	c_B	c_B	P_1	P_2	P_3	P_4	P_5	P_0
1	P_5	0	0	0	0	2	$-\frac{3}{10}$	1	30
2	P_1	40	1	1	0	2	$-\frac{1}{10}$	0	90
3	P_2	120	1	0	1	−1	$-\frac{1}{10}$	0	10
4		$z_j^{(1)} - c_j^{(1)}$		0	0	−40	8	0	4800
5			$z_j^{(2)} - c_j^{(2)}$	0	0	1	0	0	100

The economic interpretation of the elements of this matrix as the shadow prices for the production factors with respect to both objectives is analogous to the interpretation of the matrix U_3^0, but now related to the efficient solution $\mathbf{x}^{(4)}$. One additional acre of land will decrease the return revenue by €40 and increase the utilization of land by one acre. One additional Euro of capital will increase the return revenue by €8, and it will not change the value of the second objective. The elements in criterion rows 4 and 5 in Table 8.7 confirm (as the reader can see in Figure 8.2) that the solution $\mathbf{x}^{(4)}$ is an efficient solution of the primal problem (8.2)–(8.4). Entering the slack variable x_3 into the basis (in other words, decreasing acres of land used for production) would increase the return revenue and decrease the utilization of the production factor land. More precisely (and in coincidence with the above interpretation of the shadow price of land), one acre of uncultivated land will increase the return revenue by €40 and decrease the utilization by one acre. Indeed, the transformation of Table 8.7, with column P_3 being the pivot column and row 1 being the pivot row, will lead to Table 8.6 with the efficient solution $\mathbf{x}^{(3)}$ again. For this solution, 15 acres of land remain uncultivated and the return revenue has been raised by €600 (€40 $\times$ 15) to €5,400.

The reader can easily verify that U_4^0 is the efficient solution such that $C\mathbf{x}^{(4)} = U_4^0\mathbf{b}$. As can be seen from the graphical representation in Figures 8.1 and 8.2, the set of efficient solutions for the MOLP (8.2)–(8.4) is the set of all linear convex combinations of the extreme points $\mathbf{x}^{(3)}$ and $\mathbf{x}^{(4)}$, described by (8.5).

By Theorem 8.2, both problems (8.6) and (8.13) have no efficient solution if and only if at least one problem has no feasible solution. In order to find out whether for each problem of the dual pair (8.6) and (8.13) at least one efficient solution exists, the linear program (8.16) may be applied.

For the example (8.2)–(8.4), it becomes

$$\text{minimize} \quad h(\mathbf{u}) = 100u_1 + 1100u_2 + 160u_3 \tag{8.17}$$

$$\begin{aligned} \text{subject to} \quad u_1 + 10u_2 + u_3 - 40\alpha_1 - \alpha_2 &\geqq 0, \\ u_1 + 20u_2 + 4u_3 - 120\alpha_1 - \alpha_2 &\geqq 0, \end{aligned} \tag{8.18}$$

$$\begin{aligned} \alpha_1 &\geqq 1, \\ \alpha_2 &\geqq 1, \end{aligned} \tag{8.19}$$

$$\mathbf{u} \geqq \mathbf{0}, \tag{8.20}$$

with the optimal solution $u_1^0 = 0$, $u_2^0 = 2.15$, $u_3^0 = 19.5$, $\alpha_1^0 = 1$, $\alpha_2^0 = 1$ and the optimal objective function value $h(\mathbf{u}^0) = 5485$. The problem (8.17)–(8.20) is the dual problem to the following program:

$$\begin{aligned} \text{maximize} \quad & f(\mathbf{z}) = z_1 + z_2 \\ \text{subject to} \quad & x_1 + x_2 \leqq 100, \\ & 10x_1 + 20x_2 \leqq 1100, \\ & x_1 + 4x_2 \leqq 160, \\ & -40x_1 - 120x_2 + z_1 \leqq 0, \\ & -x_1 - x_2 + z_2 \leqq 0, \\ & x_1 \geqq 0, \quad x_2 \geqq 0, \quad z_1 \geqq 0, \quad z_2 \geqq 0. \end{aligned} \tag{8.21}$$

Generally, the linear program (8.16) is the dual to the problem

$$\begin{aligned}
\text{maximize} \qquad & f(\mathbf{z}) = \sum_{k=1}^{s} z_k \\
\text{subject to} \qquad & A\mathbf{x} \leqq \mathbf{b}, \\
& -C\mathbf{x} + E\mathbf{z} \leqq \mathbf{0}, \\
& \mathbf{x} \geqq \mathbf{0}, \quad \mathbf{z} \geqq \mathbf{0}.
\end{aligned} \tag{8.22}$$

Problem (8.22) can be interpreted as a special case of the parametric problem (7.17) with $\boldsymbol{\alpha} = \mathbf{e}$. In other words, the sum of the objective function values should be maximized subject to $\mathbf{x} \in K = \{\mathbf{x} | A\mathbf{x} \leqq \mathbf{b}, \mathbf{x} \geqq \mathbf{0}\}$.

The optimal solution in example (8.21) is $x_1^0 = 60$ and $x_2^0 = 25$, with the value of the first objective function $z_1^0 = 5400$ and the value of the second objective function $z_2^0 = 85$, corresponding to the efficient solution $\mathbf{x}^{(3)}$ in Figure 8.2 and Table 8.6. The variables z_1 and z_2 from problem (8.21) are the dual variables for the constraints (8.19), and x_1, x_2 are the dual variables for the constraints (8.18).

Thus, if (8.16) has an optimal solution $(\mathbf{u}^0, \boldsymbol{\alpha}^0)$, due to Theorems 7.2 and 8.2, a first efficient basic solution for (8.6) is obtained. Solving (8.17)–(8.20), we find the vector $\mathbf{u}^0 = (0; 2.15; 19.5)$ and, as a dual solution, the efficient or Pareto–Koopmans maximal point $\mathbf{x}^{(3)} = (60; 25)$ with the objective function values $\mathbf{z} = (5400; 85)$. Then due to Theorem 8.2, there exists an efficient solution for (8.13). To illustrate, we will again use the production problem (8.2)–(8.4).

According to (7.22), a vector of dual variables $\mathbf{u}$ associated with the constraints $A\mathbf{x} \leqq \mathbf{b}$ can be seen as an aggregation of the matrix U:

$$\mathbf{u}' = \boldsymbol{\alpha}' U. \tag{7.22}$$

Setting $\mathbf{u}^0, \boldsymbol{\alpha}^0$ from (8.17)–(8.20) into (7.22) leads to the following set of equations:

$$u_{11}^0 + u_{21}^0 = 0, \tag{8.23}$$

$$u_{12}^0 + u_{22}^0 = 2.15, \tag{8.24}$$

$$u_{13}^0 + u_{23}^0 = 19.5. \tag{8.25}$$

The linear vector optimization problem (8.6) is a special case of the multiobjective mathematical programming problem (7.12) such that the Kuhn–Tucker conditions from Section 7.2 can be applied. It follows from (7.20) that $U^0(A\mathbf{x}^0 - \mathbf{b}) = \mathbf{0}$ for $\mathbf{x}^0 = (60; 25)$ that both shadow prices u_{11}^0 and u_{21}^0 must equal zero. The equality $C\mathbf{x}^0 = U^0\mathbf{b}$ from Theorem 8.2 generates the next two conditions for the shadow prices u_{12}, u_{13}, u_{22}, and u_{23}:

$$1100u_{12}^0 + 160u_{13}^0 = 5400, \tag{8.26}$$

$$1100u_{22}^0 + 160u_{23}^0 = 85. \tag{8.27}$$

Solving the system of equations (8.24)–(8.27) yields $u_{12}^0 = 2$, $u_{13}^0 = 20$, $u_{22}^0 = \frac{3}{20}$, $u_{23}^0 = -\frac{1}{2}$, and therewith the same matrix of shadow prices

$$U^0 = \begin{pmatrix} 0 & 2 & 20 \\ 0 & \frac{3}{20} & -\frac{1}{2} \end{pmatrix},$$

as obtained in Table 8.6. If, however, (8.16) has no optimal solution, (8.6) and (8.13) have no efficient solution.

Changing the coefficients on the right-hand side in (8.19)—in other words, the weights for the objectives—the new efficient solution can be found. Useful information for this purpose yields the *postoptimality analysis*,[1] dealing with the changes in the optimal solution due to variation in the parameters of the model. Generally, two types of postoptimality—namely, *sensitivity analysis* and *parametric programming*—will be distinguished.

In sensitivity analysis, the question is, what is the range over which a given parameter can change without changing the optimal basis? In parametric programming, the question is, what happens if an actual change is such that it does not fall within the above range? Sensitivity analysis related to the right-hand side of (8.19) provides the following answers:[2] If $\alpha_1 \geqq 0.02$ and all other parameters remain unchanged, the same efficient point $\mathbf{x}^{(3)}$ will be obtained. If $0 \leqq \alpha_2 \leqq 40$ and all other parameters remain unchanged, the optimal solution will not change. Therefore, changing the constraint $\alpha_2 \geqq 1$ in (8.19), e.g., to $\alpha_2 \geqq 50$, the new optimal solution for problem (8.17)–(8.20) follows: $u_1^0 = 10$, $u_2^0 = 8$, $u_3^0 = 0$, $\alpha_1^0 = 1$, and $\alpha_2^0 = 50$. The optimal solution of the connected dual problem is $x_1^0 = 90$, $x_2^0 = 10$, $z_1^0 = 4800$, and $z_2^0 = 100$. The Pareto–Koopmans maximal point $\mathbf{x}^{(4)}$ from Table 8.7 has been obtained. Using relation (7.22), we can write the following:

$$u_{11}^0 + 50u_{21}^0 = 10, \tag{8.28}$$

$$u_{12}^0 + 50u_{22}^0 = 8, \tag{8.29}$$

$$u_{13}^0 + 50u_{23}^0 = 0. \tag{8.30}$$

Because of condition (7.20), $U^0(A\mathbf{x}^0 - \mathbf{b}) = \mathbf{0}$ for $\mathbf{x}^0 = (90; 10)$, the shadow prices for the production factor labor with respect to the first and second objective functions, u_{13}^0 and u_{23}^0, respectively, must be zero. The efficiency condition $C\mathbf{x}^0 = U^0\mathbf{b}$ leads then to the following equations:

$$100u_{11}^0 + 1100u_{12}^0 = 4800, \tag{8.31}$$

$$100u_{21}^0 + 1100u_{22}^0 = 100. \tag{8.32}$$

Solving the set of linear equations (8.28), (8.29), (8.31), and (8.32) yields $u_{11}^0 = -40$, $u_{12}^0 = 8$, $u_{21}^0 = 1$, $u_{22}^0 = 0$, and together with $u_{13}^0 = u_{23}^0 = 0$, the same matrix of shadow prices

$$U_4^0 = \begin{pmatrix} -40 & 8 & 0 \\ 1 & 0 & 0 \end{pmatrix},$$

---

[1] Despite the high relevance of postoptimality analysis in economic models, here the reader is referred to the literature, e.g., [5]. Most of the textbooks on linear programming contain a section devoted to postoptimality analysis.

[2] For technical details, see, e.g., [10, Chapter 8, Section 3].

as given in Table 8.7.

This very simple farmer problem illustrates how to find a pair $(\mathbf{x}^0, U^0)$ of efficient solutions for the primal and dual linear vector optimization problem (for the mathematically exact description of the vector-maximum algorithms, see [25, Chapter 9]) and what the economic meaning of the dual variables U^0 is. At the same time, the basic multiple-criteria decision-making problem arises: How do we choose, from among the numerous (not only two, as in the farmer problem above) efficient solutions, an efficient solution that will be accepted by the decision maker. In other words, there is no other efficient solution that will be preferred by the decision maker. Obviously, the acceptance or rejection of a specific efficient solution depends on the preference structure of the decision maker. In Sections 7.1.1 (dealing with welfare economics) and 7.1.2 (devoted to quantitative macroeconomic policy), we have already underlined and pointed out an interactive procedure for the problem of an "ex-ante" explicitly specified scalar-valued preference function. The interactive methods represent a very important tool for multiple-criteria decision making, and therefore the following section is devoted to the discussion of the underlying principles and the presentation of one of the interactive methods, namely, the Zionts–Wallenius algorithm, which has a clear economic interpretation and has been repeatedly applied for optimization of macroeconomic and environmental policy [27, 2, 3].

8.3 Interactive Procedures and the Zionts–Wallenius Method

The interactive methods of multiple-objective programming are methods in which the full preference structure of the decision maker is not postulated a priori, but is implicitly revealed in response to a simple question–answer procedure with the decision maker.

Interactive procedures are characterized by phases of decision making alternating with phases of computation. According to Gardiner and Steuer [9] and Stewart [26], they involve the following characteristic steps:

- A feasible (and usually efficient) solution or small number of solutions is generated according to some specified procedure and presented to the decision maker for examination.
- If the decision maker is satisfied with the solution (or one of the solutions) generated, then the process stops. Otherwise, he/she is requested to provide some local preference information (make improved judgments) in the vicinity of the solution(s) presented.
- The local information provided by the decision maker is the input to the solution procedure in which a new solution is computed, and the process returns to the first step.

The feedback between human and model enables the decision maker to explore more deeply the range of possibilities in his feasible region and how the objectives trade off against one another. In this way, the interactive procedure helps the decision maker to understand better the complex structure of the system and to learn more about the analyzed problem.

In the literature, many interactive methods can be found. They can be classified with respect to different criteria, the distinguishing feature being the relative difficulty in applying them to real problems having multiple objectives. Such methods include those of Benayoun et al. [1] (called the STEM method), Geoffrion, Dyer, and Feiberg [12], Zionts and Wallenius [28], the interval criterion weights/vector-maximum approach by Steuer [24], the visual interactive approach by Korhonen and Laakso [18], interactive goal programming by Spronk [23], and others. As a representative of this approach of multiple optimization, we introduce the Zionts–Wallenius method because of its simplicity to understand and use for those not familiar with such models, and also because of its workability in practice.

The problem considered by Zionts and Wallenius [28] is one in which each of the objective functions $f_k(\mathbf{x})$ $(k = 1, 2, \ldots, s)$ that the decision maker wants to maximize is a concave function of decision variables, and the constraint set K is a convex set. The utility function U or composite objective function is a linear function (and, more generally, a concave function) of the objective function variables $f_k(\mathbf{x})$ $(k = 1, 2, \ldots, s)$, but the precise weights in such a function are not known explicitly. The method starts with an arbitrary set of positive weights α_k satisfying $\sum_{k=1}^{s} \alpha_k = 1$, and it generates a composite objective function using these weights. The composite objective function $\sum_{k=1}^{s} \alpha_k f_k(\mathbf{x})$ is then maximized subject to the set $K = \{\mathbf{x} | A\mathbf{x} = \mathbf{b},\ \mathbf{x} \geqq \mathbf{0}\}$. The result is an efficient or Pareto–Koopmans solution in the sense that it is not possible to increase one objective function without decreasing at least one other objective function.

Returning to our numerical example (8.2)–(8.4) and setting $\alpha_1 = \alpha_2 = \frac{1}{2}$, we will solve the following problem:

$$\begin{aligned} \text{maximize} \quad U(\mathbf{x}) &= \frac{1}{2}(40x_1 + 120x_2) + \frac{1}{2}(x_1 + x_2) \\ &= 20.5x_1 + 60.5x_2 \end{aligned}$$

subject to (8.3)–(8.4). The initial simplex tableau is a slight modification of Table 8.4 (see Table 8.8).

Table 8.8. Initial tableau for problem (8.2)–(8.4) with $\alpha_1 = \alpha_z = \frac{1}{2}$.

		$\frac{1}{2}$		f_2	1	1	0	0	0	
		$\frac{1}{2}$	f_1		140	120	0	0	0	
i	B	$\boldsymbol{\alpha}' C_B$	$\mathbf{c}_B^{(1)}$	$\mathbf{c}_B^{(2)}$	P_1	P_2	P_3	P_4	P_5	P_0
1	P_3	0	0	0	1	1	1	0	0	100
2	P_4	0	0	0	10	20	0	1	0	1100
3	P_5	0	0	0	1	4	0	0	1	160
4	α_1	$\frac{1}{2}$	$z_j^{(1)} - c_j^{(1)}$		−40	−120	0	0	0	0
5	α_2	$\frac{1}{2}$		$z_j^{(2)} - c_j^{(2)}$	−1	−1	0	0	0	0
6		$z_j - c_j$			−20.5	−60.5	0	0	0	0

As the optimal solution, the Pareto–Koopmans-efficient point $\mathbf{x}^{(3)}$ will be obtained (see Table 8.9).

Table 8.9. Simplex tableau for the efficient point $\mathbf{x}^{(3)}$.

i	B	$\alpha' C_B$	$\mathbf{c}_B^{(1)}$	$\mathbf{c}_B^{(2)}$	P_1	P_2	P_3	P_4	P_5	P_0
1	P_3	0	0	0	0	0	1	$-\frac{3}{20}$	$\frac{1}{2}$	15
2	P_1	20.5	40	1	1	0	0	$\frac{1}{5}$	-1	60
3	P_2	60.5	120	1	0	1	0	$-\frac{1}{20}$	$\frac{1}{2}$	25
4	α_1	$\frac{1}{2}$	$z_j^{(1)} - c_j^{(1)}$		0	0	0	2	20	5400
5	α_2	$\frac{1}{2}$		$z_j^{(2)} - c_j^{(2)}$	0	0	0	$\frac{3}{20}$	$-\frac{1}{2}$	85
6		$z_j - c_j$			0	0	0	$\frac{43}{40}$	$\frac{39}{4}$	

After obtaining an efficient solution to the starting problem with an arbitrary set of positive weights $\alpha_k (k = 1, 2, \ldots, s)$, we can divide the set of all nonbasic variables into two subsets:

- those nonbasic variables which, when introduced into the basis, lead to efficient adjacent extreme points in criterion space,
- those nonbasic variables that, when introduced into the basis, do not lead to efficient adjacent extreme points in the space of the objective function variables $f_k(\mathbf{x})$.

Denote the first subset of variables as efficient variables and the second subset as inefficient variables. In our very simple example with two objectives, only the classification nonbasic variables into these categories is immediately visible from criterion rows 4 and 5 in Table 8.9. As already noted in the interpretation of Table 8.6, the elements in the criterion rows describe the changes in the objective functions caused by introducing the nonbasic variables into the basis. According to Table 8.9 by introducing the nonbasic variable x_4 into the basis the value of the first objective function will decrease by two units and the value of the second objective function by 0.15 unit. Consequently, the nonbasic variable x_4 is inefficient.

On the other hand, the introduction of the nonbasic variable x_5 decreases the first objective function value by 20 units and increases the second objective function value by 0.5 unit. The nonbasic variable x_5 is efficient.

To find an efficient subset of a set N of nonbasic variables in general, Zionts and Wallenius [28] proposed solving for each nonbasic variable $j \in N$ the following linear programming problem:

$$\text{minimize} \quad \sum_{k=1}^{s} (z_j^{(k)} - c_j^{(k)}) \alpha_k$$

$$\begin{aligned} \text{subject to} \quad & \sum_{k=1}^{s}(z_j^{(k)} - c_j^{(k)})\alpha_k \geqq 0, \quad j \in N, \quad j \neq k, \\ & \sum_{k=1}^{s} \alpha_k = 1, \\ & \alpha_k \geqq 0 \quad (k = 1, 2, \ldots, s), \end{aligned} \tag{8.33}$$

where $z_j^{(k)} - c_j^{(k)}$ represents the decrease in the objective function $f_k(\mathbf{x})$ due to an increase of the nonbasic variable x_j by a unit. Then

(i) if the minimum objective function value of (8.33) is negative, the variable x_j is efficient;

(ii) if the minimum objective function value of (8.33) is nonnegative, the variable x_j is not efficient.

The reader can verify that the minimum objective function value of (8.33) for the nonbasic variable x_5 is negative and for the nonbasic variable x_4 is positive.

Considering the subset of efficient nonbasic variables the decision maker is now asked, "Here is a trade: Are you willing to accept a decrease in objective function f_1 of $z_j^{(1)} - c_j^{(1)}$, a decrease in objective function f_2 of $z_j^{(2)} - c_j^{(2)}, \ldots,$ and a decrease in objective function f_s of $z_j^{(s)} - c_j^{(s)}$? Respond yes, no, or indifferent to the trade" (Zionts and Wallenius [28, p. 656]). There will be at least one positive $z_j^{(k)} - c_j^{(k)}$ and at least one negative $z_j^{(k)} - c_j^{(k)}$ for each efficient variable. According to the decision maker's responses, the choice of the weights $\boldsymbol{\alpha}$ to be used in finding a new efficient solution is restricted as follows. For each "yes" answer, an inequality of the form

$$\sum_{k=1}^{s}(z_j^{(k)} - c_j^{(k)})\alpha_k \leqq -\varepsilon, \tag{8.34}$$

where ε is a sufficiently small positive number, will be formulated. For each "no" response, an inequality of the form

$$\sum_{k=1}^{s}(z_j^{(k)} - c_j^{(k)})\alpha_k \geqq \varepsilon \tag{8.35}$$

will be constructed, and for each response of indifference, an inequality of the form

$$\sum_{k=1}^{s}(z_j^{(k)} - c_j^{(k)})\alpha_k = 0 \tag{8.36}$$

will be constructed.

As mentioned by Zionts and Wallenius [28], responses of indifference imply an unreasonably high degree of precision of judgment by the decision maker. Therefore, for practical implementation and in order to improve the convergence of the procedure, they ignored the "I don't know" responses when generating consistent multipliers.

Using linear programming a feasible solution to the set of constraints (8.34)–(8.36), and

$$\sum_{k=1}^{s} \alpha_k = 1, \quad \alpha_k \geqq \varepsilon \quad (k = 1, 2, \ldots, s)$$

is found. The relationships (8.34)–(8.36) are also used to eliminate efficient variables from subsequent question sessions. In other words, they are added to the constraints (8.33) in the next step when an efficient subset of a set N of nonbasic variables is to be found.

The resulting weights $\boldsymbol{\alpha}$ are used to generate a new composite function,

$$\sum_{k=1}^{s} \alpha_k f_k(\mathbf{x}),$$

which is then maximized under the constraints (8.3)–(8.4) to obtain a new solution, and the procedure goes to the next iteration. An efficient subset of variables is assembled as questions for the decision maker, and the process is continued until an efficient solution has been reached, where there are no remaining efficient nonbasic variables. (The reader should remember that the constraints (8.34)–(8.35) using earlier responses by the decision maker are adjoined to the constraints of problem (8.33).)

Returning to our small model (8.2)–(8.4), let us assume that the (implicit) utility function is $U(\mathbf{x}) = 0.7 f_1(\mathbf{x}) + 0.3 f_2(\mathbf{x})$, but we will only use the knowledge of this function in answering the "yes" or "no" questions (because of the absence of the "real" decision maker). The starting optimal solution is given in Table 8.9 with the nonbasic variables x_4 and x_5. The only efficient variable is x_5. We ask the decision maker whether he/she is willing to accept a decrease of 20 units of f_1 in return for an increase of 0.5 unit of f_2. To simulate a response, we compute an evaluation: $0.7 \cdot (20) + 0.3 \cdot (-0.5) > 0$. Thus there is a net decrease in the decision maker's composite function; he/she does not like the trade-off posed by x_5. Hence the solution $\mathbf{x}^{(3)}$ is optimal and the weights $\alpha_1 = 0.5$ and $\alpha_2 = 0.5$ are equivalent (for this problem) to our unknown multipliers.

Assume, for example, that the (implicit) composite function is $U(\mathbf{x}) = 0.02 f_1(\mathbf{x}) + 0.98 f_2(\mathbf{x})$. The evaluation of the efficient trade-off—a decrease of 20 units of f_1 in return for an increase of 0.5 unit of f_2—then yields $0.02 \cdot (20) + 0.98 \cdot (-0.5) < 0$. Thus there is a net increase in the decision maker's utility; he/she should like the trade-off posed by nonbasic variable x_5. Thereby an inequality of the form (8.34) is generated and a feasible solution to the set of constraints (arbitrarily setting $\varepsilon = 0.02$)

$$\begin{aligned} 20\alpha_1 - \frac{1}{2}\alpha_2 &\leqq -0.02, \\ \alpha_1 + \alpha_2 &= 1, \\ \alpha_1, \alpha_2 &\geqq 0.02 \end{aligned}$$

Table 8.10. Simplex tableau for the efficient point $\mathbf{x}^{(4)}$.

i	B	$\alpha' C_B$	$\mathbf{c}_B^{(1)}$	$\mathbf{c}_B^{(2)}$	P_1	P_2	P_3	P_4	P_5	P_0
1	P_5	0	0	0	0	0	2	$-\frac{3}{10}$	1	30
2	P_1	1.78	40	1	1	0	2	$-\frac{1}{10}$	0	90
3	P_2	3.38	120	1	0	1	-1	$\frac{1}{10}$	0	10
4	α_1	0.02	$z_j^{(1)} - c_j^{(1)}$		0	0	-40	8	0	4800
5	α_2	0.98		$z_j^{(2)} - c_j^{(2)}$	0	0	1	0	0	100
6		$z_j - c_j$			0	0	0.18	0.16	0	

is determined. A basic feasible solution is $\alpha_1 = 0.02$, $\alpha_2 = 0.98$. Using these weights, we generate a new composite function $U(\mathbf{x}) = 0.02f_1(\mathbf{x}) + 0.98f_2(\mathbf{x}) = 1.78x_1 + 3.38x_2$, which is maximized subject to the constraints (8.3)–(8.4). Applying the Jordan–Gauss elimination procedure with row 1 being the pivot row (the variable x_3 will be eliminated from the basis) and column P_5 being the pivot column (the variable x_5 will enter into the basis) to Table 8.9 (however, with the new coefficients $\alpha_1 = 0.02$ and $\alpha_2 = 0.98$), the new optimal solution $\mathbf{x}^{(4)} = (90.10)$ will be obtained (see Table 8.10).

Due to the higher weighting of the second objective, its value is increased to 100 units (all 100 acres of land are used) and the first objective function value decreased to €4,800. Hence the efficient solution $\mathbf{x}^{(4)}$ may be accomplished by using $0.02f_1(\mathbf{x}) + 0.98f_2(\mathbf{x})$ as the (implicit) utility function. As already mentioned in Section 8.1 and expressed by (8.5), the set of efficient solutions for the MOLP (8.2)–(8.4) is described as the set of all linear convex combinations of the extreme points $\mathbf{x}^{(3)}$ and $\mathbf{x}^{(4)}$. Being an extreme-point method, the Zionts–Wallenius algorithm cannot find an optimal solution if it lies in the interior of a facet.

8.4 The Leontief Pollution Model with Multiple Objectives

In Section 7.1.4, two versions of the augmented Leontief optimization model have been formulated. In the first version, the objective functions are factor cost to produce the gross national product and net pollution. Both are to be minimized for the given final demand level. The criteria in the second model are maximization of the final demand value and the minimization of net pollution under the constraint for primary input. In the general notation used in Chapter 7, the model has the following form:

$$\text{Minimize} \quad F(\mathbf{x}_1, \mathbf{x}_2, \mathbf{y}_1, \mathbf{y}_2) = \begin{pmatrix} -\mathbf{p}'\mathbf{y}_1 \\ \mathbf{y}_2 \end{pmatrix} \tag{8.37}$$

$$\text{subject to} \quad -(E - A_{11})\mathbf{x}_1 + A_{12}\mathbf{x}_2 + \mathbf{y}_1 \leqq \mathbf{0}, \tag{7.8}$$

$$A_{21}\mathbf{x}_1 - (E - A_{22})\mathbf{x}_2 - \mathbf{y}_2 \leqq \mathbf{0}, \tag{7.9}$$

$$\mathbf{v}_1'\mathbf{x}_1 + \mathbf{v}_2'\mathbf{x}_2 \leqq \overline{V}, \tag{7.10}$$

$$\mathbf{x}_1 \geqq \mathbf{0}, \quad \mathbf{x}_2 \geqq \mathbf{0}, \quad \mathbf{y}_1 \geqq \mathbf{0}, \quad \mathbf{y}_2 \geqq \mathbf{0}. \tag{7.11}$$

Let us apply the Kuhn–Tucker conditions (7.13)–(7.16) as the necessary conditions for a Pareto–Koopmans minimal point of this problem. The variables in our problem are $\mathbf{x}_1 \in R^n$, $\mathbf{x}_2 \in R^k$, $\mathbf{y}_1 \in R^n$, and $\mathbf{y}_2 \in R^k$. All variables are nonnegative. We compute

$$\nabla F(\mathbf{x}_1, \mathbf{x}_2, \mathbf{y}_1, \mathbf{y}_2) = \begin{pmatrix} \mathbf{0}' & \mathbf{0}' & -\mathbf{p}' & \mathbf{0}' \\ 0_{k\times n} & 0_{k\times k} & 0_{k\times n} & E_{k\times k} \end{pmatrix},$$

where $0_{k\times n}$, $0_{k\times k}$, and $E_{k\times k}$ denote matrices in dimensions $k \times n$ and $k \times k$, respectively,

$$\nabla G(\mathbf{x}_1, \mathbf{x}_2, \mathbf{y}_1, \mathbf{y}_2) = \begin{pmatrix} -(E - A_{11}) & A_{12} & E_{n\times n} & 0_{n\times n} \\ A_{21} & -(E - A_{22}) & 0_{k\times n} & -E_{k\times k} \\ \mathbf{v}_1' & \mathbf{v}_2' & \mathbf{0}' & \mathbf{0}' \end{pmatrix}.$$

Because of the nonnegativity constraints (7.11), the Kuhn–Tucker conditions (7.13) will be written as inequalities and complementary slackness conditions will be added. Denoting by $\mathbf{u}_1 \in R^n$ the vector of Lagrange multipliers corresponding to the constraints (7.8), by $\mathbf{u}_2 \in R^k$ the vector of Lagrange multipliers corresponding to the constraints (7.9), and by $u \in R$ the Lagrange multiplier for the constraint (7.10), we obtain

$$-\mathbf{u}_1'(E - A_{11}) + \mathbf{u}_2'A_{21} + \mathrm{u}\mathbf{v}_1' \geqq \mathbf{0}', \tag{8.38}$$

$$\mathbf{u}_1'A_{12} - \mathbf{u}_2'(E - A_{22}) + \mathrm{u}\mathbf{v}_2' \geqq \mathbf{0}', \tag{8.39}$$

$$-\alpha_1\mathbf{p}' + \mathbf{u}_1'E \geqq \mathbf{0}', \tag{8.40}$$

$$\boldsymbol{\alpha}_2'E - \mathbf{u}_2'E \geqq \mathbf{0}', \tag{8.41}$$

$$[-\mathbf{u}_1'(E - A_{11}) + \mathbf{u}_2'A_{21} + \mathrm{u}\mathbf{v}_1']\mathbf{x}_1 = 0, \tag{8.42}$$

$$[\mathbf{u}_1'A_{12} - \mathbf{u}_2'(E - A_{22}) + \mathrm{u}\mathbf{v}_2']\mathbf{x}_2 = 0, \tag{8.43}$$

$$[-\alpha_1\mathbf{p}' + \mathbf{u}_1'E]\mathbf{y}_1 = 0, \tag{8.44}$$

$$[\boldsymbol{\alpha}_2'E - \mathbf{u}_2'E]\mathbf{y}_2 = 0, \tag{8.45}$$

$$\mathbf{x}_1 \geqq \mathbf{0}, \quad \mathbf{x}_2 \geqq \mathbf{0}, \quad \mathbf{y}_1 \geqq \mathbf{0}, \quad \mathbf{y}_2 \geqq \mathbf{0}, \tag{7.11}$$

where $\alpha_1 \in R$ and $\boldsymbol{\alpha}_2 \in R^k$. Because the primary factor is perfectly transferable between the production and abatement activities, we have one (equilibrium) shadow price for this input.

The Kuhn–Tucker conditions (7.14)–(7.16) yield

$$\mathbf{u}_1'[-(E - A_{11})\mathbf{x}_1 + A_{12}\mathbf{x}_2 + \mathbf{y}_1] = 0, \tag{8.46}$$

$$\mathbf{u}_2'[A_{21}\mathbf{x}_1 - (E - A_{22})\mathbf{x}_2 - \mathbf{y}_2] = 0, \tag{8.47}$$

$$u[\mathbf{v}_1'\mathbf{x}_1 + \mathbf{v}_2'\mathbf{x}_2] = 0, \tag{8.48}$$

the feasibility constraints (7.8)–(7.10), and the nonnegativity constraints for the Lagrange multipliers:

$$\mathbf{u}_1 \geqq \mathbf{0}, \qquad \mathbf{u}_2 \geqq \mathbf{0}, \quad u \geq 0.$$

Because of the linearity of $F(\mathbf{x}_1, \mathbf{x}_2, \mathbf{y}_1, \mathbf{y}_2)$ and $G(\mathbf{x}_1, \mathbf{x}_2, \mathbf{y}_1, \mathbf{y}_2)$, the Kuhn–Tucker conditions (8.38)–(8.45) and (7.11) are necessary and sufficient for Pareto–Koopmans efficiency.

Assuming indecomposability of the input coefficients matrix A_{11}, for any non-negative vector of final demand $\mathbf{y}_1$, all goods will be produced ($\mathbf{x}_1 > \mathbf{0}$) and the conditions (8.38) are fulfilled as equalities. The Lagrange multipliers $\mathbf{u}_1$ correspond then to the prices in the Leontief augmented model. Rewriting the expression in the brackets of (8.42) as

$$\mathbf{u}_1' = \mathbf{u}_1' A_{11} + \mathbf{u}_2' A_{21} + u\mathbf{v}_1',$$

we see that the commodity prices must be such that they cover not only the cost of inputs from other sectors of the economy ($\mathbf{u}_1' A_{11}$) and the cost of primary factors ($u\mathbf{v}_1'$), but also the pollution cost ($\mathbf{u}_2' A_{21}$).

Interesting implications follow from conditions (8.40) and (8.41). The Lagrange multipliers or shadow prices u_i $(i = 1, 2, \ldots, n)$ cannot be lower than the given prices p_i $(i = 1, 2, \ldots, n)$ weighted by the factor α_1. Because of the positivity of α_1 and $\mathbf{p}$, the shadow prices $\mathbf{u}_1$ must be positive, and due to (8.46), the constraints (7.8) are fulfilled as equalities. There is no reason to have net production exceeding the final demand. If the exogenously given price p_i of the commodity i weighted by the factor α_1 is lower than the shadow price u_{1i}, the final demand y_{1i} will be zero. In other words, if the production cost u_{1i} of commodity i is higher than the final demand contribution p_i of this commodity to the social welfare described by (8.37), the commodity i is used as the input for the production of other commodities and for abatement activities only, and the final demand of this commodity y_{1i} is zero. At a Pareto–Koopmans efficient point, the positive final demand y_{1i} implies that the opportunity loss from the production of commodity i for the final demand is zero.

At the same time, condition (8.40) provides an upper bound for the factor α_1 in order to characterize a Pareto–Koopmans-efficient point of problem (8.37) under the constraints (7.8)–(7.11):

$$\alpha_1 \leqq \frac{u_{1i}}{p_i} \quad (i = 1, 2, \ldots, n),$$

$$\alpha_1 = \min_i \frac{u_{1i}}{p_i}.$$

The Kuhn–Tucker condition (8.41) yields lower bounds for the factors α_{2h} $(h = 1, 2, \ldots, k)$, in order to get a Pareto–Koopmans minimal point of problem (8.37) subject to (7.8)–(7.11):

$$\boldsymbol{\alpha}_2 \geqq \mathbf{u}_2.$$

The shadow prices $\mathbf{u}_2$ of pollutants are—similarly to the shadow prices of goods—described as the accounting value of the inputs used for the abatement activity:

$$\mathbf{u}_2' = \mathbf{u}_1' A_{12} + \mathbf{u}_2' A_{22} + u\mathbf{v}_2'. \tag{8.49}$$

Equation (8.49) determines the prices of pollutants from the abatement cost ($\mathbf{u}_1' A_{12}$), the primary cost per unit level of abatement activities ($u\mathbf{v}_2'$), and the pollution cost of the abatement activities themselves ($\mathbf{u}_2'A_{22}$). Therefore, these equations provide the economic foundation to the "polluter pays principle."

If the marginal contribution of the net pollution reduction to the social welfare α_{2h} is higher than the shadow price u_{2h} for the pollutant h—expressed as the accounting cost of the corresponding abatement activity—the net pollution will be zero. The complete abatement of the pollutant h leads to a Pareto–Koopmans-efficient point. For the positive net pollution y_{2h}, the accounting abatement cost for the pollutant h is equal to the welfare benefit resulting from its reduction by one unit.

In order to illustrate the above-mentioned results and to indicate another interesting implication of the multicriteria Leontief pollution model, let us return to the small numerical example (8.10)–(8.12). In coincidence with the formulation of the Kuhn–Tucker conditions provided above, we rewrite the problem (8.10)–(8.12) as a minimization problem

$$\text{Minimize} \quad F(\mathbf{x}_1, x_3, \mathbf{y}_1, y_3) = \begin{pmatrix} -\mathbf{p}'\mathbf{y}_1 \\ y_3 \end{pmatrix}, \tag{8.50}$$

subject to the constraints (8.11)–(8.12), where $\mathbf{p}' = (2, 4)$, $\mathbf{x}_1' = (x_1, x_2)$, $\mathbf{y}_1' = (y_1, y_2)$, x_3 describes the level of abatement activity, and y_3 denotes the net pollution. Supposing $\alpha_1 = 1$ and $\alpha_2 = 1$ and solving the above problem as the parametric problem (7.17), we obtain the following solution:

$$\begin{aligned} &x_1^0 = 47.54, \quad &&x_2^0 = 89.90, \quad &&x_3^0 = 0, \\ &y_1^0 = 0, &&y_2^0 = 72.4, &&y_3^0 = 41.95. \end{aligned}$$

Furthermore, the shadow price for commodity 1 is $u_{11}^0 = 2.14$, the shadow price for commodity 2 is $u_{12}^0 = 4$, and the shadow price for the pollutant is $u_{21}^0 = 1$. The reader can observe that $u_{11}^0 > \alpha_1 p_1 = 2$, and therefore $y_1^0 = 0$. $y_2^0 > 0$ implies $u_{12}^0 = \alpha_1 p_2 = 4$, and because of $y_3^0 > 0$, $u_{21}^0 = \alpha_2 = 1$. At the same time, $\alpha_1 = \min\{\frac{u_{11}^0}{p_1}; \frac{u_{12}^0}{p_2}\} = \min\{\frac{2.14}{2}; \frac{4}{4}\} = 1$.

Usually the coefficients α_1 and α_2 can be interpreted as the weights for the particular objectives. In our example, α_1 is the weight for the maximization of the final demand value and α_2 is the weight for the minimization of net pollution. Now suppose that the preferences of society toward higher environmental quality increase. The coefficients α_1 will decrease (e.g., $\alpha_1^* = \frac{1}{2}$), or the relative weight for the minimization of net pollution increases. Solving problem (8.50), subject to the constraints (8.11)–(8.12), as the parametric problem (7.17) with $\mathbf{p}' = (2, 4)$ and $\alpha_1^* = \frac{1}{2}, \alpha_2 = 1$ yields

$$\begin{aligned} &x_1^* = 42.96 \quad &&x_2^* = 80.56, \quad &&x_3^* = 37.59, \\ &y_1^* = 0, &&y_2^* = 57.36, &&y_3^* = 0. \end{aligned}$$

Looking at the dual solutions, the reader may observe again that all Kuhn–Tucker conditions are fulfilled: $u_{11}^* = 1.189 > \alpha_1^* p_1 = 1$, and consequently $y_1^* = 0$. Because the accounting abatement cost, expressed by the shadow price $u_{21}^* = 0.717$, is lower than the marginal contribution of the net pollution reduction to the social welfare ($\alpha_2 = 1$), the complete abatement of the pollution ($y_3^* = 0$) occurs in the Pareto–Koopmans-efficient solution.

Comparing the solutions of both examples, the following observation can be made: With the higher relative weight for the environmental goal of net pollution minimization, the production of both commodities and the final demand was decreasing ($x_1^* < x_1^0, x_2^* < x_2^0, y_2^* < y_2^0$) and the environmental quality measured by the level of net pollution was increasing ($y_3^* < y_3^0$). Is this a general result for the augmented Leontief model (8.37) under the constraints (7.8)–(7.11), or a specific example only?

As pointed out by Luptáčik and Böhm [21], the following proposition provides an answer to the question formulated above.

Proposition 8.3. *In the augmented Leontief model* (8.37) *subject to* (7.8)–(7.11), *the increasing relative weights for environmental quality will lead, in tendency, to a nonincreasing final demand and a nonincreasing net pollution.*

Proof. Using the parametric approach for solving the augmented Leontief model with objectives (8.37) (should be maximized) and the constraints (7.8)–(7.11), we obtain the following problem:

$$\begin{aligned} &\text{maximize} \quad \alpha_1 \mathbf{p}'\mathbf{y}_1 - \boldsymbol{\alpha}_2'\mathbf{y}_2 \\ &\text{subject to} \quad (7.8)\text{–}(7.11). \end{aligned} \tag{8.51}$$

Now we are looking for the change in the optimal solution of problem (8.51) caused by the changes of the coefficients α_1 and $\boldsymbol{\alpha}_2$. Due to Luptáčik and Böhm [21], the answer may be found using the Le Chatelier–Samuelson principle applied to the following pair of linear programming problems:

$$\text{maximize} \quad \mathbf{c}'\mathbf{x} \quad \text{subject to} \quad A\mathbf{x} \leqq \mathbf{b}, \quad \mathbf{x} \geqq \mathbf{0} \tag{8.52}$$

and

$$\text{minimize} \quad \mathbf{u}'\mathbf{b} \quad \text{subject to} \quad \mathbf{u}'A \geqq \mathbf{c}', \quad \mathbf{u}' \geqq \mathbf{0}, \tag{8.53}$$

where $\mathbf{x} \in R^n, \mathbf{c} \in R^n, \mathbf{b} \in R^m, \mathbf{u} \in R^m$, and $A \in R^{m \times n}$. The following theorem [22, p. 329] is very useful for our purposes.

Theorem 8.3. *Let* $\mathbf{x}^0$ *and* $\mathbf{u}^0$ *be the optimal solutions to* (8.52) *and* (8.53), *respectively. If the vectors* $\mathbf{c}$ *and* $\mathbf{b}$ *vary by* $\Delta\mathbf{c}$ *and* $\Delta\mathbf{b}$, *respectively, the corresponding change in the optimal solutions, denoted by* $\Delta\mathbf{x}$ *and* $\Delta\mathbf{u}$, *will be such that*

$$\Delta\mathbf{c}'\Delta\mathbf{x} - \Delta\mathbf{u}'\Delta\mathbf{b} \geqq 0, \tag{8.54}$$

$$\mathbf{c}'\Delta\mathbf{x} - \Delta\mathbf{u}'b \leqq \mathbf{u}_0'\Delta\mathbf{b} - \Delta\mathbf{c}'\mathbf{x}^0. \tag{8.55}$$

The increasing weights $\boldsymbol{\alpha}_2$ for the minimization of net pollution and decreasing weight α_1 for the maximization of the final demand value imply decreasing coefficients of the parametric objective function in (8.51). For the case in which only the coefficients of the objective function in (8.51) will be changed ($\Delta\mathbf{c} \neq \mathbf{0}$ and $\Delta\mathbf{b} = \mathbf{0}$), inequality (8.54) reduces to

$$\Delta\mathbf{c}'\Delta\mathbf{x} \geqq 0. \tag{8.56}$$

If $\Delta\mathbf{c} = \mathbf{0}$ and $\Delta\mathbf{b} \neq \mathbf{0}$, then we obtain

$$\Delta\mathbf{u}'\Delta\mathbf{b} \leqq 0.$$

The solution of the maximization problem *tends* to vary in the same direction (the solution of the minimization problem in the opposite direction) as the coefficients of the objective function. Direct application of the so-called Le Chatelier–Samuelson inequality (8.56) to our problem (8.51) with $\mathbf{c}' = (\mathbf{0}', \mathbf{0}', \alpha_1\mathbf{p}', -\boldsymbol{\alpha}_2)$ and $\mathbf{x}' = (\mathbf{x}_1', \mathbf{x}_2', \mathbf{y}_1', \mathbf{y}_2')$ then yields

$$\Delta\mathbf{c}'\Delta\mathbf{x} = \Delta\alpha_1\mathbf{p}'\Delta\mathbf{y}_1 + \Delta\boldsymbol{\alpha}_2\Delta\mathbf{y}_2 \geqq 0,$$

where $\Delta\alpha_1 < 0$ and $\Delta\boldsymbol{\alpha}_2 = \mathbf{0}$ (as in our numerical example). Because we want to maximize the value of final demand and to minimize net pollution, according to the Le Chatelier–Samuelson principle, the final demand $\mathbf{y}_1$ *tends* to vary in the *same* direction as the weight for the final demand, and the level of net pollution $\mathbf{y}_2$ in the *opposite* direction as the weights for the environmental goal. Summarizing, the increasing relative weights for environmental quality will lead—*in tendency*—to *nonincreasing* final demand and *nonincreasing* net pollution. The nonincreasing or decreasing gross production $\mathbf{x}_1$ and increasing level of abatement activities $\mathbf{x}_2$, as illustrated in our numerical example, follow from the structure of the Leontief model. For the decreasing final demand, lower gross production is needed, which subsequently reduces the amount of gross pollution. Furthermore, the level of net pollution is reduced by increasing abatement activity. □

These results depend crucially on the definition of the economic objectives. As shown by Luptáčik and Böhm [21] for the other version of the multicriteria augmented Leontief model described in Section 7.1.4, the opposite results have been derived.

References and Further Reading

[1] R. Benayoun, J. de Montgolfier, J. Tergny, and O. Laritchev, Linear programming with multiple objective functions: Step method (STEM), *Math. Programming*, **1**-3 (1971), 366–375.

[2] B. Böhm and P. Brandner, Optimal economic policy making with an econometric model using interactive methods of vector optimization, in J. Gruber, ed., *Econometric Decision Models: New Methods of Modeling and Applications*, Lecture Notes in Economics and Mathematical Systems vol. 366, Springer-Verlag, Berlin, 1991, 149–163.

[3] B. Böhm and M. Luptáčik, Optimisation with an environmental input–output model, in G. Boero and A. Silberston, eds., *Environmental Economics*, St. Martin's Press, New York, 1995, 21–33.

[4] J. P. Evans and R. E. Steuer, A revised simplex method for linear multiple objective programs, *Math. Programming*, **5**-1 (1973), 54–72.

[5] T. Gal, *Postoptimality Analysis, Parametric Programming, and Related Topics*, McGraw–Hill, New York, 1979.

[6] T. Gal, Th. J. Stewart, and Th. Hanne, eds., *Multicriteria Decision Making: Advances in MCDM Models, Algorithms, Theory, and Applications*, Kluwer Academic Publishers, Dordrecht, The Netherlands, 1999.

[7] D. Gale, H. W. Kuhn, and A. W. Tucker, Linear programming and the theory of games, in T. C. Koopmans, ed., *Activity Analysis of Production and Allocation*, Cowles Commission Monograph 13, Wiley, New York, 1951, Chapter XIX, 317–329.

[8] E. Galperin and P. Guerra Jimenez, Duality of nonscalarized multiobjective linear programs: Dual balance, level sets, and dual clusters of optimal vectors, *J. Optim. Theory Appl.*, **108** (2001), 109–137.

[9] L. R. Gardiner and R. E. Steuer, Unified interactive multiple objective programming, *Eur. J. Oper. Res.*, **74** (1994), 391–406.

[10] S. I. Gass, *Linear Programming: Methods and Applications*, 3rd ed., McGraw–Hill, New York, 1969.

[11] A. M. Geoffrion, Proper efficiency and the theory of vector maximization, *J. Math. Anal. Appl.*, **22** (1968), 618–630.

[12] A. M. Geoffrion, J. S. Dyer, and A. Feiberg, An interactive approach for multicriteria optimization with an application to the operation of an academic department, *Manage. Sci.*, **19**-4 (1972), 357–368.

[13] H. Isermann, Existence and duality in multiple objective linear programming, in H. Thiriez and S. Zionts, eds., *Multiple Criteria Decision Making*, Lecture Notes in Economics and Mathematical Systems vol. 130, Springer-Verlag, Berlin, 1976, 64–75.

[14] H. Isermann, Ein Algorithmus zur Lösung linear Vektormaximumprobleme, in J. Kohlas, O. Seifert, P. Stáhly, and H. J. Zimmermann, eds., *Proceedings in Operation Research*, Vol. 5, Physica-Verlag, Würtzburg, Germany, 1976, 55–65.

[15] H. Isermann, The relevance of duality in multiple objective linear programming, *TIMS Stud. Manage. Sci.*, **6** (1977), 241–262.

[16] H. Isermann, Duality in multiple objective linear programming, in S. Zionts, ed., *Multiple Criteria Problem Solving: Proceedings of a Conference, Buffalo*, 1977, Lecture Notes in Economics and Mathematical Systems vol. 155, Springer-Verlag, Berlin, 1978, 274–285.

[17] H. Isermann, On some relations between a dual pair of multiple objective linear programs, *Z. Oper. Res.*, **22** (1978), 33–41.

[18] P. Korhonen and J. Laakso, A visual interactive method for solving the multiple criteria problem, *Eur. J. Oper. Res.*, **24** (1986), 277–287.

[19] J. S. H. Kornbluth, Duality indifference and sensitivity analysis in multiple objective linear programming, *Oper. Res. Q.*, **25** (1974), 599–614.

[20] B. Kreko, *Lehrbuch der linearen Optimierung*, VEB Deutscher Verlag der Wissenschaften, Berlin, 1970.

[21] M. Luptáčik and B. Böhm, An environmental input–output model with multiple criteria, *Ann. Oper. Res.*, **54** (1994), 119–127.

[22] Y. Murata, *Mathematics for Stability and Optimization of Economic Systems*, Academic Press, New York, 1977.

[23] J. Spronk, *Interactive Multiple Goal Programming*, Nijhoff, Leiden, The Netherlands, 1981.

[24] R. E. Steuer, An interactive multiple objective linear programming procedure, *TIMS Stud. Manage. Sci.*, **6** (1977), 225–239.

[25] R. E. Steuer, *Multiple Criteria Optimization: Theory, Computation and Application*, Wiley, Singapore, 1986.

[26] Th. J. Stewart, Concepts of interactive programming, in T. Gal, Th. J. Stewart, and Th. Hanne, eds., *Multicriteria Decision Making: Advances in MCDM Models, Algorithms, Theory, and Applications*, Kluwer Academic Publishers, Boston, 1999, Chapter 10.

[27] H. Wallenius, Optimizing macroeconomic policy: A review of approaches and applications, *Eur. J. Oper. Res.*, **10** (1982), 221–228.

[28] S. Zionts and J. Wallenius, An interactive programming method for solving the multicriteria problem, *Manage. Sci.*, **22** (1976), 652–663.

9

Multiobjective Geometric Programming

As shown in Chapter 6, problems leading to geometric programming models arise very often, not only in engineering design but also in economics and management science. Referring to the introductory discussion in Chapter 7, it seems obvious to include geometric programming problems with several objectives in our consideration. A nonlinear model of environmental control from Section 7.1.6 provides one example and will be analyzed in Section 9.3.

This chapter uses the notation and results of Chapters 6 and 7 and contains four sections. Efficient and properly efficient solutions of multiobjective geometric programming problems are derived in Section 9.1 as optimal solutions of ordinary geometric programming problems. To illustrate, we present a simple production model with maximization of production and maximization of environmental quality.

In Section 9.2, duality for a multiobjective optimization problem in parametric form, as developed in Section 7.3.1, is used for multiobjective geometric programming, and as an application a nonlinear model of environmental control from Section 7.1.6 is analyzed in Section 9.3. The chapter concludes with Section 9.4, which deals with the model of a monopolistic firm maximizing revenue and profit, as formulated in Section 7.1.4.

9.1 Vector Minimization Problems in Geometric Programming

We consider the multiobjective mathematical programming problem (7.12):

$$\begin{array}{lll} \text{Minimize} & F(\mathbf{x}) = (f_1(\mathbf{x}), f_2(\mathbf{x}), \ldots, f_s(\mathbf{x})) & (9.1)\\ \text{subject to} & g_i(\mathbf{x}) - 1 \leqq 0 \qquad (i = 1, 2, \ldots, m), & (9.2) \end{array}$$

and

$$x_j > 0 \qquad (j = 1, 2, \ldots, n), \qquad (9.3)$$

M. Luptáčik, *Mathematical Optimization and Economic Analysis*,
Springer Optimization and Its Applications 36, DOI 10.1007/978-0-387-89552-9_9,

where the objective functions $f_1(\mathbf{x}), f_2(\mathbf{x}), \ldots, f_s(\mathbf{x})$ and the constraints $g_1(\mathbf{x})$, $g_2(\mathbf{x}), \ldots, g_m(\mathbf{x})$ are polynomials:

$$f_k(\mathbf{x}) = \sum_{t=1}^{S_k} c_{kt} \prod_{j=1}^{n} x_j^{a_{ktj}} \quad (k = 1, 2, \ldots, s),$$

$$g_i(\mathbf{x}) = \sum_{t=1}^{T_i} d_{it} \prod_{j=1}^{n} x_j^{b_{itj}} \quad (i = 1, 2, \ldots, m).$$

The exponents a_{ktj} and b_{itj} are arbitrary real numbers, and the coefficients c_{kt} and d_{it} are positive.

Writing the multiobjective optimization problem (9.1)–(9.3) as the parametric problem (7.17), we see that the following geometric programming problem (P_α) ensues:

$$\begin{aligned} &\text{minimize} \quad \sum_{k=1}^{s} \alpha_k f_k(\mathbf{x}) \\ &\text{subject to} \quad (9.2) \text{ and } (9.3). \end{aligned} \tag{P_α}$$

The corresponding dual problem is (by Section 6.2)

$$\text{maximize} \qquad v(\boldsymbol{\alpha}, \boldsymbol{\delta}) = \prod_{k=1}^{s} \prod_{t=1}^{S_k} \left(\frac{\alpha_k c_{kt}}{\delta_{kt}} \right)^{\delta_{kt}} \prod_{i=1}^{m} \prod_{t=1}^{T_i} \left(\frac{d_{it}}{\delta_{it}} \right)^{\delta_{it}} \prod_{i=1}^{m} \lambda_i(\delta)^{\lambda_i(\delta)},$$

$$\text{where} \qquad \lambda_i(\delta) = \sum_{t=1}^{T_i} \delta_{it} \quad (i = 1, 2, \ldots, m),$$

$$\text{subject to} \qquad \sum_{k=1}^{s} \sum_{t=1}^{S_k} \delta_{kt} = 1,$$

$$\sum_{k=1}^{s} \sum_{t=1}^{S_k} a_{ktj} \delta_{kt} + \sum_{i=1}^{m} \sum_{t=1}^{T_i} b_{itj} \delta_{it} = 0 \qquad (j = 1, 2, \ldots, n),$$

$$\delta_{kt} \geqq 0 \qquad \begin{pmatrix} k = 1, 2, \ldots, s, \\ t = 1, 2, \ldots, S_k \end{pmatrix},$$

$$\delta_{it} \geqq 0 \qquad \begin{pmatrix} i = 1, 2, \ldots, m, \\ t = 1, 2, \ldots, T_i \end{pmatrix}.$$

For fixed $\boldsymbol{\alpha} = (\alpha_1, \alpha_2, \ldots, \alpha_s)$ with $\alpha_k > 0$ $(k = 1, 2, \ldots, s)$, using geometric programming theory (see Chapter 6) we can solve problem (P_α) to provide, by Theorem 7.2, a Pareto–Koopmans-efficient solution to problem (9.1)–(9.3).

The linear combination of the objective functions with fixed α_k $(k = 1, 2, \ldots, s)$ in the parametric problem (P_α) implies—from an economic interpretation point of view—that the marginal rate of substitution between the objective function values (the marginal gain in one objective relative to the marginal loss in another objective) is constant and independent of their levels. Pascual and Ben-Israel [5] proposed a new parametric problem ($\prod \mathrm{P}_\alpha$) with a nonlinear combination of the objectives via the power function. Then the new problem is

$$\begin{aligned} &\text{minimize} \quad \prod_{k-1}^{s} f_k(\mathbf{x})^{\alpha_k} \\ &\text{subject to} \quad (9.2) \text{ and } (9.3). \end{aligned} \tag{$\prod \mathrm{P}_\alpha$}$$

The coefficients α_k can be interpreted as elasticities of the "master" or "composite" objective function $F(\mathbf{x})$ with respect to the particular objectives $f_k(\mathbf{x})$ $(k = 1, 2, \dots, s)$. In the formulation (P_α), the coefficients α_k $(k = 1, 2, \dots, s)$ express the marginal contribution of the objective k to the "master" or "composite" objective function $F(\mathbf{x})$. Problem ($\prod \mathrm{P}_\alpha$) is not necessarily a geometric program, since $\prod_{k=1}^{s} f_k(\mathbf{x})^{\alpha_k}$ need not be a polynomial. However, as shown in Section 6.4, problem ($\prod \mathrm{P}_\alpha$) can be transformed to the following geometric programming model:

$$\text{minimize} \quad \prod_{k=1}^{s} x_{n+k}^{\alpha_k}$$

subject to (9.2), (9.3), and the additional constraints

$$f_k(\mathbf{x}) x_{n+k}^{-1} \leqq 1 \quad (k = 1, 2, \dots, s)$$

and

$$x_{n+k} > 0 \quad (k = 1, 2, \dots, s).$$

The dual problem becomes

$$\text{maximize} \quad v(\boldsymbol{\delta}) = \left(\frac{1}{\delta_0}\right)^{\delta_0} \prod_{k=1}^{s} \prod_{t=1}^{S_k} \left(\frac{c_{kt}}{\delta_{kt}}\right)^{\delta_{kt}} \prod_{i-1}^{m} \prod_{t-1}^{T_i} \left(\frac{d_{it}}{\delta_{it}}\right)^{\delta_{it}} \prod_{k=1}^{s} \alpha_k^{\alpha_k} \prod_{i-1}^{m} \lambda_i^{\lambda_i} \tag{9.4}$$

subject to

$$\delta_0 = 1, \tag{9.5}$$

$$\alpha_k \delta_0 - \sum_{t=1}^{S_k} \delta_{kt} = 0 \quad (k = 1, 2, \dots, s), \tag{9.6}$$

$$\sum_{k=1}^{s} \sum_{t=1}^{S_k} a_{ktj} \delta_{kt} + \sum_{i=1}^{m} \sum_{t=1}^{T_i} b_{itj} \delta_{it} = 0 \quad (j = 1, 2, \dots, n), \tag{9.7}$$

$$\delta_{kt} \geqq 0 \quad \begin{pmatrix} k = 1, 2, \dots, s, \\ t = 1, 2, \dots, S_k \end{pmatrix}, \tag{9.8}$$

$$\delta_{it} \geqq 0 \quad \begin{pmatrix} i = 1, 2, \dots, m, \\ t = 1, 2, \dots, T_i \end{pmatrix}. \tag{9.9}$$

Here

$$\lambda_i(\boldsymbol{\delta}) = \sum_{t=1}^{T_i} \delta_{it} \quad (i = 1, 2, \ldots, m). \tag{9.10}$$

Using the duality theory of geometric programming, Pascual and Ben-Israel [5, p. 101] proved the following.

Theorem 9.1. *Let* $\alpha_k > 0$ $(k = 1, 2, \ldots, s)$ *satisfying* $\sum_{k=1}^{s} \alpha_k = 1$ *be fixed. If* $\mathbf{x}^0$ *is an optimal solution of the primal program* $(\prod \mathrm{P}_\alpha)$*, then* $\mathbf{x}^0$ *is an efficient solution of problem* (9.1)–(9.3).

To illustrate, let us consider a simple production model taking into account two objectives: maximization of production and maximization of environmental quality. Assume that the firm produces a single output in the quantity q according to the production function

$$q = f(x_1) = ax_1^{\beta},$$

where x_1 denotes the amount of the input factors (e.g., labor) used for production and $a > 0$. $\beta > 0$ is the elasticity coefficient with respect to labor. For simplicity, we consider only one production factor. The environmental quality depends on the intensity of environmental protection activities and therefore on the amount of production factor labor devoted to the preservation of environmental commodities and to the creation of natural areas. Denoting the amount of labor used for the environmental protection by x_2, the second objective function is $f_2(x_2) = bx_2^{\mu}$, where $b > 0$ and $0 < \mu \leqq 1$. It should be maximized as well. The total amount of labor is restricted to t, i.e.,

$$x_1 + x_2 \leqq t. \tag{9.11}$$

The following vector optimization problem arises:

$$\begin{aligned} &\text{Maximize} && F(\mathbf{x}) = (f_1(\mathbf{x}), f_2(\mathbf{x})) \\ &\text{subject to} && (9.11). \end{aligned}$$

The reader may verify that the formulation of the parametric problem (P_α) leads to a geometric programming with degree of difficulty one (see Section 6.2), while the parametric $(\prod \mathrm{P}_\alpha)$ is a problem with degree of difficulty zero. As shown by Pascual and Ben-Israel [5, p. 101], the degree of difficulty of problem $(\prod \mathrm{P}_\alpha)$ is smaller by $s - 1$ than the degree of difficulty of problem (P_α). Problem $(\prod \mathrm{P}_\alpha)$ for our above example becomes

$$\begin{aligned} &\text{minimize} && f_1^{-\alpha_1} f_2^{-\alpha_2} = (ax_1^{\beta})^{-\alpha_1}(bx_2^{\mu})^{-\alpha_2} = a_0 x_1^{-\alpha_1\beta} x_2^{-\alpha_2\mu} \\ &\text{subject to} && x_1 t^{-1} + x_2 t^{-1} \leqq 1, \\ &\text{where} && a_0 = a^{-\alpha_1} b^{-\alpha_2} > 0. \end{aligned}$$

Fortunately, the objective function is a posynomial and geometric programming theory (see Section 6.2) can immediately be applied. Because the degree of difficulty is zero, the dual solution is easily estimated:

$$\delta_0^0 = 1, \qquad \delta_{11}^0 = \alpha_1\beta, \quad \text{and} \quad \delta_{12}^0 = \alpha_2\mu.$$

According to (6.28), the solution of the primal problem is

$$x_1^0 = \frac{\alpha_1\beta}{\alpha_1\beta + \alpha_2\mu}t \quad \text{and} \quad x_2^0 = \frac{\alpha_2\mu}{\alpha_1\beta + \alpha_2\mu}t.$$

The ratio in which the production factor labor should be allocated between the production and environmental protection activities is uniquely determined by the weighted coefficients α_k $(k = 1, 2)$ and elasticity coefficients β and μ:

$$\frac{x_1^0}{x_2^0} = \frac{\alpha_1\beta}{\alpha_2\mu}.$$

A nonlinear model of environmental control by Mastenbroek and Nijkamp [4] (described in Section 7.1.6) represents an extension of the model discussed above considering the pollution generated by production of goods and by the consumption activities. It will be analyzed in Section 9.3.

An alternative approach for vector minimization problems with polynomials in the objectives and in the constraints will be discussed in the following section. It is based on duality for multiobjective optimization problems in parametric form, as introduced in Section 7.3.1.

9.2 Duality for Multiobjective Geometric Programming in Parametric Form

Let us return to the multiobjective geometric programming problem (9.1)–(9.3), where the objective functions $f_k(\mathbf{x})$ $(k = 1, 2, \ldots, s)$ as well as the function $g_i(\mathbf{x})$ $(i = 1, 2, \ldots, m)$ in the constraints (9.2) are posynomials. In order to transform problem (9.1)–(9.3) into a convex program, we change the variables x_j $(j = 1, 2, \ldots, n)$ by letting

$$e^{z_j} = x_j, \quad \text{or} \quad z_j = \ln x_j \quad (j = 1, 2, \ldots, n).$$

Then due to Theorem 6.1, each positive exponential function

$$h_k(\mathbf{z}) = \sum_{t=1}^{S_k} c_{kt} e^{\sum_{j=1}^n a_{ktj} z_j} \quad (k = 1, 2, \ldots, s),$$

$$l_i(\mathbf{z}) = \sum_{t=1}^{T_i} d_{it} e^{\sum_{j=1}^n b_{itj} z_j} \quad (i = 1, 2, \ldots, m)$$

is convex. S_k and T_i denote the number of terms in the objective k and the forced constraints i, respectively. Since the logarithmic function is monotone increasing, instead of the transformed problem with functions $h_k(\mathbf{z})$ $(k = 1, 2, \ldots, s)$ and $l_i(\mathbf{z})$ $(i = 1, 2, \ldots, m)$, we consider the following program:

$$\underset{\mathbf{z}}{\text{Minimize}} \quad \ln H(\mathbf{z}) \tag{9.12}$$

$$\text{subject to} \quad \ln L(\mathbf{z}) \leqq 0, \tag{9.13}$$

where

$$\ln H(\mathbf{z}) = (\ln h_1(\mathbf{z}), \ln h_2(\mathbf{z}), \ldots, \ln h_s(\mathbf{z})),$$
$$\ln L(\mathbf{z}) = (\ln l_1(\mathbf{z}), \ln l_2(\mathbf{z}), \ldots, \ln l_m(\mathbf{z})),$$

and

$$\mathbf{z} \in R^n.$$

The functions $\ln h_k(\mathbf{z})$ $(k = 1, 2, \ldots, s)$ and $\ln l_i(\mathbf{z})$ $(i = 1, 2, \ldots, m)$ are convex for arbitrary real numbers a_{ktj} and b_{itj} and positive real numbers c_{kt} and d_{it}. Thus the duality for multiobjective optimization problems from Section 7.3.1 can be applied. The dual problem to problem (9.12)–(9.13) is (see (7.38))

$$\text{maximize} \qquad \Phi(\mathbf{z}, \boldsymbol{\lambda}) = \boldsymbol{\alpha}'[\ln H(\mathbf{z})] + \boldsymbol{\lambda}'[\ln L(\mathbf{z})] \tag{9.14}$$

$$\text{subject to} \quad \boldsymbol{\alpha}'[\nabla \ln H(\mathbf{z})] + \boldsymbol{\lambda}'[\nabla L(\mathbf{z})] = \mathbf{0}, \tag{9.15}$$

$$\boldsymbol{\lambda} \geqq \mathbf{0}. \tag{9.16}$$

Differentiation of (9.14) with respect to z_j yields

$$\frac{\partial \Phi}{\partial z_j} = \sum_{k=1}^{s} \alpha_k \frac{\sum_{t=1}^{S_k} c_{kt} e^{\sum_{j=1}^{n} a_{ktj} z_j} a_{ktj}}{\sum_{t=1}^{S_k} c_{kt} e^{\sum_{j=1}^{n} a_{ktj} z_j}} + \sum_{i=1}^{m} \lambda_i \frac{\sum_{t=1}^{T_i} d_{it} e^{\sum_{j=1}^{n} b_{itj} z_j} b_{itj}}{\sum_{t=1}^{T_i} d_{it} e^{\sum_{j=1}^{n} b_{itj} z_j}}.$$

Denote

$$\delta_{kt} = \alpha_k \frac{c_{kt} e^{\sum_{j=1}^{n} a_{ktj} z_j}}{\sum_{t=1}^{S_k} c_{kt} e^{\sum_{j=1}^{n} a_{ktj} z_j}} \quad \begin{pmatrix} k = 1, 2, \ldots, s, \\ t = 1, 2, \ldots, S_k \end{pmatrix} \tag{9.17}$$

and

$$\delta_{it} = \lambda_i \frac{d_{it} e^{\sum_{j=1}^{n} b_{itj} z_j}}{\sum_{t=1}^{T_i} d_{it} e^{\sum_{j=1}^{n} b_{itj} z_j}} \quad \begin{pmatrix} i = 1, 2, \ldots, m, \\ t = 1, 2, \ldots, T_i \end{pmatrix}, \tag{9.18}$$

where δ_{kt} describes the dual variable associated with term t in the objective k and δ_{it} is the dual variable associated with term t in the forced constraint i. Then the dual constraints (9.15) become

$$\sum_{t=1}^{s}\sum_{t=1}^{S_k} a_{ktj}\delta_{kt} + \sum_{i=1}^{m}\sum_{t=1}^{T_i} b_{itj}\delta_{it} = 0 \quad (j = 1, 2, \ldots, n). \tag{9.19}$$

It follows from (9.17) that

$$\sum_{t=1}^{S_k} \delta_{kt} = \alpha_k \quad (k = 1, 2, \ldots, s) \tag{9.20}$$

and from (9.18) that

$$\sum_{t=1}^{T_i} \delta_{it} = \lambda_i \quad (i = 1, 2, \ldots, m). \tag{9.21}$$

Definition (9.17) implies

$$\delta_{kt} \geqq 0 \quad \begin{matrix}(k = 1, 2, \ldots, s),\\ (t = 1, 2, \ldots, S_k)\end{matrix} \tag{9.22}$$

and the nonnegativity constraints (9.16) and definition (9.18) lead to

$$\delta_{it} \geqq 0 \quad \begin{matrix}(i = 1, 2, \ldots, m),\\ (t = 1, 2, \ldots, T_i).\end{matrix} \tag{9.23}$$

Analogously to the procedure in Section 6.2, relation (9.17) can be rewritten as

$$\frac{\delta_{kt}}{c_{kt}} \sum_{t=1}^{S_k} c_{kt} e^{\sum_{j=1}^{n} a_{ktj} z_j} = \alpha_k e^{\sum_{j=1}^{n} a_{ktj} z_j}, \tag{9.24}$$

and relation (9.18) as

$$\frac{\delta_{it}}{d_{it}} \sum_{t=1}^{T_i} d_{it} e^{\sum_{j=1}^{n} b_{itj} z_j} = \lambda_i e^{\sum_{j=1}^{n} b_{itj} z_j}. \tag{9.25}$$

After raising (9.24) to a power by δ_{kt} and (9.25) to a power by δ_{it} and then taking a logarithm, we obtain

$$\delta_{kt} \ln \frac{\delta_{kt}}{c_{kt}} + \delta_{kt} \ln h_k(\mathbf{z}) = \delta_{kt} \ln \alpha_k + \delta_{kt} \sum_{j=1}^{n} a_{ktj} z_j \quad (k = 1, 2, \ldots, s), \tag{9.26}$$

$$\delta_{it} \ln \frac{\delta_{it}}{d_{it}} + \delta_{it} \ln l_i(\mathbf{z}) = \delta_{it} \ln \lambda_i + \delta_{it} \sum_{j=1}^{n} b_{itj} z_j \quad (i = 1, 2, \ldots, m). \tag{9.27}$$

Summing (9.26) over k and t and (9.27) over i and t and taking into account (9.20) and (9.21) gives

$$\sum_{k=1}^{s}\sum_{t=1}^{S_k}\delta_{kt}\ln\frac{\delta_{kt}}{c_{kt}}+\sum_{k=1}^{s}\alpha_k\ln h_k(\mathbf{z})=\sum_{k=1}^{s}\sum_{t=1}^{S_k}\sum_{j=1}^{n}\delta_{kt}a_{ktj}z_j+\sum_{k=1}^{s}\alpha_k\ln\alpha_k,$$

$$\sum_{v=1}^{m}\sum_{t=1}^{T_i}\delta_{it}\ln\frac{\delta_{it}}{d_{it}}+\sum_{i=1}^{m}\lambda_i\ln l_i(\mathbf{z})=\sum_{i=1}^{m}\lambda_i\ln\lambda_i+\sum_{i=1}^{m}\sum_{t=1}^{T_i}\sum_{j=1}^{n}\delta_{it}b_{itj}z_j.$$

Summing the above equations and taking into account (9.19) yields

$$\sum_{k=1}^{s}\alpha_k\ln h_k(\mathbf{z})+\sum_{i=1}^{m}\lambda_i\ln l_i(\mathbf{z})$$
$$=\sum_{k=1}^{s}\sum_{t=1}^{S_k}\delta_{kt}\ln\frac{c_{kt}}{\delta_{kt}}+\sum_{i=1}^{m}\sum_{t=1}^{T_i}\delta_{it}\ln\frac{d_{it}}{\delta_{it}}+\sum_{k=1}^{s}\alpha_k\ln\alpha_k+\sum_{i=1}^{m}\lambda_i\ln\lambda_i.$$

The dual objective function (9.14) is now expressed as a function of the dual variables δ_{kt} and δ_{it}:

$$\Phi(\mathbf{z},\boldsymbol{\lambda})=\sum_{k=1}^{s}\sum_{t=1}^{S_k}\delta_{kt}\ln\frac{c_{kt}}{\delta_{kt}}+\sum_{i=1}^{m}\sum_{t=1}^{T_i}\delta_{it}\ln\frac{d_{it}}{\delta_{it}}+\sum_{k=1}^{s}\alpha_k\ln\alpha_k+\sum_{i=1}^{m}\lambda_i\ln\lambda_i$$
$$=\ln\left(\prod_{k=1}^{s}\prod_{t=1}^{S_k}\left(\frac{c_{kt}}{\delta_{kt}}\right)^{\delta_{kt}}\prod_{i=1}^{m}\prod_{t=1}^{T_i}\left(\frac{d_{it}}{\delta_{it}}\right)^{\delta_{it}}\prod_{k=1}^{s}\alpha_k^{\alpha_k}\prod_{i=1}^{m}\lambda(\boldsymbol{\delta})^{\lambda(\boldsymbol{\delta})}\right)=\ln v(\boldsymbol{\delta}).$$

Because of the monotonicity property of the logarithmic function, $v(\boldsymbol{\delta})$ and $\ln v(\boldsymbol{\delta})$ have the same set of maximizing points. Thus the dual program (9.14)–(9.16) takes the form

$$\text{maximize}\quad v(\boldsymbol{\delta})=\prod_{k=1}^{s}\prod_{t=1}^{S_k}\left(\frac{c_{kt}}{\delta_{kt}}\right)^{\delta_{kt}}\prod_{i=1}^{m}\prod_{t=1}^{T_i}\left(\frac{d_{it}}{\delta_{it}}\right)^{\delta_{it}}\prod_{k=1}^{s}\alpha_k^{\alpha_k}\prod_{i=1}^{m}\lambda_i(\boldsymbol{\delta})^{\lambda_i(\boldsymbol{\delta})}\qquad(9.28)$$

subject to (9.19)–(9.23), which corresponds exactly to the model (9.4)–(9.10). The reason is that by taking a logarithm of the objective function in ($\prod P_\alpha$), we obtain problem (7.37). In other words, both approaches are based on the same type of "master" or "composite" objective function $F(\mathbf{x})$. It is a Cobb–Douglas type of utility function, as expressed in ($\prod P_\alpha$), with elasticity of substitution between the objectives equal to one.

Analogously to the single geometric programming problem in Chapter 6, an efficient solution for (9.1)–(9.3) can be estimated from the system of equations

$$\frac{c_{kt}x_1^{a_{kt1}}\dots x_n^{a_{ktn}}}{\sum_{t=1}^{S_k}c_{kt}x_1^{a_{kt1}}\dots x_n^{a_{ktn}}}=\frac{\delta_{kt}^0}{\alpha_k}\qquad\begin{array}{l}(k=1,2,\dots,s),\\(t=1,2,\dots,S_k),\end{array}\qquad(9.29)$$

$$d_{it} x_1^{b_{it1}} \dots x_n^{b_{itn}} = \frac{\delta_{it}^0}{\lambda_i(\delta^0)} \quad \begin{array}{l}(i = 1, 2, \dots, m),\\ (t = 1, 2, \dots, T_i),\end{array} \tag{9.30}$$

where δ^0 is an optimal solution for the dual problem (9.28) subject to (9.19)–(9.23) and i ranges over all positive integers for which $\lambda_i(\delta^0) > 0$.

The advantage of the approach in this section is that for models in which the objective function in ($\prod P_\alpha$) is not a posynomial, transformation to a geometric program is not necessary. To illustrate, let us consider the following example:

$$\text{Minimize} \quad F(\mathbf{x}) = \begin{Bmatrix} f_1(\mathbf{x}) = 2x_1 + 2x_2, \\ f_2(\mathbf{x}) = x_1^{-1} x_2^{-1} \end{Bmatrix}$$

$$\text{subject to} \quad x_1 > 0, \quad x_2 > 0.$$

This is a problem without forced constraints. According to (9.19), (9.20), and (9.22), the dual constraints become

$$\begin{aligned} \delta_{11} - \delta_{21} &= 0, \\ \delta_{12} - \delta_{21} &= 0, \\ \delta_{11} + \delta_{12} &= \alpha_1, \\ \delta_{21} &= \alpha_2 \end{aligned}$$

and yield the solution $\delta_{11}^0 = \delta_{12}^0 = \alpha_2$ and $\delta_{21}^0 = \alpha_2$ for $\alpha_1 = 2\alpha_2$. Because $\delta_{11}^0 = \delta_{12}^0$, it follows from (9.17) that $2x_1^0 = 2x_2^0$ and consequently $x_1^0 = x_2^0$. It is a well-known result that for a given girth, a rectangle maximizing its volume is a square. For a given volume, a rectangle minimizing its girth is a square.

Applying the approach by Pascual and Ben-Israel [5] from Section 9.1, the objective function

$$f_1(\mathbf{x})^{\alpha_1} f_2(\mathbf{x})^{\alpha_2} = (2x_1 + 2x_2)^{\alpha_1} x_1^{-\alpha_2} x_2^{-\alpha_2}$$

is not a posynomial. Introducing a new variable x_0, we can write the geometric programming problem

$$\begin{array}{ll} \text{minimize} & x_0^{-\alpha_1} x_1^{-\alpha_2} x_2^{-\alpha_2} \\ \text{subject to} & 2x_0^{-1} x_1 + 2x_0^{-1} x_2 \leqq 1, \\ & x_0 > 0, \quad x_1 > 0, \quad x_2 > 0. \end{array}$$

The dual constraints

$$\begin{aligned} \delta_0 &= 1, \\ -\alpha_1 \delta_0 + \delta_{11} + \delta_{12} &= 0, \\ -\alpha_2 \delta_0 + \delta_{11} &= 0, \\ -\alpha_2 \delta_0 + \delta_{12} &= 0 \end{aligned}$$

yield the solution $\delta_{11}^0 = \delta_{12}^0 = \alpha_2$ for $\alpha_1 = 2\alpha_2$. The optimal solution will be obtained by using (6.28),

$$2x_0^{-1}x_1 = \frac{\delta_{11}^0}{\delta_{11}^0 + \delta_{12}^0} = \frac{\alpha_2}{2\alpha_2} = \frac{1}{2},$$

$$2x_0^{-1}x_2 = \frac{\delta_{12}^0}{\delta_{11}^0 + \delta_{12}^0} = \frac{\alpha_2}{2\alpha_2} = \frac{1}{2},$$

which yields $x_1^0 = x_2^0$.

9.3 A Nonlinear Model of Environmental Control

The model from Section 7.1.6 proposed by Mastenbroek and Nijkamp [4] leads to the vector minimization problem

$$\text{Minimize} \quad F(c,i,p,y) = (f_1(p) = ap^{\kappa},\ f_2(c) = c^{-1}) \tag{9.31}$$

$$\text{subject to} \quad li^{-\mu}yp^{-1} + dcp^{-1} \leqq 1, \tag{9.32}$$

$$iy^{-1} + cy^{-1} \leqq 1, \tag{9.33}$$

$$c > 0, \quad i > 0, \quad p > 0, \quad y > 0, \tag{9.34}$$

where $l = b_0 i_0^{\mu} > 0$. The reader can verify that the parametric problem (P_α) related to problem (9.31)–(9.34) leads to a geometric programming problem with degree of difficulty equal to one. The corresponding parametric problem ($\prod \text{P}_\alpha$),

$$\begin{aligned} &\text{minimize} \quad (ap^{\kappa})^{\alpha_1} c^{-\alpha_2} \\ &\text{subject to} \quad (9.32)\text{–}(9.34), \end{aligned} \tag{9.35}$$

is a geometric programming problem with zero degree of difficulty. Denoting by δ_0 the dual variable related to the term in the objective function (9.35) and by δ_{it} ($i = 1, 2;\ t = 1, 2$) the dual variables related to the term in the constraints (9.32)–(9.33), the dual program takes the form

$$\text{maximize} \quad v(\delta) = \left(\frac{1}{\delta_0}\right)^{\delta_0} \left(\frac{l}{\delta_{11}}\right)^{\delta_{11}} \left(\frac{d}{\delta_{12}}\right)^{\delta_{12}} \left(\frac{1}{\delta_{21}}\right)^{\delta_{21}} \left(\frac{1}{\delta_{22}}\right)^{\delta_{22}} \tag{9.36}$$

$$\text{subject to} \quad \delta_0 = 1, \tag{9.37}$$

$$\alpha_1 \kappa \delta_0 - \delta_{11} - \delta_{12} = 0, \tag{9.38}$$

$$\delta_{11} - \delta_{21} - \delta_{22} = 0, \tag{9.39}$$

$$-\alpha_2 \delta_0 + \delta_{12} + \delta_{22} = 0, \tag{9.40}$$

$$-\mu \delta_{11} + \delta_{21} = 0, \tag{9.41}$$

$$\delta_0 \geqq 0, \quad \delta_{11} \geqq 0, \quad \delta_{12} \geqq 0, \quad \delta_{21} \geqq 0, \quad \delta_{22} \geqq 0. \tag{9.42}$$

Denoting by δ_{11}^* and δ_{21}^* the dual variables corresponding to the terms in the first and second objective functions, respectively, the dual constraints (9.20) yield $\delta_{11}^* = \alpha_1$ and $\delta_{21}^* = \alpha_2$. Consequently, the dual constraints (9.19) provide the set of equations (9.38)–(9.41) as above. The following optimal solution will be obtained:

$$\delta_0^0 = 1, \qquad \delta_{11}^0 = \frac{\alpha_1\kappa - \alpha_2}{\mu}, \qquad \delta_{12}^0 = \frac{\alpha_1\kappa(\mu - 1) + \alpha_2}{\mu},$$

$$\delta_{21}^0 = \alpha_1\kappa - \alpha_2, \qquad \text{and} \qquad \delta_{22}^0 = \frac{(\alpha_1\kappa - \alpha_2)(1 - \mu)}{\mu}.$$

Assuming that $0 < \mu < 1$ (due to the assumption in Section 7.1.6), the nonnegativity condition for δ_{11}, δ_{21}, and δ_{22} implies that

$$\frac{\alpha_2}{\alpha_1} \leqq \kappa, \quad \text{or} \quad \alpha_2 \leqq \alpha_1\kappa.$$

An upper bound for the weighting of consumption is reasonable because the model (9.31)–(9.34) does not explicitly contain constraints that take into account the scarcity of production factors. Production and consequently consumption are restricted implicitly only because of their negative environmental impact. In order to compensate the negative effect of the damage cost for the vector-valued function (9.31), the (relative) weight for the second objective can be higher if the parameter κ is higher.

On the other side, from the nonnegativity condition for δ_{12}^0 follows the lower bound for the relative weights,

$$\frac{\alpha_2}{\alpha_1} \geqq (1 - \mu)\kappa.$$

With increasing efficiency of environmental investments, described by the parameter μ, the lower bound for the relative weight α_2 can be reduced.

In order to get a nonnegative solution of the dual problem (9.36)–(9.41), the coefficients α_1 and α_2 (or the weights for the objectives) are restricted to

$$(1 - \mu)\kappa \leqq \frac{\alpha_2}{\alpha_1} \leqq \kappa. \tag{9.43}$$

Due to the results of Section 6.3.3 dealing with the economic interpretation of dual variables in geometric programming, $\lambda_1^0 = \sum_t \delta_{1t}^0$ indicates the effect of a change in the pollution level (p) on the social disutility function (9.31). Because

$$\lambda_1^0 = \delta_{11}^0 + \delta_{12}^0 = \alpha_1\kappa > 0, \tag{9.44}$$

the constraint (9.32) for the optimal solution δ^0 must be fulfilled, for any α_1, α_2 satisfying (9.43), as equality. In the other case, due to the minimization of (9.31), the level of pollution can be reduced without violating any constraint. It follows from (9.44) that for fixed α_1, a higher coefficient κ (describing the influence of pollution for the damage cost) implies a stronger impact of the pollution level for the social disutility (9.31).

The coefficient $\lambda_2^0 = \sum_t \delta_{2t}^0$ describes the impact of a change in the production of good (y) for the social disutility (9.31). We obtain

$$\lambda_2^0 = \delta_{21}^0 + \delta_{22}^0 = \frac{\alpha_1\kappa - \alpha_2}{\mu} \geqq 0 \tag{9.45}$$

because of (9.43). For $\frac{\alpha_2}{\alpha_1} < \kappa$, the dual variables δ_{21}^0 and δ_{22}^0 are positive, and therewith λ_2^0 is positive as well. The constraint (9.33) is fulfilled as equality, the produced good can be used either for environmental investments (i)—decreasing the emission coefficient b and consequently the damage cost—or for consumption. It follows from (9.45) that for fixed α_1 and α_2, a lower coefficient κ and/or higher efficiency of environmental investments described by parameter μ diminish the negative impact of production for the social welfare combining minimization of pollution cost and maximization of consumption.

On the other side, for given parameters κ and μ, the (relative) higher weighting of the consumption maximization diminishes the negative effect of production. For α_2, at the upper bound of (9.43) is $\lambda_2^0 = 0$. The negative effect of increasing production causing higher environmental cost by production (reduced by decreasing emission coefficient b) and consumption is compensated by the positive effect from increasing consumption. If $\alpha_2 < \alpha_1\kappa$ holds, the weighting—or, using the terminology of Lange [3], *social significance*—of the consumption in the welfare function is not enough to outweigh the negative effect of the increasing production for the environment.

Similarly, the higher weighting of the pollution cost in the social welfare strengthens the impact of pollution level (p) for the composite welfare

Knowing the optimal solution of the dual problem (9.36)–(9.42), we can find the optimal solution of the primal geometric problem (9.31)–(9.34). According to (9.25)—analogously with (6.28)—each optimal solution c^0, i^0, p^0, and y^0 for the primal problem (9.31)–(9.34) satisfies the system of equations

$$li^{-\mu}yp^{-1} = \frac{\delta_{11}^0}{\lambda_1^0} = \frac{\alpha_1\kappa - \alpha_2}{\alpha_1\kappa\mu}, \tag{9.46}$$

$$dcp^{-1} = \frac{\delta_{12}^0}{\lambda_1^0} = \frac{\alpha_1\kappa(\mu - 1) + \alpha_2}{\alpha_1\kappa\mu}, \tag{9.47}$$

$$iy^{-1} = \frac{\delta_{21}^0}{\lambda_2^0} = \mu, \tag{9.48}$$

$$cy^{-1} = \frac{\delta_{22}^0}{\lambda_2^0} = 1 - \mu, \tag{9.49}$$

where $\lambda_1(\boldsymbol{\delta}^o)$ and $\lambda_2(\boldsymbol{\delta}^o)$ must be positive. Due to (9.44) $\lambda_1(\boldsymbol{\delta})$ is positive for any $\alpha_1 > 0$ and $\kappa > 0$. In order to ensure the positivity of $\lambda_2(\boldsymbol{\delta})$, it is assumed in what follows that $\alpha_2 < \alpha_1\kappa$. From (9.48) and (9.49), we have the following qualitative result:

$$\frac{i^0}{c^0} = \frac{\mu}{1 - \mu}.$$

The optimal ratio in which the output y should be allocated between environmental investment i^0 and consumption c^0 is uniquely determined by the efficiency parameter μ and is independent of the other model parameters, including the social significance of the objectives.

Substituting $i = \mu y$ from (9.48) into (9.46) and $c = (1 - \mu)y$ from (9.49) into (9.47), two equations with two variables y and p will be obtained. As in the general system (6.28), taking the logarithm of both sides of each equation, a system of linear equations with the variables $\ln y$ and $\ln p$ arises that can be solved analytically. After some transformations, the optimal solution results:

$$y^0 = \left(\frac{k_1}{k_2}\frac{\delta_{12}^0}{\delta_{11}^0}\right)^{\frac{1}{\mu}} = \left(\frac{k_1[\alpha_1\kappa(\mu - 1) + \alpha_2]}{k_2[\alpha_1\kappa - \alpha_2]}\right)^{\frac{1}{\mu}},$$

$$p^0 = k_2\frac{\alpha_1\kappa\mu}{\alpha_1\kappa(\mu - 1) + \alpha_2}y^0,$$

where $k_1 = l\mu^{-\mu}$ and $k_2 = d(1 - \mu)$. Because of the positivity condition for the primal variables y and p, the dual variable δ_{12} must be positive. Thereupon the following strict inequality is required:

$$\alpha_1\kappa(\mu - 1) + \alpha_2 > 0, \quad \text{or} \quad \frac{\alpha_2}{\alpha_1} > (1 - \mu)\kappa.$$

To obtain an efficient solution of problem (9.31)–(9.34), condition (9.43) for the weighting coefficients α_1 and α_2 should be reinforced as follows:

$$(1 - \mu)\kappa < \frac{\alpha_2}{\alpha_1} < \kappa.$$

Under these strict inequalities, all primal and dual constraints are positive and both forced constraints are fulfilled as equalities. The optimal solution of problem (9.35) yields an efficient solution of problem (9.31)–(9.34). Moreover, using the duality theory of geometric programming can provide useful insights into the qualitative features of the efficient solution.

9.4 Optimal Behavior of a Monopolist Facing a Bicriteria Objective Function

Another application of the program $(\prod \mathrm{P}_\alpha)$ and its dual (9.4)–(9.10) or (9.12)–(9.13) and (9.14)–(9.16), respectively, offers the model of a monopolistic firm maximizing revenue and profit from Section 7.1.4. Assuming the revenue function $R(q) = aq^\beta$, with $a > 0$, $\beta > 0$, implying positive marginal revenue and the cost function $C(q) = bq^2$, with $b > 0$ (yielding positive marginal cost), the profit function $\pi(q)$ becomes $\pi(q) = aq^\beta - bq^2$. The following multiobjective optimization problem appears:

$$\text{Maximize} \quad F(q) = \begin{Bmatrix} f_1(q) = aq^\beta, \\ f_2(q) = \pi(q) \end{Bmatrix}. \tag{9.50}$$

Unfortunately, the profit function $\pi(q)$ is not a posynomial, but as described in Section 6.4, simple transformation of the profit function $\pi(q)$ allows the formulation of problem (9.50) as a multiobjective geometric programming problem. Denoting the profit $\pi(q)$ by the new variable y and introducing the constraint

$$aq^{\beta} - bq^2 \geqq y,$$

problem (9.50) is equivalent to the following problem:

$$\text{Minimize} \qquad \overline{F}(q, y) = \begin{Bmatrix} f_1^{-1}(q) = \frac{1}{a} q^{-\beta}, \\ f_2(y) = y^{-1} \end{Bmatrix} \tag{9.51}$$

$$\text{subject to} \qquad \frac{1}{a} y q^{-\beta} + \frac{b}{a} q^{2-\beta} \leqq 1,$$

$$q > 0, \quad y > 0.$$

According to (9.19) and (9.20), the dual constraints are

$$\begin{aligned} -\beta\delta_{11}^{o} - \beta\delta_{11} + (2-\beta)\delta_{12} &= 0, \\ -\delta_{21}^{o} + \delta_{11} &= 0, \\ \delta_{11}^{o} &= \alpha_1, \\ \delta_{21}^{o} &= \alpha_2, \end{aligned} \tag{9.52}$$

where δ_{11}^{o} and δ_{21}^{o} denote the dual variables corresponding to the terms in the first and second objective function, respectively. Fortunately, we obtained a system of equations with zero degree of difficulty such that the optimal solution of the dual problem is easily estimated:

$$\delta_{11}^{o\,*} = \alpha_1, \qquad \delta_{21}^{o\,*} = \alpha_2, \qquad \delta_{11}^{*} = \alpha_2, \qquad \delta_{12}^{*} = \frac{\beta}{2-\beta}(\alpha_1 + \alpha_2),$$

and

$$\lambda_1^{*} = \delta_{11}^{*} + \delta_{12}^{*} = \frac{2\alpha_2 + \alpha_1\beta}{2-\beta}.$$

Taking into account the nonnegativity condition for the dual variables δ_{1t} $(t = 1, 2)$, the parameter β must be smaller than two. It is easy to show that this condition is fulfilled for any negative-sloped elastic demand function. Assume that the inverse demand function takes the form

$$p = h^{-1}(q) = aq^{-\gamma}, \quad \text{with } a > 0 \quad \text{and} \quad \gamma > 0.$$

The price elasticity of demand is then $\epsilon_p = -\frac{1}{\gamma}$. Herewith $\gamma < 1$ implies price elastic demand and, due to the Amoroso–Robinson form

$$\text{MR} = \frac{dR}{dq} = \frac{1}{p}\left[\frac{dp}{dq}\frac{q}{p} + 1\right],$$

positive marginal revenue (MR), which coincides with the assumption of positive β in the revenue function $R(q) = aq^{\beta}$. Using the above inverse demand function, the revenue function can be written as

$$R(q) = p(q)q = aq^{-\gamma}q = aq^{1-\gamma}.$$

Letting $\beta = 1 - \gamma$, the condition $\beta < 2$ then implies $\gamma > -1$, which is obviously fulfilled.

Using the form (9.30), we can derive the efficient solution of the primal problem:

$$\frac{\delta_{12}^*}{\lambda_1^*} = \frac{\beta(\alpha_1 + \alpha_2)}{2\alpha_2 + \alpha_1\beta} = \frac{b}{a}q^{2-\beta},$$

$$q^* = \left[\frac{a\beta(\alpha_1 + \alpha_2)}{b(2\alpha_2 + \alpha_1\beta)}\right]^{\frac{1}{2-\beta}}, \tag{9.53}$$

$$\frac{\delta_{11}^*}{\lambda_1^*} = \frac{\alpha_2(2-\beta)}{2\alpha_2 + \alpha_1\beta} = \frac{1}{a}yq^{-\beta}. \tag{9.54}$$

Substituting (9.53) for q in (9.54) yields

$$y^* = \frac{a\alpha_2(2-\beta)}{2\alpha_2 + \alpha_1\beta}\left[\frac{a\beta(\alpha_1 + \alpha_2)}{b(2\alpha_2 + \alpha_1\beta)}\right]^{\frac{\beta}{2-\beta}}.$$

For $\alpha_1 = 0$ and $\alpha_2 = 1$, the solution of the profit-maximizing firm is easily obtained from (9.53):

$$q_{\mathrm{PM}}^* = \left[\frac{a\beta}{2b}\right]^{\frac{1}{2-\beta}}.$$

The reader can verify that for $\beta < 2$ (fulfilled for any price elastic demand function with negative slope),

$$q_{\mathrm{PM}}^* < q^*.$$

The output of a monopolist maximizing profit *and* revenue is higher than the output of the profit-maximization monopolist (see [1]). The difference between the output levels q_{PM}^* and q^* depends, for the given parameters a, b, and β, on the weighting of both of the objectives expressed by the coefficients α_1 and α_2, respectively. It follows from (9.53) that

$$\frac{\partial q^*}{\partial \alpha_1} > 0 \quad \text{and} \quad \frac{\partial q^*}{\partial \alpha_2} < 0.$$

With increasing relevance of the revenue maximization, the output of a monopolistic firm will increase; with increasing relevance of the profit maximization, it will decrease.

As in the model of a profit-maximizing monopolist, the condition for competitive output, price equal to the marginal cost, is violated for a monopolist maximizing a utility function (with revenue and profit as arguments). The product price exceeds marginal cost, and therefore the question concerning the effects of changes in profits and turnover taxes on optimal output arises. To derive the effects of change in profits tax (ρ) on optimal output, model (9.51) is modified as

$$\text{Minimize} \qquad \overline{F}(g, y(\rho)) = \begin{Bmatrix} f_1^{-1}(q) = \frac{1}{a} q^{-\beta}, \\ f_2(y(\rho)) = \frac{1}{(1-\rho)} y^{-1} \end{Bmatrix}$$

$$\text{subject to} \quad \frac{1}{a} y q^{-\beta} + \frac{b}{a} q^{2-\beta} \leqq 1,$$

$$q > 0, \quad y > 0.$$

Because the change in the profits tax (ρ) results in changes of the coefficients but not the exponents in the posynomials, the dual constraints (9.52) remain unchanged, and therewith (due to the degree of difficulty being zero) the dual solution does as well. According to (9.30), the optimal output q^* as determined by (9.53)—and consequently the profit y—remain unchanged. For the Cobb–Douglas type of utility ("master" or "composite" objective) function, a change in the profit tax (ρ) does not affect the behavior of a monopolist maximizing revenue and profit (see [2] for an alternative derivation of this result).

What will be the change in the behavior of a monopolist facing a bicriteria objective function with respect to a change in turnover tax? In order to answer this question, we rewrite model (9.51) as $\overline{R}(q) = aq^{\beta}(1-t)$, where $0 < t < 1$ denotes a turnover tax. The profit function then becomes

$$\pi(q) = aq^{\beta}(1-t) - bq^2$$

and model (9.51) is

$$\text{Minimize} \quad \widetilde{F}(q, y) = \begin{Bmatrix} f_1^{-1}(q) = \frac{1}{a(1-t)} q^{-\beta}, \\ f_2(y) = y^{-1} \end{Bmatrix}$$

$$\text{subject to} \qquad \frac{1}{a(1-t)} y q^{-\beta} + \frac{b}{a(1-t)} q^{2-\beta} \leqq 1,$$

$$q > 0, \quad y > 0.$$

As in the previous model with the profit tax (ρ), due to the same exponents in the posynomials, the dual constraints (9.52) remain unchanged, and therefore a change in the turnover tax (t) does not change the dual solution. The solution of the primal problem will be obtained by using (9.30):

$$\begin{aligned} \frac{\delta_{12}^*}{\lambda_1^*} &= \frac{\beta(\alpha_1 + \alpha_2)}{2\alpha_2 + \alpha_1 \beta} = \frac{b}{a(1-t)} q^{2-\beta}, \\ \widetilde{q} &= \left[\frac{a\beta(1-t)(\alpha_1 + \alpha_2)}{b(2\alpha_2 + \alpha_1 \beta)} \right]^{\frac{1}{2-\beta}}. \end{aligned} \tag{9.55}$$

Comparison of $\widetilde{q}$ from (9.55) with q^* from (9.53) yields

$$\widetilde{q} < q^*.$$

A monopolist maximizing the revenue *and* profit will reduce his output if the turnover tax (t) increases.

References and Further Reading

[1] M. Brown and N. Revankar, A generalized theory of the firm: An integration of the sales and profit maximization hypothesis, *Kyklos*, **24**-3 (1971), 427–443

[2] F. Hof and M. Luptáčik, Optimal behaviour of a monopolist facing a bicriteria objective function: A reconsideration, *Politická Ekon.* (*Political Econ.*) XLI, **5** (1993), 662–679.

[3] O. Lange, The foundations of welfare economics, *Econometrica*, **10** (1942), 215–228.

[4] A. P. Mastenbroek and P. Nijkamp, A spatial environmental model for an optimal allocation of investments, in P. Nijkamp, ed., *Environmental Economics, Vol.* 2: *Methods*, Nijhoff Social Sciences, Leiden, The Netherlands, 1976.

[5] L. D. Pascual and A. Ben-Israel, Vector-valued criteria in geometric programming, *Oper. Res.*, **19** (1971), 98–104.

Index

M. Luptáčik, *Mathematical Optimization and Economic Analysis*,
Springer Optimization and Its Applications 36, DOI 10.1007/978-0-387-89552-9,

FSC
www.fsc.org
MIX
Papier | Fördert
gute Waldnutzung
FSC® C083411

Kolibri 360 GmbH
Ferdinand-Jühlke-Straße 7
99095 Erfurt, Deutschland
produktsicherheit@kolibri360.de